Constructing East Asia

Constructing East Asia

Technology, Ideology, and Empire
in Japan's Wartime Era, 1931–1945

Aaron Stephen Moore

Stanford University Press

Stanford, California

Stanford University Press
Stanford, California

Printed in the United States of America on acid-free, archival-quality paper

Library of Congress Cataloging-in-Publication Data

Moore, Aaron Stephen, 1972- author.
    Constructing East Asia : technology, ideology, and empire in Japan's wartime
era, 1931–1945 / Aaron Stephen Moore.
        pages   cm
    Includes bibliographical references and index.
    ISBN 978-0-8047-8539-6 (cloth : alk. paper)
    ISBN 978-0-8047-9724-5 (pbk. : alk. paper)
    1. Technology—Political aspects—Japan—History—20th century.
2. Technology and state—Japan—History—20th century.
3. Japan—Colonies—Asia—History—20th century.   4. Public works—East
Asia—History—20th century.   5. Fascism—Japan—History—20th
century.   6. Japan—History—1926–1945.   7. World War, 1939–1945—
Japan.   I. Title.
    T27.J3M65 2013
    303.48'3095209043—dc23   2012050736

ISBN 978-0-8047-8669-0 (electronic)

Typeset by Westchester Publishing Services in 11/13.5 Adobe Garamond

To my parents, Lisa Chung Moore and Stephen William Moore

Contents

Figures

Acknowledgments

THIS BOOK BEGAN MORE than ten years ago at Cornell University and in many formal and informal research groups in Japan. Originally, I aspired to write an intellectual history of the concept of "technology" (*gijutsu*) in early twentieth-century Japan, but the project evolved greatly since then as I learned how the concept was appropriated by different social actors and took shape in large-scale infrastructure projects. It is impossible to list all of the friends and colleagues who have helped me along the way. First and foremost, I'd like to thank my intellectual mentors at Cornell University who helped stimulate my thoughts on technology: J. Victor Koschmann, Naoki Sakai, Michael Steinberg, and Frederick Neuhouser. Discussions with my peers who read and commented on various parts of my work also contributed to the development of this project: Adelheid Voskuhl, Ben Middleton, John Kim, Trent Maxey, Anna Parkinson, Sheetal Majithia, Doreen Lee, Chi-ming Yang, Mihara Yoshiaki, Sven Brandenburg, Kasai Hirotaka, and Kimoto Takeshi.

The Tokyo University of Foreign Studies was my intellectual home as I researched this project over the years. Uemura Tadao, Iwasaki Minoru, Nakano Toshio, and Yonetani Masafumi all welcomed me into their scholarly community, commented on my work, and sponsored my research in various capacities. My close colleagues in Japan—Tomotsune Tsutomu, Shidama Shinri, Suyama Daiichirō, and Kikuchi Naoko—also greatly stimulated my thoughts on technology with their engaging critiques and discussions. A special thanks goes to Yamane Nobuhiro and Tanigawa Ryūichi, who have greatly aided the history of technology portion of my work through various discussions and cooperative research endeavors. I am also grateful to Hirose Teizō and the Nishimatsu Construction Company for allowing me to examine documents that were not publicly available.

This book would have never taken off without the help of research and travel grants from the Fulbright Association; the German Academic Exchange Service (DAAD); the East Asia Program at Cornell University; the School for Historical, Philosophical, and Religious Studies and the Institute for Humanities Research at Arizona State University; the Library of Congress; and the Terasaki Center for Japanese Studies at UCLA. The fellowship I was awarded at UCLA's Terasaki Center in 2008 enabled me to conduct research for the section in this book on Japan's large-scale colonial infrastructure projects. Teaching, organizing an international conference, and presenting at various forums at UCLA greatly helped my development into a historian of science and technology through the various people I met there. Special thanks to Sharon Traweek, Nakayama Shigeru, Sooraya de Chandarevian, and Hiromi Mizuno.

Portions of this manuscript have been presented at many conference panels, colloquia, and public talks, and the comments and criticisms I received at these events have been invaluable. Special thanks to James Bartholomew, John Dower, Louise Young, Ruth Rogaski, John DiMoia, Tae-Ho Kim, Janis Mimura, Michael Shin, Joan Fujimura, Daqing Yang, Max Ward, Steven Levine, Kobayashi Hideo, Eric Dinmore, Eric Schatzberg, and Seung-joon Lee. Many thanks to the anonymous reviewers of my work and others who have read various portions of the manuscript, particularly Stephen MacKinnon, James Rush, Wesley Sasaki-Uemura, and Lisa Onaga, all of whom helped to greatly sharpen my arguments. The hard work of Stacy Wagner and the editorial staff at Stanford University Press, as well as the production editor, Melody Negron, and the staff at Westchester Publishing Services, have made the publication of my manuscript possible.

This book was also greatly aided by my friends Ahilan Kadirgamar, Cenan Pirani, and Lawrence Surendra, who have kept me intellectually engaged with different fields and parts of the world other than Japan, thereby enabling me to gain a broader perspective on the issues presented in this book. My wife, Nilanjana Bhattacharjya, has encouraged me throughout the many years of research and writing, and helped clarify my ideas through her thorough critiques. Finally, this book would not have existed without the loving support and patience of my parents, Lisa Chung Moore and Stephen William Moore.

Constructing East Asia

Introduction
The Technological Imaginary of Imperial Japan

Japan as Techno-Superpower

In December 1990, Japan's Science and Technology Agency and the National Institute of Science and Technology Policy published a report titled *Historical Review of Japanese Science and Technology Policy*—a "postwar comprehensive history of Japan's science and technology policies." The report's purpose was ambitious: to educate the world about how Japan's science and technology policy had played an essential role in its economic and social development and to reflect on how Japan could adopt policies "aimed at not only creating a wealthy nation but a wealthy world as well."[1] The report was written during the 1980s "economic bubble" era, when Japan was viewed as the global leader in technology and technical innovation in such areas as consumer electronics, automobiles, semiconductors, manufacturing technology, and robotics. Numerous books with such sensational titles as *The Technopolis Strategy: Japan, High Technology, and the Control of the Twenty-First Century* and *Japan as a Scientific and Technological Superpower* detailing Japan's unique approach to economic development appeared during this time.[2] Kodama Fumio, dean and professor of engineering management at the Shibaura Institute of Technology, described Japan's model of promoting technological innovation as one that represented a global "techno-paradigm shift" and went so far as to credit the Japanese cassette tape recorder, videocassette recorder, and fax machine for making possible the Iranian and Philippine revolutions and the Tiananmen uprising.[3] Thus, in the 1980s and early 1990s, Japanese technology and technology policy were widely seen as a progressive force for social development and economic prosperity—and in some cases, even democratic values.[4]

Japan became the world's largest foreign aid donor by 1989 and kept that position for the first decade of the post–Cold War era. In 1995, Japan's overseas development assistance spending reached $14.5 billion, almost double the U.S. figure for that year.[5] Much of this aid consisted of technical assistance not only in the form of goods but also in technical knowledge and personnel. These technical assistance programs began soon after the U.S. occupation ended in 1952 and first took the form of wartime reparations agreements with formerly occupied Southeast Asian countries. Typically, the Japanese government subsidized large-scale hydropower, transportation, or industrial projects, which in turn provided lucrative contracts to domestic construction and manufacturing firms, thereby providing a boost to the Japanese economy. In the 1960s, Japanese scholars even developed and publicized a stage theory of economic development based on the promotion of technology—the "flying geese model"—which governed Japan's foreign policy in Asia through the 1990s. According to this model, in the first stage an underdeveloped country imports manufactured goods from a more advanced nation (i.e., Japan), which promotes industrial development. Import substitution constitutes the second stage, and when that country begins to produce manufactured goods for export in surplus, it enters the third and final stage. The history of Asia's development seemed to prove this theory of industrial growth through technology transfer and import substitution. Longtime receivers of Japanese technical assistance South Korea, Taiwan, Hong Kong, and Singapore became known as the "Four Tigers," followed by "secondary emerging markets," such as the Philippines, India, Indonesia, Thailand, and China.[6] Thus, the aggressive promotion of technology not only was essential to Japan domestically but also has played a prominent role in its foreign policy.

For the Japanese government and its admirers, technology and technology policy represented a modernizing, progressive force that was essential to Japan's national development and security throughout its modern history, as well as Asia's recent economic success. Technology was something to be instrumentally used and promoted to achieve national prosperity, innovation, and productivity. This familiar narrative of Japan's farsighted development of technology and a new "techno-nationalist" paradigm constituted a pillar of Japan's modernization narrative built up by Japanese and Western scholars alike.[7]

Reconceptualizing Technology in Modern Japan

Constructing East Asia critiques this conventional narrative of postwar technological modernization by tracing its origins in the "technological imaginary" of wartime Japan in order to draw attention to some of the many ways that technology has operated as an ideology and a system of power throughout much of the twentieth century. By "technological imaginary," I mean the ways that different groups invested the term "technology" (*gijutsu*) with ideological meaning and vision. Rather than offering an overarching model or definition of technology, I examine how the discourse surrounding technology in wartime Japan developed and changed according to who was discussing it or what political objective that person or group had. This book expands the conventional modernization narrative of technology as an abstract, universal force for progress and prosperity by analyzing how a technological imaginary was formulated not only in relation to domestic capitalist development and wartime mobilization but also in close relation to colonial expansion and rule—an arena that has often been viewed as separate from the "main" trajectory of Japan's historical development.[8] The emergence of a discourse on technology among Japan's elites was also a process of "imagineering," a blending of creative imagination and technical expertise in the formulation of wartime and colonial policies, as well as the construction of numerous large-scale infrastructure projects designed to incorporate the hopes and desires of various peoples.[9] This book's main premise is that technology's employment as a mobilizing force during the wartime period through various policies and projects of imagineering continued as Japan transformed itself into a global technological superpower and an influential nation in international development circles after 1945. At the popular level, these continuities have been repressed by the conventional narrative of postwar Japan's economic miracle and rise into a major force for peace on the global stage.

Modernization theorists have typically depicted the wartime 1930s and 1940s as a "dark valley" of irrationality, spiritualism, and reactionary politics in Japan's modern history.[10] Postwar Japanese democratization, they have argued, rested largely on the gains of the U.S. occupation period

(1945–52). Since the 1980s, however, a number of works have demonstrated how the postwar Japanese democratic system in fact rested on many statist components developed during the wartime era. Some works have also shown how wartime authoritarianism and militarism were not simply ultraconservative reactions to liberal democracy but contained many modern and progressive characteristics as well. For example, Chalmers Johnson traces the origins of Japan's postwar "economic miracle" to the various techniques of industrial policy and planning formulated by economic bureaucrats in the prewar and wartime eras. John Dower provocatively calls the Asia-Pacific War a "useful war" in its development of key postwar institutions for high-speed growth, such as the economic bureaucracy, semi-monopolistic business combinations, and the Japanese system of management and "cooperative" labor relations. Sheldon Garon argues that values of modernization, such as "progress, science, and rationality," allowed the state to develop techniques of "moral suasion" in the prewar era to mobilize civil society into a more authoritarian, managed social system during and after the war. *Constructing East Asia* focuses these earlier analyses of the relationship between "modernization" and authoritarianism by examining how modernization's most visible product—technology—operated as an ideology for wartime mobilization and colonial rule, which in turn shaped the course of postwar democratic Japan's history.[11]

To illuminate the relationship between technology and power, and thereby question the conventional, instrumentalist view of technology, this book borrows from the Frankfurt School's rich body of work on technology's political nature. Max Weber, for example, argued that the emergence of a Protestant ethic of discipline, calculation, and rationality in modern capitalism created a disenchanted order whereby people were "bound to the technical and economic conditions of machine production which to-day determine the lives of all the individuals who are born into this mechanism . . . with irresistible force."[12] Purposive and instrumental forms of activity, organization, and technology became embodied in large bureaucracies and administrations, building an "iron cage" of reason whereby people were transformed into "specialists without spirit, sensualists without heart." Formal systems of rationality that optimize calculability and control and were concerned with "efficiency of means" rather than "choice of ends" dominated people's everyday lives.[13] Thus, for Weber the formation of a technological society was not so much a linear march of progress as modernization theorists have argued but a dehumanizing and inescapable process of rationalization. *Constructing East Asia* incorporates Weber's argument by examining how technology constituted a widespread force of rationalization that in turn shaped the nature of power in wartime Japan.

Technology in wartime Japan did not simply represent an oppressive force of rationalization but actively mobilized the people for state goals as well. The work of Herbert Marcuse and Jürgen Habermas is important to this book's argument that technology is a system of power, as they have suggested certain ways that technology has dynamically incorporated people's hopes and desires into mechanisms of social control. Marcuse, for example, linked the spread of technical rationality to the naturalization of capitalist relations of domination. For him, technology "provides the great rationalization of the unfreedom of man and demonstrates the 'technical' impossibility of being autonomous, of determining one's own life." Whereas Weber emphasized an oppressive "iron cage of reason," Marcuse described capitalist domination as "submission to the technical apparatus which enlarges the comforts of life and increases the productivity of labor."[14] Habermas elaborated on Marcuse's formulations by noting how a pervasive "technocratic consciousness" or logic of "purposive-rational action" expands outside the realm of economic activity and reproduces itself at the level of social systems into which people are functionally integrated.[15] "[Politics] is oriented toward the elimination of dysfunctions and the avoidance of risks that threaten the system: not . . . toward the *realization of practical goals* but toward the *solution of technical problems*," he argued.[16] As a result, the public sphere had become depoliticized and concerned more with the system's proper functioning than with any practical vision of the "good life."[17] Technology for Habermas and Marcuse represented more than physical technology; it represented specific techniques of power and mobilization. Their theoretical conclusions are significant because they help capture an important political dynamic at work in wartime Japan when new definitions of technology emerged and became predominant in the public discourse.

Constructing East Asia demonstrates this political dynamic through an analysis of the influential groups and actors who shaped Japan's technological imaginary—intellectuals, technology bureaucrats, engineers, and state planners. Such Marxists as Aikawa Haruki articulated a notion of a technologized system society whereby the people's economic, political, social, and cultural lives were mobilized for radical social transformation—including those in Japan's expanding empire. Technology bureaucrats like Miyamoto Takenosuke insisted on the importance of "social engineers" who incorporated technical expertise into national policy making, and put forth the notion of "technologies for Asian development" (*kōa gijutsu*) as a guiding vision for engineers to modernize and therefore "liberate" Asia from Western imperialism. Such engineers as Naoki Rintarō, Haraguchi Chūjirō, and Kubota Yutaka traveled to Japan's empire to escape bureaucratic red tape in Japan and developed concepts of "comprehensive technology" (*sōgō gijutsu*) and "national land planning" (*kokudo keikaku*)—coordinating and integrating such technical projects as urban planning, dam construction, flood control, and industrial development to bring about mutually sustaining relationships and benefits. Reform bureaucrats (*kakushin kanryō*—literally, "renovationist bureaucrats") like Mōri Hideoto formulated such notions as "economic technology" (*keizai gijutsu*) or policies designed by bold "economic technicians" to integrate Japan and its empire into an organic mechanism based on voluntarist "life organizations." As "technology" became a prominent word in Japanese public discourse during the 1930s, influential elites appropriated the term and expanded its conventional meaning as physical artifacts to include techniques of social organization and transformation.

Technology and Japanese Fascism

Technology in wartime Japan meant much more than simply advanced machinery and infrastructure; it included a subjective, ethical, and visionary dimension. As in Europe and elsewhere, from the early twentieth century, technology in Japan began to represent certain forms of creative thinking, acting, or being, as well as values of rationality, cooperation, and efficiency. Technology also lent itself easily to utopian visions of an egalitarian society without ethnic or class conflict. Particularly during the 1930s, as Japan was shifting from a light to a heavy industrial wartime economy, elites developed a more subjective view of technology as increasingly permeating and altering every aspect of life. This more subjective, practical, and mobilizing view of technology—the "technological imaginary"—guided a whole range of social actors (or "imagineers") from bureaucrats designing Japan's wartime managed economy to engineers planning and constructing massive colonial infrastructure projects, from Marxists struggling to make sense of Japanese capitalist development and the possibility of revolution to cultural critics advocating the cultivation of a "neorealist" technological aesthetic in film and mass media.

Wartime Japan's technological imaginary represented a form of fascist ideology that employed familiar tropes of modernity and rationality rather than relying primarily on cultural appeals to spiritualism or ultranationalism. The technological imaginary left a particularly strong legacy on postwar Japanese society and foreign policy. Several scholars have examined the connections between technology and fascism in other contexts. Most notably, Jeffrey Herf has analyzed how such German "reactionary modernists" as Oswald Spengler, Ernst Jünger, Martin Heidegger, and Werner Sombart appropriated technical reason to pathological, irrational, and romantic ends of "community, blood, will, self, form, productivity, and finally race."[18] But Japanese elites did not merely "pervert" technology's inherent rationality by infusing it with irrationality and romanticism. Rather, in varying ways, they articulated a practical, a political, and an inventive notion of technology whereby different areas of life were rationally planned and mobilized to exhibit their maximum potential and creativity.

Instead of searching for some particular notion of "Japanese" technology, *Constructing East Asia* examines how Japanese elites actively incorporated utopian notions of technology from other contexts into their fascist ideologies. Charles Maier has made the connection between technology and fascism in the West by tracing how Taylorism in the early twentieth century spread beyond rationalizing factory work techniques to become a powerful political ideology of industrial management and social reorganization. "Scientific management" lent itself to visions of overcoming class conflict on both the left and the right in the early twentieth-century United States and Europe. According to these visions, society would be reorganized along the lines of a "coherent system" of "efficiency, optimality, enhanced productivity and expanded output." For example, in the United States during the Progressive Era, Charles Ferguson and Thorstein Veblen put forth the engineer as the ideal person to "impose optimality upon society" and end capitalist waste and class conflict. In France, "Saint Simonianism embodied a proto-technocratic ideology that rejected traditional class divisions in favor of the unity of all 'productive' and 'industrious' elements, bourgeois, peasant, and proletarian." In Italy, the Futurists envisioned the fascist state as a "dynamo," and therefore "more than a state." In the Soviet Union, communists celebrated technology's potential to facilitate social revolution. Finally, in Germany, industrialist-engineers, such as Walther Rathenau and Wichard von Moellendorf, employed technological paradigms in pushing for a "planned economy" (*Planwirtschaft*) that would eliminate competition and transform capitalists into public employees. Thus, technology became a powerful signifier of social harmony, innovation, and efficiency all over the industrialized world, especially in the face of the crisis of capitalism and growing labor unrest during the Great Depression. Japanese elites did not reject these notions of technology and social management but incorporated them into their own fascist ideological programs.[19]

This book defines fascism as an ideology and mode of power translated globally into various national contexts that combined antimodern and modern elements for the revolutionary transformation and mobilization of society.[20] Although "fascism" has been a contentious term for English-language scholars writing on wartime Japan, those who have used it have often borrowed from Maruyama Masao's conception of "fascism from above." Maruyama believed that Japanese fascism in the end was spread not by a mass movement "from below" like in many European cases but by the state's various organs. Furthermore, Japanese fascism was "particular" in its emphasis on emperor-centered familialism, antimodern agrarianism, and emancipatory pan-Asianism.[21] Although English-language scholars have reformulated various points of Maruyama's thesis or have refused to use the term altogether, many have continued to emphasize the antimodern, authoritarian, and spiritualist or communitarian elements of Japanese fascism more than its rational, modernizing components.[22] In their frameworks, Europe constitutes fascism's original model according to which Japan always appears particular. For example, the lack of a charismatic leader or a mass fascist-style party or the continuity between Meiji institutions and those of the 1930s was seen as providing enough evidence to "prove" that Japan was not fascist.[23] Instead of deriving a standard model from the German or Italian experience, *Constructing East Asia* views fascism as a common set of ideas and programs that were translated into different national contexts.[24] Focusing solely on the particularities and minutia of a so-called pure model of fascism ignores fascism's importance as a broader historical force that developed simultaneously in different places. More important, a focus on fascist particularity in Japan overlooks common processes of modernization within fascism, such as rationalization, social reorganization, and the construction of a technological imaginary as a form of power and mobilization that continued to have important effects after the war.

In her recent book on notions of "the scientific" among wartime Japanese elites, Hiromi Mizuno avoids the term "fascism" in favor of "scientific nationalism"—an ideology whereby "science and technology are the most urgent and important assets for the integrity, survival, and progress of the nation."[25] In her work on Japan's wartime reform bureaucrats, Janis Mimura aptly describes their ideologies as "techno-fascism"—a fusion of technical rationality, comprehensive planning, and modern values of productivity and efficiency with ethnic nationalism and right-wing ideologies of organicism. In various ways, she shows how techno-fascism aimed to transcend traditional Japanese political divisions and incorporate them within a larger politics of technocratic planning, which she briefly describes as a new "mode of power."[26] *Constructing East Asia* expands on Mizuno's notion of "scientific nationalism" and Mimura's suggestion about techno-fascism by arguing that the technological imaginary represented something more than a politics of nationalism and technocratic planning. In their techno-fascist or scientific nationalist ideologies, we also see the contours of another mode of power, one that was based more on harnessing the creativity and vitality of human subjects than solely on repression and violence. Within the ideas and policies of Japan's elites, power was not simply something that organized society from above, but dynamically shaped it from within through the productive practices of a whole array of institutions and people. Fascism was more than the existence of a totalitarian state; it also created a form of "molecular or micropolitical power" throughout everyday life that sought to preserve capitalism "without all of its consequences for class conflict, its alienating effects, instability, and cultural and economic unevenness," as Harry Harootunian argues.[27] *Constructing East Asia* examines how the technological imaginary articulated such a fascist mode of power.

Japan's Technological Imaginary

The figures and groups discussed in this book were at the vanguard of a wide range of Japanese elites who began to articulate a more subjective, utopian notion of technology from the 1920s. For example, American scientific management ideology was translated into the Japanese context and developed into the world-famous modern Japanese management system after 1945. A number of engineers, managers, and bureaucrats in the prewar and wartime eras promoted the state-sponsored efficiency movement and the industrial rationalization movement (*sangyō gōrika undō*).[28] The notion of technocracy—rule by technical experts—also became popular during the early twentieth century. Technocracy's proponents included such heavy chemical industrial combine (*zaibatsu*) leaders as Nissan's Ayukawa Yoshisuke, who promoted the idea of multilateral "public holding companies" over private corporations, and Ōkochi Masatoshi of the Physical and Chemical Research Institute (*rikagaku kenkyūjo*; hereafter Riken), proponent of the philosophy of "scientific industry," and such reform bureaucrats as Mōri Hideoto, Okumura Kiwao, and Kishi Nobusuke, who developed a conception of technology as the efficient management of the economy and society. Engineers organized themselves into the Japan Engineers' Club (*Nihon kōjin kurabu*; hereafter, Kōjin Club) in 1920—becoming the Japan Technology Association (*Nihon gijutsu kyōkai*) in 1935—and began asserting that technology formed the basis of national culture and ethics. Heavily influenced by the New Deal in the United States and Nazi economic policies in Germany, they pushed an agenda of encouraging cooperation between labor and management, improving administrative and bureaucratic efficiency, increasing engineers' involvement in national policy-making positions, and intensifying East Asia's colonization. Their leader, Miyamoto Takenosuke, who became assistant director of the powerful Cabinet Planning Board, played a key role in drafting such important plans as the 1941 Outline for a New Order of Science and Technology.[29]

After World War I and the advent of total war, the military became a locus for new notions of technology and society. "Control officers" called for the establishment of a "national defense economy" that efficiently utilized natural resources and optimized industrial production for war. In alliance with reform bureaucrats and engineers, they proposed policies to organize society based on such principles of technology as rationalization and efficiency.[30] For example, Tada Reikichi, an officer who headed the Army Science Laboratory and later the Army Technology Bureau in the 1930s, viewed science and technology as essential components of the organic Japanese "national body" that the state had to actively develop to achieve victory in the current stage of evolutionary struggle among nations. He even envisioned Japan as an organic "electronic fortress" equipped with advanced radar technologies (the state's eyes) as well as remote control and guidance systems (the state's limbs) that reacted promptly to any foreign military threat. To attain this advanced state, the government's technology bureaucracy had to be centralized and the nation's research apparatus integrated to rapidly develop the necessary technological innovations.[31]

Broader utopian conceptions of technology permeated the social sciences as well. In sociology, Matsumoto Junichirō and Hayase Toshio introduced to Japan the ideas of the technocracy movement and their importance for the New Deal in the United States, socialism in the Soviet Union, and fascism in Germany.[32] In economics, Ōkuma Nobuo emphasized the study of techniques related to the reproduction of human labor as well as material production, and Ōkochi Kazuo argued for the introduction of policies to promote private consumption as well as production. These studies crystallized into a wider discipline of the "life sciences," which helped increase the scope of state technocratic control for wartime mobilization.[33] In political science, Rōyama Masamichi defined technology as the "tactics of managing human life" and applied technology to administrative reform. Rōyama argued that through the adoption of rational management techniques in administration, technological consciousness and method would begin to hold sway in administrative conduct and eventually spread to local government and the numerous organizations governing daily life.[34]

Philosophers conceptualized technology as praxis, imagination, and creation ("subjective technology") in order to articulate new potentialities of sensation and subjectivity within modern life. From the 1920s, the philosopher Nishida Kitarō used the term "technology" as a synonym for *poiesis* or what he called "acting intuition" (*kōiteki chokkan*), which concerned the simultaneous self-formation of the subject and the formation of the world.[35] Along these lines, the philosopher Miki Kiyoshi wrote in a 1938 essay, "Technology is the act of making things. The common essence of technology is to make things, whatever they may be, whether they are tools, machines, mental and bodily forms, social systems or ideas."[36] Thus, he equated technology with the production of all areas of life. The philosopher of science Shimomura Toratarō viewed the human body as "an organism that in some fashion uses machines as its own organs" and criticized other Japanist philosophers involved in the famous 1942 Overcoming

Modernity symposium who simplistically dismissed science and technology as "Western" and therefore "inauthentically Japanese."[37] The Marxist philosopher Tosaka Jun defined technology as a dynamic "mass intelligence," comprising the innumerable skills, techniques, and practices that were not confined to the factory but emerged within modern everyday life.[38] The aesthetics philosopher Nakai Masakazu incorporated the language of cinema into his analyses of media technology's effects on subjectivity. He viewed the modern human body as "similar to a palace filled with mirrors that infinitely reflect and project various things such as light, sound and words off of each other" and articulated a notion of "technological time" of mass invention and creativity.[39] Japanese society's increasing permeation by various technologies undermined contemporary attempts by other prominent philosophers, such as Watsuji Tetsurō and Kuki Shūzō, to posit an "authentic" Japanese subjectivity supposedly untouched by "Western" modernity.

From the 1920s, technology in the form of radio, film, mass-circulating magazines, journals, and newspapers also radically transformed cultural expression and people's subjective experience. The spread of mass media technologies signified the formation of a technological culture full of new aesthetic sensations and possibilities. For many, modernity's most visible product—the machine—infused all areas of life, yet not in an alienating manner as labor unions or right-wing ideologues often proclaimed. Cultural commentators celebrated the new experiences of "speed, shock, sensation, and spectacle" that new mass media technologies embodied. The Marxist critic Hirabayashi Hatsunosuke explored the specific ways that cinema, radio, and mass detective novels brought art and culture closer to the masses, and generated possibilities for the emergence of a "people's culture" imbued with a scientific, critical attitude.[40] The film critic Imamura Shōhei wrote that the emerging documentary film aesthetic of the 1930s ("culture films") possessed "a fresh, original perception of the life of the machine, a poetic originality with regard to the machine, a new yearning for the machine."[41] The avant-garde artist Murayama Tomoyoshi argued for the necessity of "loving the beauty of the bluntly courageous machine," and his MAVO movement consciously employed the technically manufactured materials of modern industrial society in their artistic work to break down barriers between art and everyday life.[42] Miriam Silverberg examines how the emergence of mass-mediated culture in the

1920s facilitated the creation of a range of "consumer subjects," such as the café waitress, housewife, modern girl and boy, salaryman, and lumpen proletarian, who employed shifting strategies of "erotic grotesque nonsense" to challenge statist discourses of rationalization until the late 1930s.[43] Thus, the technological imaginary emerged out of a society that was already very much familiar with and even saturated by industrial and media technologies.

Although scholars have emphasized the importance of technocratic management from above in their studies on technology or "technofascism" during the wartime era, another aspect of technology frequently has been overlooked: the ability of technology to mobilize, create, innovate, and organize something new.[44] As Silverberg notes, the newly empowered "consumer subjects" of the 1920s and 1930s developed a culture characterized by "enormous energy, the urge to create, and acerbic challenges to the status quo," such as the state's rationalization and morality campaigns.[45] This mass culture of "erotic grotesque nonsense," she argues, largely died by the end of the 1930s as Japan fully turned toward militarism and totalitarianism. *Constructing East Asia* addresses the question of how statist elites attempted to control this dynamic mass culture of innovation and experimentation by explicitly appealing to technology. The "technological imaginary" signified much more than an instrumental deployment by technocrats of rational means-ends technology for social management. It also represented the formation of a new mode of power that sought to operate at the level of people's hopes and desires and direct them toward national objectives. Broadly speaking, the technological imaginary envisioned society as an organic system constituted by a whole series of economic, scientific, cultural, intellectual, and administrative technologies. According to many of these emerging visions, every member of society had a productive role in the operation of the social system, which was dedicated to constructing a "New Order in East Asia." Technology began to take on the meaning of a vast technical system similar to what Yamanouchi Yasushi has described as the "autopoiesis characteristic of organic life."[46] The incorporation and systematization of all areas of life through the technological imaginary during Japan's wartime era serves as a compelling paradigm to analyze certain fascist tendencies that have continued well into postwar Japan, and might be somewhat inherent to the process of technological modernization itself.

Technology and Japanese Imperialism

The technological imaginary was not limited to Japan proper but extended to its empire as well. When Japan acquired Taiwan as its first colony in 1895, Gotō Shinpei, Taiwan's head of civilian affairs and its governor from 1898, developed the concept of "scientific colonialism." Largely influenced by German thought on colonial administration, scientific colonialism meant a "systematic and research-oriented approach" to development, whereby Taiwan became a laboratory for a wide range of such policies as urban planning, hygiene improvement, infrastructure construction, and the introduction of civil institutions. Similar to a biological organism, the colonies had their own particular laws and practices that required systematic study by Japanese experts as the basis for "modernizing" and "civilizing" them.[47] Gotō became the first director of the South Manchurian Railway (*Minami Manshū tetsudō gaisha*; hereafter, Mantetsu) in 1906 and rose to become Japan's home minister—the ministry responsible for infrastructure development—in 1916 and 1923. Scientific colonialism inspired the first wave of technical bureaucrats, scientists, doctors, and engineers sent to Japan's formal and informal empire in Korea, Taiwan, the Guandong Leased Territories, and Mantetsu's railway zones. There they built the roads, railways, ports, and cities for Japan's commercial empire, which was then largely rooted in agricultural commodities. Although scientific colonialism led to a wide range of technological achievements, it referred more to the "useful application of knowledge, most of which was derived from Western science" as a means to justify Japanese imperial rule rather than to any kind of systematic conceptualization of technology's relationship to empire.[48] It was not until the establishment of Manchukuo in 1932 and the shift toward a militarized bloc economy that a new wave of bureaucrats and engineers emerged to articulate conceptions of an integrated technological empire.

Under the banner of "constructing East Asia" (*tōa kensetsu*), several thousand idealistic engineers flocked to Korea, Taiwan, Manchukuo, and China primarily during the 1930s to construct roads, canals, ports, dams, cities, irrigation, sewage and water works, and electrical and communications networks. Under their direction, those colonies developed the technological imaginary, or a new mode of power that envisioned society as an organic system of productive processes and mobilized citizens in close relationship to colonial planning and management. Bureaucrats and engineers were able to formulate and realize many of their grand visions in urban planning, industrial development, and comprehensive river basin management in the colonies because of the stronger centralized institutions, lack of red tape, and decreased factionalism there.[49]

The technological imaginary was not simply an intellectual construct among Japanese elites but was formulated in close relation to actual technical projects. Bureaucrats and engineers also formed their conceptions on technology by getting their hands dirty on the ground. As Daqing Yang notes, "the material means of either building Japan's empire or holding it together are still largely taken for granted rather than being thoroughly investigated."[50] Through an analysis of several specific wartime projects—river basin planning in Manchuria on the Liao River, urban planning in Beijing, regional planning on the border between Korea and Manchuria, and the construction of the Fengman (Hōman) and Sup'ung (Suihō) Dams in Manchuria and Korea—this book demonstrates how the technological imaginary also developed in constant negotiation with a variety of institutions, people, and forces in the colonies. Existing work on technology in the Japanese empire has largely affirmed the political scientist James Scott's influential discussion of massive, state-sponsored public works projects during the mid-twentieth century as examples of an authoritarian "high-modernist" ideology—a strong "self-confidence about scientific and technical progress, the expansion of production, the growing satisfaction of human needs, the mastery of nature . . . , and, above all, the rational design of social order commensurate with the scientific understanding of natural laws."[51] Whereas Scott has claimed that "monotonic schemes of centralized rationality," such as large, multipurpose dams, simplified or "straitjacketed" the human and natural worlds, and thereby suppressed the full range of variation and complexity of local interests, knowledge, and conditions to conform to its uniform visions, I argue that conflict and negotiations with such variation lay at the very heart of the formation of technocratic ideology and the engineering process.[52] As Martin Reuss has noted, "successful engineering requires more than the application of scientific rationalization" in accordance with the dictates of high-modernist ideology but also necessitates various "social inputs" from different groups and difficult interactions with unfamiliar environments.[53] Rather than eliminating social complexity and reiterating the high-modernist intellectual constructs of wartime technocratic elites, I analyze how war campaigns, state mobilization plans, environmental forces, business interests, bureaucratic conflict, resistance by local residents, and technical limitations all played a role in shaping these particular projects and larger visions of technology. More broadly, instead of focusing on Japanese imperialism's particularities—for example, its policy of assimilation in Korea and Taiwan or the ideology of emperor worship as the basis for empire—this study examines how colonial ideology and power operated through more universal tropes of comprehensive technical planning and development prevalent throughout the world at the time. By understanding how the technological imaginary operated in the colonial context, we are then also able to see important continuities with postwar Japan's influential overseas development policy and technical assistance projects in Asia.[54]

Scope of the Book

Constructing East Asia is about the relationship between technology and power within Japanese fascism and imperialism and covers some of the technological imaginary's range and depth through the examination of different social actors who had major roles in shaping wartime discourse and national policy: intellectuals, technology bureaucrats, engineers, and state planners. The analysis of several large-scale urban and regional planning, river basin control, and dam construction projects in the Japanese empire demonstrates how the technological imaginary was not merely an intellectual construct but developed extensively on the ground as well. Chapter 1, "Revolutionary Technologies of Life," introduces some of the main issues involved in theorizing technology's meaning in Japanese society as leftist intellectuals first developed its meaning from the early 1930s. Through a study of Aikawa Haruki, one of the most prominent theorists of technology, this chapter demonstrates how a view of technology as representing broader values of rationality, efficiency, and creativity not only fell within the purview of government officials and engineers who mobilized the country for war but also was enthusiastically embraced and elaborated upon by leftist intellectuals exploring technology's role in society's revolutionary transformation. Contrary to the postwar myth of leftist resistance to or unwilling complicity with the wartime state, it was Marxist social scientists who in fact shaped some of the most radical notions of mobilizing society through technology and even imagined ways to make the state's wartime mobilization programs more efficient and effective—all of which foreshadowed techniques that were later adopted in postwar Japanese society.

Chapter 2, "Technologies of Asian Development," examines the discourse around technology among state engineers who became a strong force in the 1930s. An analysis of the writings and policies of the high-level technology bureaucrat and spokesperson for the engineers' movement, Miyamoto Takenosuke, reveals how state engineers shaped technology's meanings. While leftist intellectuals outlined a view of technology as infusing all areas of life with a spirit of creativity and revolutionary transformation, technology bureaucrats asserted a more technocratic notion of technology as comprehensive social planning through such concepts as "comprehensive technology" (*sōgō gijutsu*) and "technologies for constructing Asia" (*kōa gijutsu*). Chapter 2 thus illuminates how wartime ideology worked through tropes of modernization and technical development—as opposed to backward-looking, spiritualist discourses of emperor worship or Japanese racial superiority, as conventional narratives hold.

Chapter 3, "Constructing the Continent," shifts the scene to civil engineers and urban planners in the colonial context, where they planned and constructed a wide range of infrastructure projects. Whereas the earlier chapters focus primarily on conceptions of technology among actors in Japan, this chapter examines the manifestation of those conceptions in the form of the emerging trope of "comprehensive technology" projects in Japan's wartime empire. Using the various conceptions of comprehensiveness among project engineers as its point of departure, it illustrates how three prominent technology projects were planned and implemented in relation to various competing interests, institutions, and forces in the colonies: the Liao River Improvement Project in southern Manchuria, Beijing urban planning in China, and the Dadong port coastal industrial zone along the Yalu River in eastern Manchuria on the border with Korea. Predominant conceptions of technology are analyzed in terms of their attempted embodiment in large-scale colonial projects.

Chapter 4, "Damming the Empire," examines the most prominent examples of Japanese colonial infrastructure: the construction of two of the world's largest dams, Fengman and Sup'ung. This chapter focuses on the various power relations involved in producing the technical knowledge that transformed the Songhua River region in northern Manchuria and harnessed hydropower from the Yalu River on the border between Korea and Manchukuo, as well as their effects on the colonies. In sum, I argue that it was the overlapping forms of political, legal, scientific, managerial, police, and military power in the colonies that made Japanese expertise and the discourse of modernizing East Asia through Japanese technology possible.

Chapter 5, "Designing the Social Mechanism," analyzes the thought and actions of an influential group of wartime policymakers who actively patronized the idealistic technology bureaucrats and engineers. The "reform bureaucrats" were the experts who designed Japan's managed economy policies for wartime mobilization both in the colonies and at home. By analyzing the wartime concepts and policies of Mōri Hideoto, their chief ideologue, this chapter explores how discourses of technology lent themselves to the formation of a new mode of fascist power—one that situated technology as a productive and creative rather than coercive and repressive force within people's everyday lives. An analysis of the reform bureaucrats' discourse complicates common understandings of Japanese fascism as something either dominated by a "deliberate irrationality" or as "corrupting" technology's otherwise rational and progressive nature.[55] It also reveals that wartime technocratic ideology signified something more than a rule by experts—a form of power that combined spiritualist and rational techniques to mobilize the population for total war.

The epilogue discusses some of the continuing effects of notions of technology developed during the wartime era on postwar Japan's transformation into a technological superpower. In the midst of reinventing the Japanese political system and the institutions of high-speed economic growth, "technology" again operated as social, cultural, and political mechanisms designed to incorporate people's hopes and desires both at home and abroad. Thus, the epilogue examines some of the postwar legacies of wartime discourses on technology, particularly their implications for social control and mobilization.

The analysis of the wartime technological imaginary illuminates how ideology and power operated in Japan and its empire by seeking to mobilize people through the trope of technology. Technology was actively defined as practical, transformative, and thoroughly political in working to bring about a utopian New Order in East Asia. Each group of intellectuals, engineers, and bureaucrats defined technology for their own agendas and interests, and thereby sought to radically transform society within the contours of the wartime Japanese state's overarching goals.[56] In articulating the technological imaginary, however, they rationalized a domestic and colonial fascism that instead created a dystopian order that thoroughly exploited Asia's land, resources, and people. For example, more than ten thousand Chinese and other Asians were processed as "logs" (*maruta*) and subjected to gruesome human experimentation by Unit 731 to develop Japan's biological and chemical weapons program, and hundreds of thousands more were killed in field experiments. Ishii Shirō, the scientist who headed the unit, rationalized this research in terms of the universal scientific "search for truth" and the national importance of developing advanced weapons against the enemy.[57] Several million Chinese were systematically press-ganged, classified, and processed as forced laborers for Japan's mission to "construct East Asia." Furumi Tadayuki, a Manchukuo reform bureaucrat, testified in the 1954 Chinese war crimes trial that their goal was to transform "coolies" into "machinic extensions of the Japanese Imperial Army; nonhuman automatons absolutely obedient." They then often ended up in mass graves or what the Chinese describe as "pits with ten thousand corpses" (*wan ren keng*), which were often located near Japanese industrial and technical projects, such as Fengman and Sup'ung Dams.[58] Furumi also described how reform bureaucrats, military officers, and criminal elements systematically and with a "refined rationality" created a market that plied millions of Chinese with opium, heroin, and morphine to weaken them physically and fund Japan's total war infrastructure.[59] Two hundred thousand or more Korean, Chinese, and other women largely from Asia were pressured, tricked, or forced to become "comfort women" or sexual slaves who were made to service thirty to forty Japanese soldiers a day in the name of nurturing their "fighting spirit," prevent the male's "natural" inclination to rape civilian women, and check the spread of venereal disease within their ranks. These women were often also referred to as "sanitary public toilets."[60] The brutality and exploitation in the colonies and war front fed back into Japan, where all men, women, and children were increasingly asked to sacrifice every aspect of their lives for the war effort as soldiers or factory workers. This culminated in the gruesome 1945 slogan of *ichioku gyokusai* or "the shattering of a hundred million like a beautiful jewel," whereby all Japanese were asked to become like the *kamikaze*—the young men who "volunteered" to smash their airplanes into American naval vessels to defend Japan. Contrary to its promise of empowerment and development, the technological imaginary rationalized wartime forced labor, rape, starvation, death, and resource exploitation, and arguably made them brutally systematic and efficient.

Scholarship on Japanese war crimes often focuses on the pervasive racism or authoritarian culture within the military and the rest of society in accounting for such brutality.[61] Yet we may also partially understand them in relation to the technological imaginary's attempt to completely harness the vital subjectivities and lives of people within Japan and its empire. People ultimately became expendable for the "greater good" of achieving the modern, technological "new order" throughout Asia—their lives "processed" in the various labor camps, factories, laboratories, "comfort stations," and battlefields throughout Asia.[62] We should not therefore simply dismiss the technological imaginary's transformative, participatory messages that appealed to people's hopes and desires for a better life as empty ideology, as it also created powerful modernist development initiatives and an array of institutions that called for innumerable sacrifices to be made in its name while entrenching the interests of the elites who articulated that ideology. It is this new mode of power and mobilization that survived relatively intact after the war, pushed by many of the same bureaucrats, engineers, and intellectuals as the recipe for postwar prosperity and development throughout Asia. Instead of viewing the technological imaginary as a "positive" building block for the postwar Japanese or Asian economic miracles, *Constructing East Asia* examines the wartime and colonial origins of technology as an ideological system of power that continued to have fatal effects in the postwar era.[63] As Japan has continued to invest significant resources in technology to promote postwar economic growth and social development—for example, in the formation of the "Japanese style of management"; in the formation of the "construction state" (*doken kokka*); in Japan's influential overseas development assistance programs; in postindustrial visions of the "information society" (*jōhōka shakai*); and in Japan's strong commitment to nuclear energy—the issues of power, mobilization, and social transformation through "technology" first extensively articulated during the war remain very important today. For example, in the aftermath of the recent Fukushima nuclear disaster, Japanese are questioning the state's strong commitment to nuclear energy since the 1950s, as well as the energy industry's longtime promise of clean, abundant, and safe power—both the results of an uncritical approach to technology and powerful institutions that have continued to emphasize technology's importance to Japan's national progress often without truly incorporating the people's interests.

Chapter 1

Revolutionary Technologies of Life

The Intellectual Appropriation of Technology

In his 1941 book *Ideas on Technology*, the philosopher of science Saigusa Hiroto observed that the term "technology" (*gijutsu*) had become prominent in Japan's public discourse only from the mid-1930s as the state began to mobilize engineers, technicians, and skilled workers for the "construction of Asia."[1] "Today, a new world is being *made*," he emphasized, and "it is only natural for technical people to be sought after as new things are being made in industry, economics, politics and even in ideology, literature, and the arts."[2] In this wartime context of political, economic, and cultural "construction," technology was associated more with creation and production in general rather than solely with physical machinery and artifacts. The enormous literature on technology in the public sphere and the predominance of such terms as "technological spirit," "technological culture," "technological science," and "technological mobilization" attested to the emergence of a distinct technological imaginary as well as the contested nature of the term in wartime Japanese society.[3]

Throughout Japan's 1930s and 1940s, debates over the meaning of technology raged across the political spectrum, particularly among bureaucrats, intellectuals, and engineers. On the one hand, far right-wing ideologues and politicians pushing for a Shōwa Restoration viewed technology as something that was steadily eroding Japan's spiritual vigor, as well as traditional emperor-centered values of community and agrarianism, or tried to formulate a unique "Japanese Science and Technology."[4]

On the other hand, many engineers, bureaucrats, and businessmen viewed technology's spread throughout all areas of life as a key to resolving worsening social ills, and they campaigned vigorously for the introduction of rational techniques of management and administration throughout society.[5] Along with "culture" (*bunka*) and "nation" (*minzoku*), "technology" was an important lens through which Japanese elites defined Japan's modernity during a period of total war and empire. An important characteristic of this discourse was that for many, technology was not just accepted as the "value-neutral" machines and productive mechanisms of society, but rather, technology's very nature was questioned and redefined. In fact, technology was more and more equated with the production of all of society, not only of its laws, institutions, ideologies, social organization, and economic structure but of its citizens and subjects as well. As Victor Koschmann points out, it was interpreted more and more "in performative or existential terms, as signifying certain ways of thinking, acting, or being, or even as representing certain qualitative virtues, such as rationality, creativity, or an ethic of responsibility."[6] In sum, technology became a signifier through which Japanese intellectuals worked out solutions to some of the pressing problems of capitalist modernity, such as social inequality, spiritual alienation, and the structure and future shape of Japan's economy.

Although Marxist and leftist intellectuals became largely peripheral figures during the war because of increased state repression, their role in shaping the social scientific discourse behind state policy has been widely recognized.[7] They were the first to introduce the main issues regarding technology's meaning and role in modern capitalist society to a wide range of academics, engineers, bureaucrats, and the general reading public. They launched the "Debate over the Theory of Technology" (1932–35), which centered on whether technology was primarily "objective" as opposed to "subjective." Did technology primarily consist of instrumental tools, machinery, and infrastructure, they asked, or did it also significantly involve subjective will, imagination, and ethics? The debate introduced people to the subjective, creative aspects of technology and made parallels between technology and other processes of "making" in the realms of politics, education, and the arts.[8] At stake for Marxist and leftist intellectuals, however, was the relationship between technology and social transformation—how could technology become truly integrated into people's lives as a force for revolutionary change and human development instead of existing as an external force of spiritual alienation, unemployment, and exploitation under modern capitalism? The Soviet Union's "socialist construction" campaigns to empower the proletariat through technical education and the promotion of science and technology in everyday life offered a beacon of hope to Japanese Marxists and prompted them to examine this question in terms of Japan's particular capitalist conditions.

This chapter examines the technological imaginary among Japanese leftists, primarily through the work of Aikawa Haruki (Yanami Hisao), a leading theorist of technology during the war. Although it is more understandable for bureaucrats and engineers to have supported and articulated a view of technology as representing values of productivity, rationality, and creativity given the wartime requirements of mobilization and Japan's avowed mission to "construct East Asia," it is less clear why leftists critical of the state developed similar ideas. In fact, before his arrest in 1937, Aikawa stuck to a strictly materialist definition of technology as the physical system of the means of production and analyzed its role in exacerbating the contradictions of Japan's "semifeudal" capitalism. His primary goal back then was to understand technology and capitalist reproduction in the interests of "liberating" the means of production from capitalist control, which had transformed technology into a force for alienation and exploitation. Why did such Marxists as Aikawa abandon their more limited, materialistic view of technology as the capitalist means of production in favor of a broader idea of technology as the economic, social, and cultural processes (i.e., technologies) that organized and produced all aspects of life? What was the allure of this type of thinking among leftists such that they even began articulating similar ideas put forth by wartime bureaucrats and engineers of transforming society into a type of rationalized, productive social mechanism that involved the willful, active participation of all of its members? Answering this question helps us understand the powerful appeal of the technological imaginary and the values associated with it, which continued to grow in strength during Japan's postwar high-speed economic growth era.

These questions gained particular importance in the immediate postwar era when newly empowered leftist intellectuals debated the nature of "free and democratic" subjectivity in response to what they perceived as an "immature" or "irrational" subjectivity that they argued was responsible for Japan's descent into totalitarianism and war. In the realm of theory of technology, the physicist Taketani Mitsuo and his followers put forth the well-known definition of technology as the "conscious application of rule-governedness in human (productive) praxis" in response to what they saw as "fascist" wartime theories of technology that emphasized blind subservience to values of productivity.[9] Taketani's theory, which carved out an autonomous space for human subjectivity and praxis, seemed relevant to the democratic struggles taking place among scientists in Japan's laboratories and universities, and more broadly, to Japan's rapid technological transformation in the late 1950s and 1960s. According to Taketani, the rise of a spirit of independent and rational scientific inquiry among engineers and the general populace would bring about the advancement of the productive forces, which would ultimately conflict with the outmoded, irrational relations of production, resulting in increased class struggle and ultimately socialist revolution. Yet as Koschmann notes, such a conception of technology as fundamentally rational would not necessarily prevent the opposite effect, namely, the advancing productive relations becoming the basis for the increasing technical rationalization and systematization of society and thereby the incorporation of class and social conflict.[10] By dismissing the war merely as a period of irrationality, atavism, and spiritualism, postwar leftist intellectuals missed some of the specific ways that technology could be used to mobilize "free and democratic" subjectivity, and instead celebrated what they saw as a distinctly "rational" postwar technological development.

As discussed in the book's introduction, the emerging discourse on technology during the war among bureaucrats, engineers, and intellectuals must be understood as part of a larger process of change in how power operated and was articulated in modern Japan rather than simply as a tool of an irrational totalitarian regime. As Yamanouchi Yasushi argues, during this period of intense social, political, and cultural mobilization for total war beginning in 1937, the nature of power shifted from being something wielded repressively from above to being resituated and systematized into various institutions and people within society. Borrowing a term from Talcott Parsons, Yamanouchi argues that Japan shifted from being a class society of clearly defined realms of state and society to more of a "system society," like many other modern societies at the time. Class and other types of social conflict were continuously subject to technical rules and institutionalized as part of an idealized social mechanism. Politics became more about resolving technical problems and mobilizing subjects for state goals than posing alternative, conflicting visions of society. More important, this emerging wartime "system society" continued in various forms—for example, in the system of institutions that co-opted the strong public sympathy for the antinuclear movement into support for nuclear power—during the postwar period of high-speed economic growth.[11]

In line with contemporary trends in Europe and the United States, Japanese bureaucrats, technical experts, and intellectuals from across the political spectrum played an important role in conceiving a systematized Japan during the 1930s and 1940s, specifically by extending the meaning of technology to include the production of all aspects of life. Leftists played an especially active role in shaping the technological imaginary primarily because they had already developed a large body of research dedicated to a systematic understanding of the Japanese economy and therefore easily adapted their methodology to other realms of society. During the war, Marxist intellectuals, such as Aikawa, abandoned class as the primary lens to analyze and organize society in favor of an idea of society as a complex mechanism of actively mobilized subjects and institutions that revolutionized all areas of life. In this way, they began sharing a belief with government technocrats in the transformative power of state planning combined with corporatist mass mobilization. In the same manner that technocratic thinking was easily incorporated into different national regimes and ideologies around the world, Japanese Marxists articulated a notion of a technologized wartime system society to achieve their own specific objectives of socialist modernization and revolution. But in the end, their critical vision of a systematized society rooted in the energies of the people lent itself to the state's technological imaginary of creating a more efficient social machine for wartime mobilization and empire.

Examining Aikawa's voluminous and diverse body of work on technology provides insight into the extent to which Marxist intellectuals envisioned a modern "system society" and how it might operate. His vision of a fully technologized society challenges the popular image of wartime Japan as primarily rooted in authoritarian violence, spiritualist ultranationalism, and a pervasive atavism. More important, it reveals some of the rational techniques of power and mobilization taking shape at the time in the form of the technological imaginary, which postwar leftist intellectuals overlooked within their simplistic narratives of the wartime period as "irrational" and their self-promotion as the leaders of a movement for a more rational, democratic, and prosperous Japan. Whereas wartime Marxists possessed a naïve belief in the technical systematization of society, many postwar Marxists largely ignored the issue and claimed that technology could be controlled through the development of a strong "rational subjectivity" rooted in humanism and democracy.[12] An analysis of Aikawa's notions of the technicized society, the "technological economy" in Asia, and "cultural technologies" reveals how such a free and rational subjectivity could also be mobilized for the pursuit of war, empire, and other state objectives that undermined the development of a democratic civil society.

The Origins of Theory of Technology

Japanese intellectuals were well aware of the debates on the nature of technology occurring in Europe and the United States in the early twentieth century. With the spread of the machine throughout all areas of life and the construction of large and complex technological systems in the late nineteenth and early twentieth centuries, intellectuals throughout the industrialized world began to grapple with the meaning of technology and technological development. Unlike in the nineteenth century, when the discourse was largely characterized by a romantic rejection or an enthusiastic welcoming of specific technological artifacts, the early twentieth-century debate centered on the larger project of technological development itself and the nature of life within a world saturated by technology.[13] In Germany, for example, such intellectuals as Oswald Spengler, Werner Sombart, and Max Weber and such businessmen as Walter Rathenau attempted to "assimilate" modern technology into more familiar discourses of German *Kultur* or state economic planning (*Planwirtschaft*) in order to come to terms with some of technology's harmful effects, such as devastating warfare, massive unemployment, and spiritual alienation.[14] In the United States and Europe, Taylorist ideas of technical rationality in the factory lent themselves to utopian visions of overcoming class conflict by reorganizing society along the lines of a coherent system of efficiency, optimality, and productivity. This manifested itself in various forms throughout the world, such as the technocracy movement in the United States, St. Simonianism in France, Stakhanovism in the Soviet Union, and futurism in Italy—all of which proposed differing versions of an optimal social mechanism managed by experts and run by highly skilled, creative workers dedicated to overcoming class conflict and social inequality.[15] In sum, with technology's rapid proliferation and development throughout all areas of life, commentators began to formulate new concepts to capture that experience of social transformation in accordance with their specific historical or cultural context. By actively "appropriating" technology, intellectuals incorporated it into wider "discourses of modernity," thereby shaping the modernization process itself.[16] Technology was no longer new or novel, nor was it something that could simply be isolated or romantically rejected as in the nineteenth century.

In fact, in the United States—widely viewed as the pinnacle of modern technological civilization in the early twentieth century—the term "technology" was not even widely used as a general term for artifacts, machines, and technical systems until after World War I or perhaps not even until the Great Depression, according to Leo Marx and Eric Schatzberg.[17] Other terms, such as "useful arts," "manufacturing," "industry," "invention," "applied science," and "machine," were used instead to describe what is now generally subsumed under "technology." As Germany rapidly industrialized in the late nineteenth century, a sophisticated discourse on *Technik* arose among engineers; however, it only developed into a wider debate in the early twentieth century, when engineers and intellectuals attempted to define the relationship between *Technik* and *Kultur* and understand the relationship among *Technik*, *Wirtschaft* (economy), and *Kultur*.[18] Thus, the term "technology" not only was more widely used in the early twentieth century but also began to be defined more broadly and in less material or artifactual terms. This coincided with the proliferation of large and complex technological systems throughout the industrialized world, which blurred the boundaries between the artifactual and other components, such as the "conceptual, institutional, and human."[19] In this context, intellectuals began to define "technology" in more subjective or metaphysical terms and to expand its realm into the fields of economics, administration, social policy, and culture.

Theory of Technology in Japan

The term for technology in Japanese (*gijutsu*) was a classical Chinese term used in premodern and early modern Japan. Iida Kenichi notes, for example, that in the Tokugawa Era (1603–1868), *gijutsu* meant everything that constituted a basic samurai education: manners, music, archery, horsemanship, writing, and arithmetic. With the Meiji Restoration in 1868, however, and the full-fledged introduction of Western institutions and science and technology, the philosopher Nishi Amane used *gijutsu* to translate the English term "mechanical arts" in his 1870 work *Hyakugaku renkan* (The Chain of 100 Sciences). In 1871, the Ministry of Public Works (*kōbusho*), which supervised the introduction of industrial technology from Europe and the United States and the construction of Japan's railroad and communications networks, began to use *gijutsu* in all official documents, and the proponent of Western enlightenment and learning, Fukuzawa Yukichi, actively employed the term. However, in industrial encyclopedias and dictionaries, it continued to have a strong association with "art" or "technique" rather than simply artifacts.[20]

It was only when Japan began transforming itself from a light to heavy industrial economy after World War I that *gijutsu* really began to signify material or artifactual technology in the popular imagination. Like elsewhere, however, with the rise and expansion of the metallurgical, chemical, and electrical industries; the institutionalization of science and technology in universities and research institutes; the proliferation of communications, electrical, and transport systems; and the spread of mass production after World War I, technology began to take on other new immaterial meanings, particularly among engineers who viewed their work through technology as "culturally creative." Engineers and technology bureaucrats, for example, began articulating what they called the "standpoint of technology" (*gijutsu no tachiba*) from which society should be rationally planned and coordinated. Japanese intellectuals closely followed the debates in Europe and the United States over the cultural meanings of technology and the course of technological development.[21] But it was not until 1932, in the aftermath of the Great Depression, that debates over the nature of technology really took prominence among social scientists and academics. From then on, intellectuals began defining technology as more than machines and technical systems, as also including specific ways of thinking or acting in the world or such virtues as creativity, efficiency, and social responsibility.[22]

The "Debate over the Theory of Technology" in the Marxist journal *Yuibutsuron kenkyū* (Studies on Materialism) between 1932 and 1935 set the tone for contemporary theoretical treatments of technology among intellectuals.[23] Marxism originally entered Japan in the early twentieth century as a framework to examine "social issues." Its systematic, universal, and critical character appealed to social scientists in particular.[24] Thus, the journal maintained a following among the wider intellectual community as a leading forum on dialectical materialism. The debate itself emerged against the backdrop of a broader struggle between two Marxist factions—the *Kōza-ha* (Lectures Faction) and the *Rōnō-ha* (Labor-Farmer Faction). The *Kōza-ha* focused on the so-called warped development of Japanese capitalism—Japan had a semifeudal political system centered around the emperor and an uneven economy consisting of a "highly industrial, militarist-monopolist sector in the cities . . . atop an economic foundation consisting of semi-feudal land ownership and a semi-serflike pattern of petty farming."[25] They believed that Japan had to properly go through a bourgeois democratic revolution before socialist revolution, whereas the *Rōnō-ha* argued that Japan already had the trappings of a bourgeois capitalist society.

Aikawa and others framed the theory of technology debate as a response to the various "idealist" notions of technology that arose worldwide in response to the Great Depression. For example, they noted the rise of more spiritual definitions of technology in Germany and the technocracy movement in the United States. Following the influential arguments of the Soviet delegation to the 1931 London International Congress for the History of Science and Technology, Aikawa insisted that a theory of technology must examine technology's nature under capitalism in order to overcome its contradictions.[26] At the congress, Modest Rubinstein argued that capitalist relations of production restricted the advancing forces of production, thereby making technology the source of increasing exploitation, unemployment, alienation, and devastating warfare. Under "socialist construction," however, capitalist relations would be smashed and science and technology utilized to their full potential. The antagonism between intellectual and physical labor would be eliminated, as workers were encouraged to acquire technical knowledge and actively participate in increasing productivity and innovation. Thus, "reserves of energy, initiative, and inventiveness" become opened up as each worker contributed to "the improvement of production processes, to the development of technique, and consequently, to the development of science."[27] In consultation with the proletariat, scientists and engineers would no longer serve research organs subordinated to "financial capital" but become part of a centralized network of institutions dedicated to resolving the technical tasks of socialist construction, such as mechanizing agriculture for collective farms, developing natural resources in outlying regions, and building a centralized electricity network. In sum, socialist relations of production would liberate the forces of production and labor to the point where they dialectically advanced each other.

With this Soviet vision in the debate's background, Aikawa countered such other intellectuals as Tosaka Jun, who insisted that modern technology included subjective aspects, such as skill, intellect, and invention, by asserting that technology had to be understood in a strictly historical-materialist manner as the "organization of the means of labor" within a particular historical mode of production.[28] He borrowed heavily from *Kōza-ha* leader Yamada Moritarō's work, *Analysis of Japanese Capitalism.* Yamada sought to "concretize" Marx's theory of reproduction in the second volume of *Capital* within Japanese capitalism. He believed that capitalist reproduction achieved a "coherent completion" when production in the capital-goods sector achieved an equilibrium with the consumer commodities sector. Between 1897 and 1907, Japan rapidly developed a heavy industrial, militarist-monopolist sector in conjunction with the pursuit of imperialist wars abroad, which provided natural resources, markets, and capital for its expansion.[29] At the same time, this sector developed on the basis of a hefty national land tax combined with an exploitative semifeudal land rent system and absolutism in the countryside. Japanese capitalism focused primarily on the "production of the means of production" within military heavy industries instead of simultaneously developing a strong consumer and commodity-producing sector through capitalist agriculture and light industry, which characterized a more expansive form of capitalist reproduction.[30] The countryside's semifeudal, exploitative conditions hindered the full development of the productive forces and a more skilled workforce. Thus, following Yamada, Aikawa's insistence on defining technology as the "organization of the means of labor" in the debate was an attempt to understand its role in capitalist reproduction, which would then reveal fundamental contradictions in its material structure. By precisely understanding these contradictions, Marxists would then be able to formulate proper revolutionary strategy.

Some Marxists in the debate disagreed with Aikawa's strict materialist interpretation of technology and argued for the incorporation of subjective elements in line with changes in the nature of industrial society and labor. For example, Tosaka included immaterial forms of production, such as intellectual work and technical skills, under the rubric of technology. He conceived of technology as a "dynamic, mutual transaction of subjectivity and objectivity" and argued that technology increasingly took on the form of a dynamic "mass intelligence" in contemporary capitalist society. Oka Kunio accused Aikawa of downplaying "living labor" as an essential element in the development of technology and the forces of production. For Tosaka and Oka, Aikawa's materialist definition of technology was too schematic and dogmatic, unable to account for the myriad ways that technology dynamically infused subjectivity and everyday modern life.[31] In this manner, Marxists were some of the first intellectuals to introduce technology's subjective, creative meanings into social scientific discourse, and Aikawa would develop some of these notions in his later wartime theories.

The Practical Theory of Technology: Fusing Technology with Society and Life

Aikawa's theory of technology shifted to total support for Japanese militarism at the beginning of the war with China in 1937. In June 1936, Aikawa was arrested along with thirty-two others for ostensibly attempting to reestablish the banned Japanese Communist Party. Aikawa was already quite active in communist circles during the 1930s, having participated in the "debates over the nature of Japanese capitalism" for the *Kōza-ha* and being arrested several times during various police roundups of leftist intellectuals. In his 1936 police statement, he confessed to illegal organizing and renounced all political activity, although he refused to repudiate Marxist economic theory. He also rejected his earlier materialist definition of technology. It is not clear why Aikawa subsequently converted to statism. In his 1940 work *Modern Theory of Technology*, Aikawa credited his state-organized trip to China in 1938 as awakening him to the centrality of "cultural, intellectual, and scientific construction" in building a new order in East Asia. As "construction" became a keyword in mobilizing the populace for war, the term "technology" also gained prominence as the state expanded heavy industrial and military production, mobilized engineers and skilled workers, encouraged technical education and innovation, and rationalized management and labor practices. In this context, Aikawa wrote, it was therefore necessary to clarify technology's essence and root it in concrete human "praxis" and production in line with Japan's "world-historical mission" of building a new order in East Asia.[32]

Aikawa's wartime writings on technology, however, were not a total break with his earlier socialist theory of technology. During the earlier debates with his fellow Marxists, "technology" was generally viewed either as an alienating force causing massive unemployment, destructive warfare, and spiritual desolation or as an idealist panacea for all of humanity's ills. For Aikawa and others, the Soviet Union's program of rapid, heavy industrialization and infrastructure building, mass education of workers and peasants, and promotion of science and technology throughout life provided a model for the proper integration of technology with society. Through widespread accounts of the Five-Year Plans' successes and the influential films and literature of socialist realism, the Soviet Union offered Japanese Marxists a compelling example of technological development for the proletariat. In his 1942 work *Introduction to a Theory of Technology*, he cited the need to make technology a "living concept" within everyday life in order to overcome the widespread "spiritual crisis" whereby technology was widely seen as conflicting with the human spirit.[33] Although his rationale for making technology a "living concept" was explicitly wartime mobilization at home and empire abroad, he clearly did not give up the project of revolutionizing society through technology.

For Aikawa, technology was essentially rooted in praxis and production, and it found concrete expression in "life activity" (*seikatsu*). He continued to subscribe to Marx's first thesis on Feuerbach describing subjectivity as practice or "sensuous human activity" in the world rather than as spiritual or *Homo sapiens*.[34] As practical subjects shaping their worlds, human beings immediately made technologies to realize their goals. Thus, technology was the objective means for realizing practical life goals—"life technologies" (*seikatsu gijutsu*).[35] Yet although Aikawa continued to define technology as "means," he repudiated his earlier materialist definition of technology as the "organization of the means of production" divested entirely of subjectivity. He did so by granting a particular ontological status to "means" that was in between the ideal and the real. Citing Hegel, Aikawa noted that "means" were never dead, instrumental objects but always already infused with human purpose and values. Technology represented the realization of the rational or actualized ideals; therefore, Hegel granted it an important status in his *Science of Logic*.[36] Technology was not just objective means, then, but a "living means" infused with human goals, values, and ideals. However, in line with Marx's critique of Hegelian idealism, Aikawa asserted that technology was essentially intertwined with "human, sensuous activity" in the world or praxis, out of which ideals and objectives arose. It was not simply the "realization of Spirit," as Hegel had argued.

Aikawa continued to emphasize economic production as the principal form of expression of technology's practical nature, as it formed the basis of all social relations and the mode of production. Technology always took on a specific historical form as the economic structure's "technical constitution" (*gijutsu kōsei*). The technologies of heavy industrial production based on machine systems, chemical processes, and electrical power in union with the organization of labor and the natural resources of Japan and its empire formed the technical constitution of wartime Japan's economic structure. According to Aikawa's new definition of technology, however, the technical constitution was not only objective and material but also thoroughly infused with practical value and intent. During the war, it was the national mission of wartime mobilization and constructing a new order in East Asia that provided Japanese technology with "sociality," or practical purpose and value in everyday life.[37] The overall state discourse of "construction" and social reorganization was in tune with the essence of technology as fundamentally practical, creative, and transformative. On this point of attributing state ideals to technology, his fellow Marxists, such as Oka and Tosaka, accused Aikawa of idealism or even mysticism.[38]

Aikawa, however, borrowed from earlier Marxist reflections on technology's subjective dimensions from the debate and in turn expanded his notion of a nation's "technical constitution" to include a whole array of technologies, or what he now called "practical processes," that produced and reproduced all areas of life. In his first wartime works on theory of technology, he developed a diagram of the various technologies that constituted modern society (see Figure 1). Economic processes formed the foundation and consisted of technologies for all aspects of economic life. For example, consumer technologies consisted of household consumption and processes of reproduction linked to health and hygiene technologies. Distribution technologies were made up of "financial technologies" and "management technologies," which were connected to "production technologies" in industry, mining, and agriculture. Above this foundation were political and cultural processes or technologies. "National thought technologies" were cultural and consisted of "artistic technologies," "moral technologies," and "educational technologies," for example. Political technologies were not only military technologies but also "administrative technologies" of legislating, policing, and managing health, culture, labor, and the economy. In short, society was an integrated system of technologies,

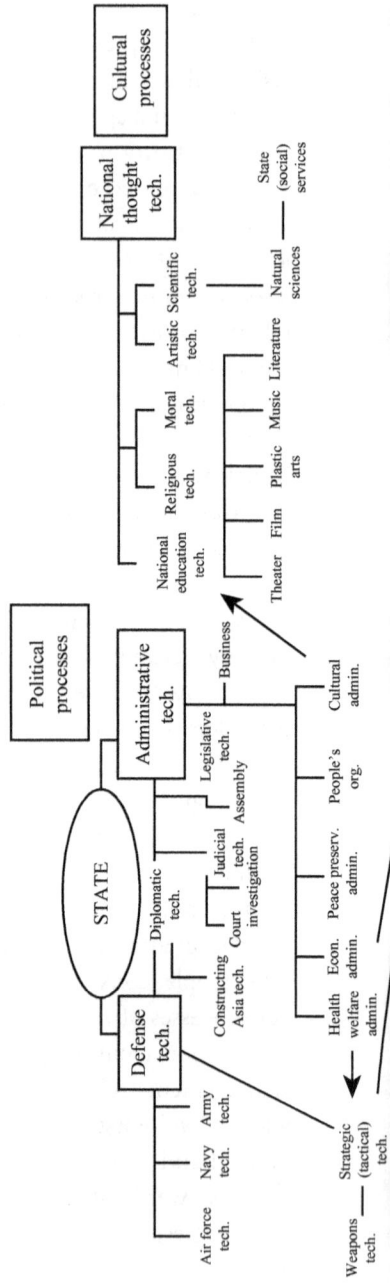

Cultural processes

National thought tech.

Political processes

Administrative tech.

STATE

Defense tech.

Scientific tech.

Artistic tech.

Natural sciences

State (social) services

Moral tech.

Religious tech.

National education tech.

Theater Film Plastic arts Music Literature

Business

Legislative tech.

Assembly

Judicial tech.

Court investigation

Diplomatic tech.

Constructing Asia tech.

Cultural admin.

People's org.

Peace preserv. admin.

Econ. admin.

Health welfare admin.

Air force tech.

Navy tech.

Army tech.

Strategic (tactical) tech.

Weapons tech.

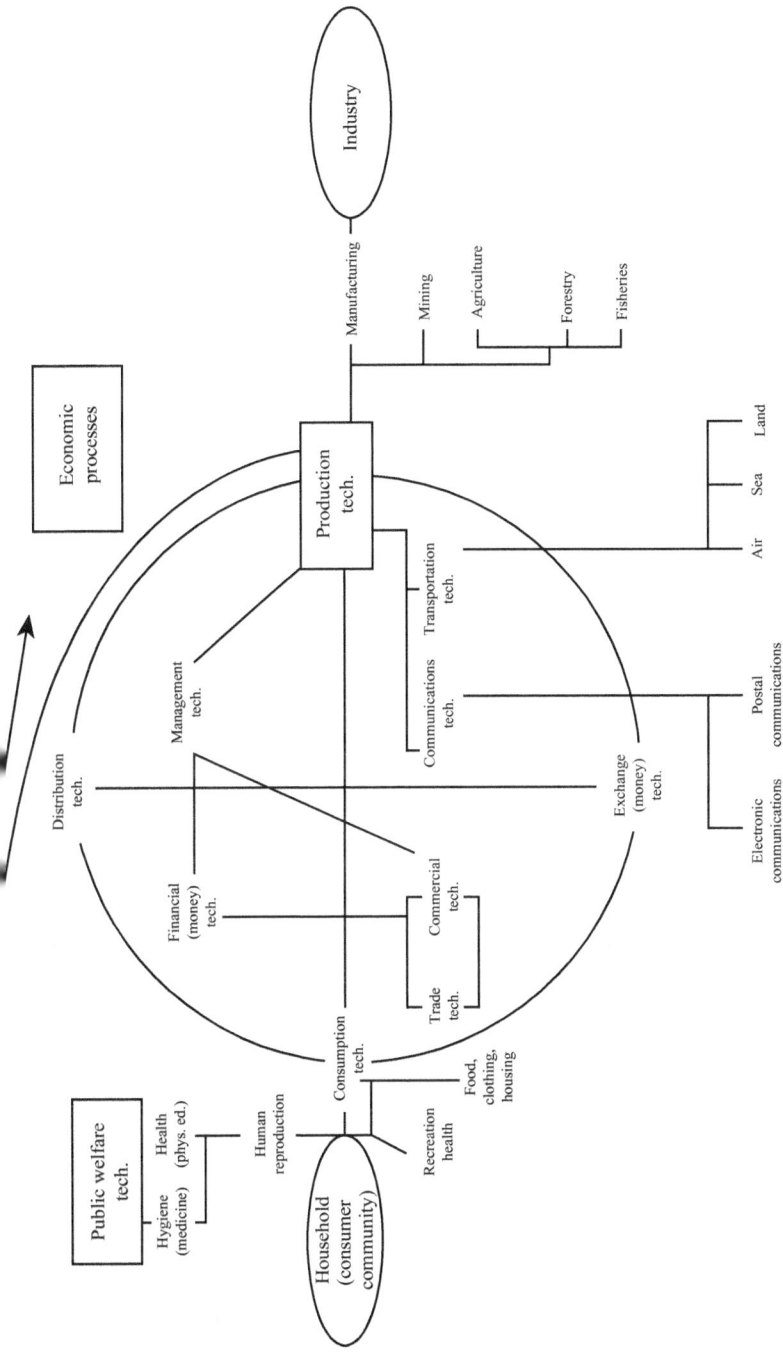

Figure 1 "The Life-Activity Levels of Technology." Source: Aikawa, *Gijutsuron nyūmon*, 85.

or "practical processes." Japan's world-historical goals of constructing the new order in East Asia provided the unifying ideals that made this a "living" system of technologies as opposed to a mere mechanism. Technology would therefore be more in tune with its fundamental nature of producing and transforming the world for human beings rather than an alienating, objective force under liberal capitalism; it would be infused with a universal ethos of resolving modernity's ills and liberating Asia from Western imperialism.[39]

According to Aikawa, a proper conceptual understanding of technology's essence as the organization of the means of production united with transformative state goals was necessary to guide Japan's cultural, economic, and political policies for both the short-term goal of winning the war and the long-term one of constructing the new order in which technology would be united with life. "Technology is the external means or complex, organization, and system of those means within activity processes that are conscious of human socio-historical goals," he wrote.[40] In this way, he attempted to root technology in more general categories of life and subjectivity rather than in the "proletariat" who seized technology and skill from the capitalists for socialist construction. The wartime state's ideals of celebrating "action," "praxis," and "construction" provided Marxists, such as Aikawa, with terms they could appropriate for their own revolutionary goals. But in the same way that other Marxist intellectuals pushed for the spread of scientific values in society, agricultural reform in Manchukuo, or the liberation of Asia from Western imperialism, Aikawa's ideas and proposals were subsumed within the Japanese state's wartime goals. By proposing new techniques for mobilization, Marxists often made their pursuit more efficient and effective, and statist elites continued to be somewhat receptive to their ideas during the war, which were expressed in various journals and roundtables for prominent think tanks.[41]

Theory and Policy: Political Technologies of Modernization and Mobilization

How did Aikawa's theory of technology as "practical processes" infused with "constructive" state goals translate into wartime policy recommendations and studies of specific problems related to science and technology? His emphasis on technology within economic production as technology's most practical, actualized form clearly privileged state priorities of increasing productivity and mobilizing the populace for war. He published several studies as proposed applications of his theory: *Theory and Policy of Technology* (1942), *Industrial Technology* (1942), *The Resources and Technology of Southeast Asia* (1944), and *Technology and Skill Management: The Shift to Mass Production* (1944). In these studies, he demonstrated how an understanding of technology's essence as equally subjective *and* objective was the key to constructing a new order in which knowledge, creativity, skill, and imagination organically fused with advanced mass production systems, complex chemical processes, and vast electrical, communications, and transportation networks. In short, Japan's "Advanced National Defense State" and "New Order in East Asia" would overcome modernity's "spiritual crisis" or the perceived conflict between technology and culture, which liberal capitalism had produced and exacerbated. The wartime new order would fulfill the Marxist goal of bringing technology and production back into the sphere of life. It would also eliminate all of the semifeudal elements within Japan's economic structure, thereby paving the way for future socialist revolution. In this way, statist discourse on technology with its overall emphasis on increasing productivity, eliminating Western-style capitalism, and constructing a new order in East Asia was even able to incorporate the hopes of those ultimately dedicated to the state's overthrow.[42]

According to Aikawa, Japan's wartime mobilization brought out technology's practical, productive nature with its emphasis on establishing a total war system through such laws as the Outline for a New Order for Science and Technology (1941), the New Order for the Economy (1940), the yearly Materials Mobilization Plans (*butsudō keikaku*) and Production Expansion Plans (*seisan kakujū keikaku*), and the Important Industries Organization Law (1941). He joined the editorial staff of the Japan Technology Association's journal, *Gijutsu hyōron* (Technology Review), in 1941, took part in study meetings of the National Policy Research Association (*kokusaku kenkyūkai*), a think tank of statist reform bureaucrats, and wrote numerous articles for such other technocratic journals as the Engineers' Leadership Association's (*kōseikai*) *Kōgyō kokusaku* (Industrial State Policy) and Riken's *Kagakushugi kōgyō* (Scientific Industry), under the title of "technology critic." These organizations were the most powerful interest groups for engineers and technology bureaucrats during the war. He was a core member of a Japan Technology Association subcommittee that submitted proposals to the cabinet for the establishment of a mass organization of engineers and technicians as part of the New Order for Science and Technology. Thus, Aikawa was very much involved in their campaign to improve the status of engineers in government, promote state policies from the "standpoint of technology," develop the colonies, expand technical rationalization in the workplace, and establish new orders to mobilize the economy and promote technical innovation.[43]

Although Aikawa lent his full support to the aims of engineers along with the overall state goals of wartime mobilization and imperial expansion, he added his own distinctive criticisms, which were designed to make state policy more efficient and effective. In line with his theory, he insisted on policies that fused technology's subjective and objective aspects—not only advanced, physical technology but also the spontaneity, creativity, and skill that shaped and operated it. His criticisms always pushed the state to further intensify the technical mobilization of all areas of life. In fact, he noted that although wartime managed economy policies in the aftermath of the Manchurian Incident of 1931 and Prime Minister Konoe Fumimaro's New Order programs (1940–41) were reforming Japan's dual economic structure of modern industrial monopolies and small, semifeudal manufacturing workshops and landlord-tenant farming, new institutions were still needed to radically transform existing structures.[44] His Marxist methodology of analyzing Japanese capitalism's "warped" nature fused with the state's program of wartime mobilization. State wartime reforms gave Marxists like Aikawa hope that Japan would finally be able to correct its semifeudal capitalist conditions and therefore advance further toward socialism.

The Outline for the Establishment of a New Order for Science and Technology (passed in May 1941) was the culmination of a widespread campaign among engineers and technology bureaucrats to realize their goal of centralizing science and technology policy, promoting research for wartime mobilization, developing colonial resources, and increasing the people's "scientific spirit." It opened with a call for the "establishment of a Japanese type of science and technology based on the autonomous resources of the Greater East Asia Co-Prosperity Sphere."[45] It formed the Technology Board (*gijutsuin*), which was designed to centralize the state's science and technology administration scattered throughout the various ministries; allocate engineers and technicians to priority industries; promote innovative research in strategic areas and enforce the sharing of patents; and encourage more scientific and technical training in schools, universities, factories, and social facilities. The outline also planned for a Science-Technology Research Organization and Science-Technology Council. The research organization was never established, whereas the council became a policy advisory organization consisting of academics, intellectuals, and bureaucrats from various priority areas, such as metallurgy, shipbuilding, and agriculture. In the end, although the government supported the outline's overall goals to improve Japan's wartime mobilization system, the New Order proposals largely failed because the military and other government ministries jealously guarded their jurisdictions and businesses remained unwilling to share their research with competitors. However, the outline did form a prototype for the centralization of postwar Japan's technology administration.[46]

Aikawa supported the state's efforts to centralize science and technology policy, create an independent, innovative "Japanese" technology based on colonial resources, and promote technical education among the populace. Yet he consistently criticized the New Order for Science and Technology for privileging scientific research in the laboratory rather than technical research at the factory level. For him, the outline was grounded in a view of technology as merely the application of scientific knowledge rather than as practical, actual production. It artificially separated technology from its real essence as production and praxis by emphasizing scientific research outside the arena of production. Technology, Aikawa argued, must be united with the people's economic life, not shut away in laboratories. Such large German heavy industrial giants as Krupp, Siemens, and Daimler-Benz best represented the proper understanding of technology as the unity of science, production, and state goals with their combination of applied innovative research, mass production, and advancement of state campaigns to increase production and efficiency.[47]

Aikawa hailed the fact that technology was being "politicized" through "political technologies," such as the New Order for Science and Technology.[48] But ultimately, the goal of technological policy was not only to win the war, he reminded everyone, but also to overcome the Western capitalist system and thereby bring technology back into a nonalienated relationship with economic life. The new order not only should focus on improving the quality and quantity of military technology but also needed to incorporate a larger, "subjective" component so that technology would truly be rooted in people's lives and constantly develop through their participation and mobilization. Such institutions as the Technology Board were only the first stage of forming a "nerve center of the entire system of science and technology" to direct technical policy and research. The next step would be to organize managers, engineers, and workers into "vocational organizations of science and technology" (*kagaku gijutsu shokunō dantai*). These factory organizations would encourage efficiency, promote invention and innovation, increase cooperative participation, and instill a sense of mission and responsibility. Only through such organizations would technology then become more than the instrumental means of production and regain its "practical nature" of being unified with human life activity. State directives would also be "operationalized" through such organizations in a way that would preserve and encourage the people's creativity, spontaneity, and innovation. The new order's final stage was to "make national life scientific" (*kokumin seikatsu no kagakuka*), whereby technical values of rationality, creativity, and efficiency as well as technical knowledge would take root in the rest of society through propagation by schools, training centers, community centers, and the media.[49] Thus, technology for Aikawa no longer meant just the organization of the means of production but a whole integrated system of political, social, and cultural technologies infused with the state's goals of wartime mobilization and the transformation of Japan into an advanced heavy industrial power that would overcome capitalism's ills, such as class conflict and spiritual alienation. Not only would the new order produce more machines, weapons, and resources but highly educated, creative, and productive citizens as well—citizens who for Aikawa and other Marxists could conceivably constitute a critical force for future revolution and socialist construction.[50]

For Aikawa, the workplace organizations at several Hitachi Manufacturing factories in Ibaraki Prefecture served as concrete models of "vocational organizations of science and technology" that would become the mobilized units on the ground for the New Order of Science and Technology. Aikawa visited several Hitachi plants in 1941 to help write the film scenario for a state-sponsored documentary, *The Present Battle* (*Konnichi no tatakai*, 1942), and published his reports for the Japan Technology Association's journal. The factories provided Aikawa with a concrete model of the type of mobilization that the new order should strive for. Management emphasized the importance of technology and technical methods of administration; innovation was promoted from top to bottom; worker creativity was respected and encouraged; and significant investment was put into research, training, and welfare facilities for workers and their families. Apparently Morikawa Kakuzō, head of the Cabinet Planning Board's Technology Section and one of the new order's drafters, visited the plants several times and viewed them as national models. For Aikawa, only the type of mobilization in Hitachi's factory that he witnessed could truly bring out technology's practical, creative nature in a way that fused organically with life.[51]

Hitachi was Japan's response to Siemens and General Electric's global dominance of heavy industrial machinery and power sources. Its fortunes turned for the better with the beginning of Japanese expansion into Manchuria and China from 1931. By 1941 it was a "living model of the New Order for Science and Technology," ballooning into a large, heavy industrial concern capitalized at 350 million yen. Hitachi was renowned for stressing "production over profit" and for being "worksite oriented" rather than "office oriented." It was an interconnected complex of manufacturing sites, research laboratories, test and training factories, schools, worker dormitories, and amusement facilities. From its production of high technology to its integrated technical organization, Hitachi represented technology's cutting edge in Japan.[52]

The central component of the "Hitachi spirit" of technology was the invention incentive system that resulted in the creation of 838 patents, 3,239 "applied ideas," and numerous world-class hydropower turbines, including several for Fengman Dam in Manchukuo (see Chapter 4). According to Aikawa, Hitachi's management and engineers had a systematic policy of guiding, encouraging, and crediting workers for their ideas and inventions. Over thirty ideas and designs per month were announced in the factory bulletin, and three hundred invention prizes and ten thousand letters of recognition were awarded each year. Cash bonuses were given to workers for inventions, with further bonuses awarded after three years, if the invention maintained profitability. Inventions were not necessarily the product of scientific education and genius, Aikawa noted, but arose through everyday interaction with technology as well as worker cooperation on the factory floor. Examples of inventions included an automatic hammer installed from the ceiling for hammering hundreds of thin metal sheets together and a special device for the planer to finish the precise wings of a turbine's water wheel. Creativity and innovation extended to finding ways to eliminate factory waste. The invention incentive system's introduction encouraged the formation of a "progressive," cooperative culture of technology that respected and appreciated worker creativity and action. In fact, Aikawa argued, such a system brought out technology's true spirit as practical, imaginative, and social. The technical division of labor no longer determined worker organization as in capitalism; instead, worker mobilization brought out a more dynamic, nonalienating sense of technology as a "living means" that was always in a process of development in relation to the factory workers and state goals. Such a technical organization of work was more effective than "bureaucratism" and imposing a "police system" from above.[53]

Aikawa was also impressed by the workplace assemblies (*shokuba jōkai*), which were the nucleus of Hitachi's organizational system and their inculcation of a "technological spirit" of rationality, creativity, and efficiency. He studied the assembly newsletters and described many worker initiatives to improve efficiency, productivity, and factory life in general. After several months of these meetings, workers apparently began to "spontaneously" look for waste of materials, time, and work. They set monthly waste elimination targets. They also set objectives for reducing the number of defective products and rewarded workers for attaining them. All kinds of innovations, such as improved machine tools, work methods and designs; successful technology transfers; and the discovery of mistakes, received some form of award or recognition. In sum, the management technique of introducing worker assemblies had the desired effect of increasing worker responsibility and cooperation without the necessity of resorting to top-down, autocratic measures.[54]

Such labor management techniques, Aikawa argued, had to be combined with other changes in the larger "technical constitution" and industrial structure outside the factory. Especially with the beginning of war with the United States in 1941, Japan had to more thoroughly transform its economy into a mass production system. Wage systems had to be modernized from contract labor systems based on paying labor bosses (*oyakata*) for temporary workers or payments of living allowances arbitrarily determined by management. The Andō-style management system, which paid workers an hourly wage plus a bonus for beating production targets based on a fixed formula, represented a system that provided a living wage yet encouraged productivity.[55] Old-fashioned, all-purpose skilled workers had to be trained in more advanced, specialized skills in line with the technical demands of aircraft and machinery production. The "keyman" or "cadre" (*kikankō*) who not only possessed advanced technical skill but also understood the entire industrial process represented the new type of worker for the age of mass production. Knowledge, intuition, skill, and technique would replace mechanical, physical labor. As engineers left their planning rooms and came down to the factory floor to supervise and reorganize the assembly line system, the class barrier between the new "management workers" and "production technology intelligentsia" would disappear. The state needed to recognize the inseparable relationship between skill, knowledge, and technology—only then would technology realize its practical, creative essence and form the basis for a new Japanese mass production system that would ultimately transform the entire society, eliminating the older, inefficient elements.[56]

The physical industrial structure also had to be transformed in order for life to be truly integrated with technology during the war. The New Order for Science and Technology needed to be closely related to the New Order for the Economy (1940), which organized key industries into control associations (*tōseikai*) managed by business to plan production, separated capital from management, reorganized small factories and subcontractors, and introduced production incentive systems. Firms should be organized not only into control associations based on such traditional industrial categories as steel, machine tools, textiles, and coal but also in accordance with the rising chemical industries that fused the latest machine, electrical, and chemical technologies to produce such wartime essentials as fertilizers, metal alloys, pharmaceuticals, synthetic fuels and materials, dyes, and explosives. Huge chemical trusts, such as IG Farben in Germany, DuPont in the United States, and ICI in Great Britain, represented the future of industrial organization with their foundation in advanced technology and their formation of research and production synergies. These trusts produced everything from basic chemical compounds, such as sulfuric acid and soda, to more complex products, such as synthetic fuels and alloys.[57] In Japan, the new zaibatsu, such as Ōkochi Masatoshi's Riken, Ayukawa Yoshisuke's Nissan, Nakano Tomonori's Nihon Sōda, and Noguchi Jun's Nihon Chisso, represented the opening steps toward establishing such megatrusts and away from the heavy industrial giant model of Mitsubishi and Mitsui based on such traditional industries as mining, shipping, and textiles.[58]

Finally, technical innovation needed to be encouraged among businesses through such measures as patent sharing, cooperative research, and joint production. In order to improve Japan's overall technological standard during the Allied economic blockade, patent sharing had to go beyond the superficial level of observing each other's production processes and joint research exchanges and carry out joint research projects, such as the Steel Control Association's organization of a research group and construction of a test smelter to process pig iron.[59] Thus, in line with his theory of technology, Aikawa called for measures to develop industrial technology in coordination with efforts to encourage technology's more subjective factors through programs to mobilize creativity, participation, and productivity. The key to technological development was to revolutionize all elements of the "technical constitution" in an integrated manner: natural resources, means of production, and labor.

The 1930s and 1940s saw a flurry of laws that promoted various aspects of technology for the war effort as well as the rise of a discourse of "constructing" a new order in Japan and East Asia. The state's emphasis on technology led to a number of discussions among engineers, intellectuals, government officials, and businessmen, whereby each group invested the increasingly prominent term "technology" with their own hopes and objectives. Marxists, such as Aikawa, who were dedicated to Japan's revolutionary transformation were no exception to this increasing engagement with technology. In Aikawa's proposals to intensify the state's wartime mobilization system through bringing out technology's "practical" sense among the people, there was a strong continuity with his earlier Marxist agenda of modernizing what he saw as Japan's persistent semifeudal economy. In fact, as noted previously, he frequently discussed the Japanese economy's "warped development" in his wartime proposals. For example, he praised the 1942 Production Expansion Plan's pursuit of "technological rationalization." The state's rational planning of resources, labor, and technology would correct Japan's inefficient industrial structure of small subcontractors who did not employ the latest technologies and still relied on physical labor power by weeding them out, transforming them into producers of priority industrial materials, and introducing newer, more standardized technologies and Fordist mass production techniques.[60] Agricultural mechanization and cooperativization more suited to Japan's geographical conditions would increase productivity and correct the semifeudal, labor-intensive, and small-scale nature of Japanese agriculture.[61] In this way, the state's discourse on technology mobilized Marxist hopes because it promised to fully modernize "semifeudal" Japan and eliminate capitalism's inequalities and ills. Technology was absolutely central to the state's wartime plans, not only for material reasons of improving Japan's productivity but also for spiritual ones of transforming its citizens into active, mobilized subjects dedicated to building a new order in East Asia. Although Aikawa tried to appropriate the state's wartime discourse on technology for "socialist construction," in the end his proposals to dynamically integrate technology into people's lives merely strengthened that discourse and promised to make Japan's wartime mobilization system all the more efficient and effective. We see a similar dynamic with regard to Aikawa's proposals for Japan's expanding empire.

The Technological Economy of Greater East Asia

For Aikawa, the war was more than just about modernizing Japan's "semifeudal" economy through establishing a managed economy and popular mobilization. It also aimed at "constructing" a new order in East Asia free of Western imperialism. Similar to the technology bureaucrats and military officers discussed in later chapters, Aikawa envisioned the incorporation of the colonies into a "technological economy" with Japan as the heavy industrial core and the colonies as centers for natural resource extraction, processing, and light manufacturing. Yet in his study for Mantetsu's East Asian Economic Investigation Bureau (*tōa keizai chōsakyoku*), titled *Technology and Southeast Asian Resources*, he provided a detailed blueprint for colonial industrialization that closely mirrored Marxist theories of economic development.[62] As Japanese troops expanded across Southeast Asia in 1941, the military and Cabinet Planning Board intensified their studies of its strategic natural resources, such as tin, rubber, nickel, and oil.[63] In this context, Aikawa's study put forth plans for rapid modernization, development, and mobilization that were consistent with his Marxist program of correcting capitalism's "warped development" in the colonized world in preparation for future nationalist and socialist revolution. Once again, although Aikawa sought to appropriate the state's technocratic discourse of the colonies as an "untapped" natural resource to be developed by a "superior" Japanese science and technology, in the end his proposals provided a vision of Japan's empire as an optimal, efficient extractive machine that exceeded the plans of Japan's leaders.

In Southeast Asia, whose heavy industrialization was obstructed by Western colonialism's singular focus on resource extraction, Japan had to first focus on building the economic base or introducing "production technologies" of heavy manufacturing, resource exploration and excavation, power production, and labor management, according to Aikawa. As in his proposals to revolutionize Japan's "technical constitution," production technologies combined three elements that represented technology's subjective and objective components: (1) the material means of production, (2) technical skills and engineering capability, and (3) natural resources. Japan's construction of a "comprehensive technological planning system" in Southeast Asia had to address all three elements in order to be comprehensive—manufacturing machinery and facilities, technical training and expertise, and quality and type of natural resources. The state of all three components determined the precise organization of production technology in Southeast Asia. As in his analysis of the technologies that constituted Japanese society, production technology formed the material foundation for such other technologies as political, social, and cultural technologies—the totality of which would be infused with the larger sociohistoric goals of meeting Asia's (i.e., Japan's) immediate war needs and building a productivist, self-sufficient "Asia for Asians" liberated from Western imperialism. Japan would be the external force that put Southeast Asian nations on the "proper" course toward independence and modernization. In this manner, Aikawa ultimately supported the formation of an "East Asian League" of independent states working together to construct a common "East Asian cultural sphere."[64]

Aikawa's study primarily focused on the contemporary state of Southeast Asia's technical infrastructure and natural resource production (factors 1 and 3 above) in the late 1930s under Western colonialism. He left the question of labor management to a future work, simply providing some general guidelines that included planning labor migration, studying the occupational tendencies and "physical, spiritual, and intellectual" capabilities of each ethnicity, and introducing technical training facilities for workers in close relation to local economic conditions. The foundation of the Southeast Asian technological economy would be the development of its rich natural resources, which had been essential to fueling the West's heavy industrialization rather than Southeast Asia's. Specifically, the location of ore processing facilities, the condition of transport networks between processing centers and mines, and the location of carbon-based and hydropower sources had to be closely studied and developed. Although Western colonial powers focused on resource extraction and power production, they did not invest very heavily in local processing and natural resource refining. Most of the processing was done in heavy chemical factories and refineries in the European metropole, with low-tech assembly and light processing conducted in the colonies. Japan, however, would focus on building the "technological system" or infrastructure necessary for further developing natural resources and processing them into higher-value industrial materials or goods—very similar to its Manchukuo industrialization programs occurring at the time. Providing for Japan's total war needs and Asian "industrial self-sufficiency" or "liberation" would be the twin ideological pillars of East Asia's new technological economy. Aikawa's study thus countered the Japanese leadership's "Greater East Asia Co-Prosperity Sphere" proposals to make Southeast Asia primarily into a provider of natural resources while concentrating industry and processing in Japan (including Taiwan and Korea) and Manchukuo.[65]

Much of the study used publicly available statistics to analyze the existing colonial economic foundation on top of which Japan would help establish an industrialized Southeast Asia. Aikawa began by examining three economic "pyramids" for tin, rubber, and oil—three key natural resources for the colonial powers. Castigating studies that did not take a "comprehensive" viewpoint or use precise statistics, he analyzed each resource's location, production figures, and quality; the technologies used for excavation, planting, and refining; the existing processing and manufacturing facilities; the transportation networks and electricity facilities; and finally, the technical infrastructure of Singapore, which was at each pyramid's apex because it was the processing and re-export center for most Southeast Asian natural resources bound for the West.

For example, the "rubber pyramid" constituted a typical colonial "trade pyramid" with its focus on maximizing raw rubber production through the intense exploitation of physical labor in huge estates, little investment in high-tech processing facilities or production techniques, and final reexport to the metropole. In contrast, Japan needed to focus on introducing high-tech production facilities (building on already existing facilities, such as Goodyear's plant in the East Indies), developing necessary infrastructure, and sending experts trained in tropical agricultural and chemical-mechanical processing as well as providing knowledge of "management" and "administrative" technologies to "efficiently utilize" hundreds of Southeast Asian workers.[66] In this way, the Southeast Asian rubber pyramid would be transformed from a "trade pyramid" geared toward colonial extraction into a "production pyramid" oriented toward "national defense" and self-sufficient, heavy industrial development.

The "tin pyramid," on the other hand—centered on Great Britain's development of Malay's tin resources in the Kinta River Valley region—was closest to what Japan should be doing for Southeast Asian development. Because of the relative ease of processing the high-quality tin and overall profitability, Britain introduced dredging technology; advanced tin smelting plants and reverberating furnaces; constructed the Chenderoh hydropower plant; and built transport and port facilities at Penang. Instead of raw ore, high-quality processed tin was produced and exported all over the world. Japan needed to build on this model; however, unlike Great Britain, it would promote more extensive hydropower over carbon-based power and concentrate more on building industrial cities than on building commercial ones. This focus on hydropower was based on Japan's dam-building experience in East Asia (see Chapter 4). According to Aikawa, Western experts were often pessimistic about Southeast Asian industrialization because of its lack of quality coking coal resources for steel production, which formed the basis of Western industrial civilization. Rather than following the so-called Western model of industrialization through steam power followed by electric power, however, Japan would immediately bring about industrialization through large-scale hydropower projects along the lines of the Tennessee Valley Authority (TVA), the Dnieprostroy dam project in the Soviet Union, and Japan's heavy industrial dam projects in Manchukuo. East and Southeast Asia would largely skip the steam era and pursue an "electric power–led form of industrialization" based on the development of what Lewis Mumford called the twin pillars of neotechnic civilization—electricity and alloy metal production.[67]

Inspired by the above-mentioned comprehensive development projects, Aikawa stated that the main priority of long-term development in Southeast Asia was to "organically" unify resource development with on-site local manufacturing and processing. A "three-dimensional" industrial plan in which resources were sent to large urban manufacturing centers and factories were built near the rural resource areas was essential to constructing an expanding "technological economy." This was opposed to a "two-dimensional plan," which merely connected resource areas to urban centers. A three-dimensional plan balanced "centripetal" forces of resources destined for heavy industrial centers with "centrifugal" forces of developing processing plants in rural resource areas. Along these lines, Aikawa targeted six different resource groups to be linked to electrification and expanding industrial development: iron, bauxite, tin, coal, chrome and nickel, and manganese and antimony. For each resource, he examined the conditions of specific resource areas and their potential to be developed into expanding industrialization zones.[68]

Aikawa provided a rough map of the different industrialization zones centered on particular natural resources, energy sources, and urban centers (see Figure 2). Aside from the Kinta River valley tin mining area, he examined the technological conditions for bauxite development in northern Sumatra around the Asahan River and the city of Medan; coal development in the Tonkin Delta around Hanoi and Haiphong in Indochina; copper, zinc, and lead development in the Baldwin mine region around Mandalay in Burma; chrome development in the Zambales region in the Philippines near Manila; and nickel development around Laguna Lake, also near Manila, among other areas. Each region would be an integrated complex of mines, transportation networks, processing and manufacturing industries, power stations, and urban centers. In addition to analyzing the existing conditions, he detailed the developmental difficulties that Japan would have to overcome—the lack of good transportation infrastructure, the high expense of labor, the scarcity of essential natural resources nearby that were necessary for advanced manufacturing, the low level of mechanization and technical facilities, and insufficient power supplies, among other things. Japan would build on the existing resource areas, power stations, manufacturing facilities, and the West's failed development plans; however, as Western colonial powers were largely trapped by an old-fashioned colonial mindset centered on resource extraction and steam power, Japan was on the verge of building a neotechnic order based on comprehensive zones of heavy industrial production and more efficient hydropower.[69]

At the center of each industrial region would be large, multiethnic "industrial cities." Aikawa analyzed the characteristics of thirty-five main cities with populations of fifty thousand or more. Southeast Asian cities had a "colonial character" as political and commercial centers for facilitating the reexport of raw materials—development focused primarily on military, transport, and commercial facilities rather than industrial ones. Urban populations largely consisted of migrant laborer or trader populations, whose lives were based on monocultural agriculture and mining, a

Figure 2 "Outline of Southeast Asia Industrialization Regions." The dotted circles indicate the industrialization zones centered on specific natural resources and major cities (marked by small circles). The various forms of cross-hatching and lines indicate the locations of important natural resources such as oil, iron ore, tin, and coal (other natural resources are indicated by their chemical symbols). "X" designates planned power projects. Source: Aikawa, *Tōna Ajia no shigen to gijutsu*, 412.

dual economy of semifeudal rural areas and modern cities, trade and com-merce, and light manufacturing. Japan's goal was to transform these colo-nial cities into productive metropolitan areas in line with the comprehen-sive industrial zones outlined above. Laboring and trader populations would become working-class populations. Aikawa divided colonial cities into three types and examined their existing technical infrastructure: ag-ricultural cities were centers for agricultural products and had such basic processing facilities as sugar refineries or rubber processing plants, as well as small electricity plants (e.g., Mandalay, Bandung, and Saigon); the metallurgical cities were centers for natural ores and had such facilities as oil processing, simple chemical and building materials factories, and large power stations (e.g., Palembang, Padang, and Hanoi); and comprehensive cities combined both types and included large ports and industrial plants (e.g., Singapore, Jakarta, and Surabaya). Japan's goal would be to develop cities as "mediating" centers for the development, processing, and transport of the natural resources within each industrialization zone. For example, Medan would be the center for the development of the Eastern Sumatra /Malacca Straits bauxite-based industrialization zone; Manila for its non-ferrous metals and alloys-based industrialization zone; Penang and Kuala Lumpur for their tin- and rubber-based industrialization zones. Thus, re-gional industrial development would be linked to urbanization within the new Greater East Asia productivist order centered on Japan.[70]

Southeast Asian industrialization was to be based on a network of massive hydropower development projects and an intensified mobilization of labor, which Aikawa called the "Asiatic Energy System." Whereas the Western colonial powers largely relied on wasteful coal and oil resources and technology for their extractive resource economies and commercial form of urbanization, Japan would build on and develop dams, for ex-ample, on the Asahan River in Sumatra, the Larona River in Sulawesi, and Lanao Lake in Mindanao, to fuel heavy industrialization and urban-ization. He was particularly influenced by German *Energiewirtschaft* the-ory, which believed in comprehensively recording all of a nation's energy resources in order to maintain a "developmental balance" necessary to create an expanding industrial economy.[71] Extending this, he even claimed that "Western" energy theories were reductive in that they did not take into account human energy, which was at the root of all other energy forms. "The national powers of one billion East Asians signify an enormous im-mediate energy reserve, and a massive storehouse of igniting and burning blood gasoline," he wrote.[72] The Asian Energy System must harness not only such material energy sources as oil, coal, wood, and water but also human biomaterial power. In pure technocratic fashion, he even conducted a brief comparison of caloric intake and food prices between Asian and Western countries in order to determine the "food structure of human energy."[73]

Thus, Aikawa's blueprint presented an integrated system of strategic natural resources, regional processing and manufacturing plants, large-scale hydropower projects, industrial cities, and optimally mobilized labor populations all geared toward maximum productivity. His proposals complicate the conventional view of Japanese imperialist ideology, whereby the colonies were largely seen as providers of resources for the more advanced Japanese metropole. Through Aikawa's study, we see how Japanese leftists were attracted to and actively appropriated the state's imperialist discourse of "Asian construction" to emphasize development, modernization, and technical values of systematization and connectivity in order to correct the "deformities" of Western colonialism, thereby implicitly preparing the structural conditions for socialist revolution. Ultimately, this strengthened the "technologies of Asian construction" discourse put forth by state engineers to develop comprehensive infrastructures on the Asian continent (see Chapters 3 and 4). With its detailed plans to establish interconnected comprehensive industrial zones centered on strategic natural resources and hydropower facilities (as well as Asian "blood" power), Aikawa's study in fact made their plans much more intensive and integrated, ultimately justifying the total exploitation of Asia's life and resources in the name of "liberation" and "Asian construction." In this way, state technocratic visions of an optimal, rationalized colonial system guided by technical elites (as opposed to "irrational" military officers or law bureaucrats) incorporated the socialist dream of eliminating uneven development and constructing a proletarian society.

Cultural Technologies of Film

Aikawa's work on technology went beyond the spheres of economic production and policy on which state bureaucrats and engineers primarily focused their attention. He also wrote extensively on film and the mass media, or what he referred to as "cultural technologies." Mirroring other contemporary technology theorists, such as Lewis Mumford and Werner

Sombart, he noted that Japan had become a "culture of electricity" during the 1920s. Electrification had "fundamentally reorganized the veins and nerves of private and state monopolies."[74] Industries expanded into rural areas, the "chemical revolution" enabled the mass production of synthetic materials, transportation networks brought regions into closer contact with each other, and wired and wireless communications technologies brought news and images to Japan from around the world within the newly emerging mass media of radio, newspapers, magazines, and cinema.[75] For Aikawa and such other leftist intellectuals writing about technology as Nakai Masakazu, Tosaka Jun, and Miki Kiyoshi, mass media brought about changes in the very nature of human sensation—a shift from the age of individual subjectivity to mass subjectivity. Mass media technologies problematized the common perception of technology as something alienated from human life or as "external means," as they were inextricably tied up with the people's subjective makeup. Once again, Aikawa delved more deeply into technology's nature than state engineers or bureaucrats did by rooting it at the level of mass subjectivity and culture. He intensified and expanded government statements, similar to his prescriptions on state technology and imperial policy, that promoted science and technology for the wartime effort.

The main characteristic of contemporary technological culture, Aikawa noted, was the predominance of "reproductive art" (*fukusei geijutsu*). Radio culture had replaced the culture of prose and poetry, film had overtaken theater, photography had challenged painting, and the record had replaced the live symphony orchestra. However, similar to his German contemporary Walter Benjamin, instead of romantically lamenting the decline of classical culture and the replacement of "originality" with an inferior "reproduction," Aikawa praised the "originality" of the "new genres of mechanical reproductive art."[76] As other Marxist film critics, such as Gonda Yasunosuke and Imamura Taihei, had recognized, these new genres possessed a "popular mass nature" and a "circulatory character that can bring art closer to the mass psyche."[77] Mechanical reproductive art emphasized artistic form more than content, which in turn created new possibilities for content. Aikawa bemoaned the fact that many artists were unable to take advantage of these technologies to generate new art and were stuck in their old romantic, individualistic ways. Art was not "genius" or spiritual creativity but the employment of specific methods, processes, and techniques of mass "expressive technologies" (*hyōgen gijutsu*) with a "multitude of rich forms," such as film, radio, recordings, and pho-tography.[78] "New *sake* for a new flask," Aikawa implored all artists.[79] For him and other Marxists, reproductive art did not degrade artistic content (or represent "false" capitalist ideology) but enriched and multiplied its possibilities.

Film was a prime example of mechanical reproductive art's new po-tential. "Not only has film produced a form with mass transmissibility, it has begun to develop a dynamic potential to express the sensuous content of a new age," Aikawa wrote.[80] Film production itself embodied the me-dium's mass nature. It was an immense production process that synthe-sized "literary (scenario), theatrical and aesthetic (filming), and musical (recording) elements for reproduction onto the screen through movie pro-jectors and sound mechanisms."[81] Innovations in artistic form made pos-sible by film also generated new content. For example, films like Charlie Chaplin's *Modern Times* (1936) did not go beyond a romantic caricatur-ing of industrial rationalization and technology as dehumanizing and therefore suppressive of creativity, but animations like *Popeye* overcame the old "technology vs. art" dichotomy to instead create new technologi-cal sensations.[82] Aikawa wrote of *Popeye*:

Among Popeye's unusual powers are a winding motor, the destructive force of a can-nonball, and a heart like a continuous electric engine. His astounding activity repre-sents a burst of mechanical energy, and the music emits a metallic dissonance. Why does the conception of Popeye completely match the dynamic optical technology that supports its cinematic form? Moreover, why is it strangely accompanied by a sense of reality and freshness? Perhaps if this metallic and optical sensation aesthetically unites with real image content, film's aesthetic character is further developed.[83]

Classical art forms, such as music and poetry, that assumed a split be-tween culture and technology could not yet achieve such an aesthetic fu-sion of technology and life, which Aikawa described as having a "fresh modern sensation." Instead of spiritual alienation, he felt a potential dy-namism, power, and life in media technology's saturation of society. Their mass transmissibility and ability to stimulate new sensations should be actively embraced and developed by artists rather than rejected for a ro-manticized past.[84]

As noted earlier, Aikawa viewed the war as the primary means to establish a Japan that would eliminate "feudal remnants" and usher in a new order in which technology formed an integral part of society and life rather than being a force of alienation and instrumentality. In the cultural realm, his support for the war manifested itself in his activities as a theorist and scriptwriter of documentary or "culture films" (*bunka eiga*). Wartime filmmakers and critics regarded the culture film as the period's preeminent aesthetic form. Even theatrical or fiction films (*geki eiga*) began to take on a realistic documentary style and quality. Documentary films also incorporated theatrical forms and techniques, creating the culture film genre. This genre blurring was the result of a number of factors. The 1939 Film Law, which established the "New Order for Cinema," created an inspection system and enforced the showing of a minimum of 250 meters of nonfiction film in any movie program. Documentaries also often escaped censorship laws because they often did not have scenarios, thereby making their production much easier. State inspection fees were even waived for nonfiction films, which further encouraged their production. Such state promotion of culture films led to an explosion of them, with the Ministry of Education approving 4,460 documentaries in 1940 alone. As a result, documentary film styles quickly made themselves felt throughout the wartime cinematic world. The historical exigencies of the war and battlefront, audience popularity, and increased attention from film critics contributed to the spread of the documentary film form and the resulting preeminence of the documentary aesthetic.[85]

The term "culture film" originated from the import of *Kulturfilme* produced by the German film giant UFA beginning in 1930. Culture films were originally films describing the achievements and processes of modern science and technology.[86] The Ministry of Education, however, soon began using the term "culture film" to refer to all documentary films produced in Japan.[87] With the war's beginning, culture films took on more nationalistic overtones, and the term designated films that served to "enlighten," "modernize," and mobilize the people for war and empire. Films ranged from describing the spiritual vigor of elite air force pilot trainees (Young Soldiers of the Sky, *Sora no shōnenhei*, 1942) to portraying the rural poverty and cooperative spirit of rural improvement among peasants in northern Japan (Snow Country, *Yukiguni*, 1939) to capturing the lives of railway workers (Locomotive C-57, *Kikansha C57*, 1941).[88] The term "culture" in "culture film" had multiple meanings depending on the particular film. Almost all culture films had nationalist overtones of valuing discipline, cooperation, self-sacrifice, and hard work. Many films celebrated the trappings of modern technology—the factory, the machine, management techniques, and innovation. Many also sought to correct the people's "superstitious" customs and beliefs. For Aikawa, the culture film was at the vanguard of representing all aspects of modern technological society and of mobilizing people to construct a new "technological culture" throughout Japan and East Asia.

The Theory of the Culture Film (1944) was primarily a collection of Aikawa's essays on documentary film during the 1940s with some additional material and revisions. It was the culmination of his experiences writing and studying documentary film at the Art Film Company (*geijutsu eigasha*), a haven for leftist filmmakers. Along with the film critic Imamura Taihei's book *Theory of the Documentary Film*, it was one of the few full-length theoretical treatments of the documentary in prewar and wartime Japan.[89] Like many film commentators, Aikawa admired the documentary's scientific and high-tech quality. Film in general had a "particular technological structure." "Films are possible worlds conditioned by creative and projective mechanisms that are the integrated product of physical, chemical, and electrical applications such as modern mechanics, optics, photo-chemistry, acoustics, and so on," he wrote.[90] Unlike classical artistic production, which was primarily individual and artisan-like, film production had a "modern manufacturing structure."[91]

Aside from documentary film production's "technological structure," Aikawa admired culture films for their "principal standpoint of discovering rational constructions within social and natural scientific reality." In their very structure and organization, culture films "bring out knowledge against feeling, concepts against intuition, documentation against imagination, fact against fiction." Thus, they represented progress over theatrical films, which primarily emphasized feeling, intuition, imagination, and fiction. The scientific elements within the film's "technological system . . . awaken the scientific elements asleep within the producer's subject matter, and permeate the film's planning, construction, production, and in the end, even the way the film perceives, understands, and represents objects."[92] Science and technology therefore thoroughly infused the subjectivities and actions of those cooperating in the documentary's production. As such, culture films constituted a vital supplement to the state's wartime campaigns to generate a scientific spirit.

In line with his theory of technology, which united technology's subjective and objective components, Aikawa asserted that the culture film's scientific qualities, which emerged out of cinematic technologies and the cinematic form of production, were inseparable from its more emotive, aesthetic qualities. The image itself was a subjectively created "form of expression" and inevitably involved some aesthetic element; however, the subjective, aesthetic factors also inhered in the objective cinematic technology itself. Different cameras, lenses, sound equipment, lighting technology, and shooting techniques had their own aesthetic qualities, and any improvement in them could transform the film's overall qualities.[93] In sum, for Aikawa, the wartime documentary film's "cultural technologies" constituted the fusion of rationality with sensation, scientificity with aesthetics. Culture films were a "living means" of cultural production intertwined with subjectivity rather than external objects to be stoically viewed from a distance or produced in a purely instrumental manner.

Aikawa and others called the culture film's technological aesthetic "neo-realism." He was inspired by Soviet avant-garde filmmakers like Sergei Eisenstein and Fridrikh Ermler, and this was evident in his criticisms of "documentarianism"—the idea that the camera should only objectively portray reality—and his emphasis on "the spirit of flow, the spirit of process, and the spirit of montage" as cinema's essence. Because the culture film's primary goal was not just to record reality but also to bring out national ideals latent within everyday life, it had to utilize its very own techno-aesthetic or "constitutive power" (kōseiryoku) of continuity and flow through such various techniques as editing, camerawork, acting, music, and so on. In this way, the culture film fused the objectively oriented documentary and the subjectively oriented theatrical film into a higher cultural form that united both. "Neo-realism" synthesized aesthetic feeling and scientific conceptuality in order to bring out hitherto unseen "truths" within reality that documentarianism and romanticism by themselves were unable to capture.[94]

As mentioned earlier, Aikawa developed his theories on the cultural technologies of film while working on the production of a culture film titled *The Present Battle* (1942). In 1940, the Cabinet Information Bureau commissioned the Art Film Company to make an epic film on "national solidarity." The filmmakers immediately seized on the heavy industrial factory as an excellent choice for concretely demonstrating solidarity between people from different classes and backgrounds and went on to produce a series of three films on factory life.[95] Aikawa was asked to help write the script, and in preparation, he conducted study trips to several large Hitachi electric generator plants near Tokyo in 1941 and 1942, one of which became the film site. The film is interesting not only for its illustration of cinematic techniques of mobilization or "cultural technology" but also for its object of representation: one of Japan's most technologically advanced factories.

As mentioned earlier, Aikawa was most interested in Hitachi's invention promotion and workplace assembly programs, and these formed a key part of the film's story. According to the script, *The Present Battle* depicted a dysfunctional electric generator factory of frequent accidents and delays, lazy workers, tension between younger and older and urban and rural workers, and broken machine parts. The manager decided to introduce a worker assembly system to let workers vent their frustrations, work out problems, and express their spontaneity in the form of onsite improvements. This, he thought, would generate unity among the different workers and help them feel more responsibility for the state's goal of producing more generators. Things gradually improved, and the film ended with the construction and test run of a ten-thousand-kilowatt generator destined for the occupied Dutch East Indies. All of the factory's problems were resolved, and the workers became filled with a feeling that their labor was "constructing the New Order in East Asia." Their "spark-emitting" machine tools exuded creativity, the generator would power the new empire, and every 1/100th of a millimeter shaved off a precision engine part contributed to East Asia's cultural construction. In many ways, the film depicted the old industrial structure being overcome by the new system of mass, hi-tech production—a goal shared by Marxists and the state during the war.[96]

In *Theory of the Culture Film*, Aikawa discussed many of the problems that the filmmakers encountered and reflected on the effectiveness of cinematic technologies and techniques. For example, the film apparently ended with a dry series of shots of a dam; water overflowing the dam's spillway; water flowing into a turbine to produce electricity; a factory; a house; and soldiers marching to the front. For Aikawa, this was too "documentarian" and did not burn the important ideas of national solidarity through commitment to one's vocation into the audience's minds. The filmmakers therefore cut the scene but again ended abruptly in an unsatisfying manner with a short, generic shot of the rising sun shining over the globe. Cinematic "cultural technologies" should not simply remain at the level of dry scientific representation or a "boring national policy lecture," Aikawa argued, but needed to work at the level of aesthetics and feeling as well. Technology should not only represent but also create national thought, utilizing the full array of editing, lighting, sound effects, and camerawork. In the end, Aikawa felt that the film failed to dynamically synthesize the ideals of "constructing East Asia" and national solidarity, presenting a challenge to future culture films.[97]

One of the main reasons for choosing a heavy industrial factory was that it served as a microcosm of national solidarity and the functional integration of a variety of people. Yet in the film, Aikawa wanted to extend this microcosm of technical integration to the rest of society. He tried to make the filmmakers come up with ways to represent society as somewhat similar to the integrated factory, so they inserted a scene where an alienated worker returned to the countryside only to witness his home village prospering from the introduction of agricultural machinery and working cooperatively to produce more food for the nation. There he realized that his lathe operator job had national meaning, and he returned to the factory. Aikawa wanted to make even more links to consumer organizations and neighborhood associations. The filmmakers even tried filming what he called a "household of technology" followed by the factory scenes in an attempt to represent the technological links between the two. In the end, they cut these scenes because they were too diversionary. Thus, another "cultural technology" of film was its ability to integrate different contexts and spaces, thereby creating more of a sense of connectivity between various people.[98]

Aikawa always argued that the cinematic flow of images or the film's tempo was more important than the actual content. Editing was the most important "technology of representation" because such cinematic techniques as the montage created different film tempos, which immediately influenced people's feelings and sensations. Effective editing boosted the film's narrative, which went from a negative beginning to a positive ending. Editing also enabled the insertion of such different temporalities as the sudden occurrence of historical events in the workers' lives. For example, Japan's attack on Pearl Harbor happened during the film's shooting, and the filmmakers wanted to capture the immediate mobilizing effect of war with the United States among the workers. They therefore inserted a scene of a mass worker rally and constantly referred to the urgent mission of constructing East Asia throughout the film. Aikawa also favored letting images speak for themselves rather than inserting too much narration and employing music to bring out the cinematic flow. Rather than using cinema as a mere instrument for portraying reality, he urged filmmakers to actively create different sensations made possible by newer "technologies of representation." Technology, he argued, needed to actively transform national policy into image.[99]

For Aikawa, "reproductive art" revolutionized human sensation and held multiple possibilities for popularizing culture and enabling mass expression. Culture films in particular represented one of the highest forms of cultural expression in modern technological society. Their mixture of scientific precision and rationality along with their incorporation of creative aesthetic techniques of editing, lighting, and sound helped create a dynamic technological sensation among the people and encouraged technical values of organization, innovation, and scientific rationality. Through the continuous development of newer cinematic technologies and innovative aesthetic techniques, "neorealism" would become the predominant aesthetic and help mobilize the people's vital energies for the war effort and the "cultural construction of East Asia." Like the other technologies that Aikawa viewed as constitutive of society, "neorealist" cultural technologies, such as culture films, fused subjective, ethical intent with objective processes and practices. For example, *The Present Battle* combined the spiritual goals of encouraging national values of responsibility, creativity, and integration with actual technologies of filming, editing, script writing, and acting in order to proliferate those very values among the people. Thus, although he recognized some of cinema's revolutionary potential, his political horizon (at least explicitly) was no longer the proletariat as an agent of socialist revolution but rather the nation incorporated into the state's goals of wartime mobilization and increasing productivity. Yet similar to his calls to intensify technical mobilization in the economic realm, his strident criticisms of culture films suggested a broader, unspoken political objective: his earlier socialist dream of uniting technology with life, sensation, and subjectivity. Cultural technologies and aesthetics played an essential role in realizing this vision—artists and filmmakers had to become "engineers of the human soul," as Soviet socialist realism had proclaimed, in order to lay the foundation for the dictatorship of the proletariat. But in the end, Aikawa became more of a cheerleader for the government's wartime mobilization campaigns and encouraged (or "engineered") Japan's subjects to sacrifice themselves spontaneously and willingly for the nation.[100]

Conclusion: Aikawa's Postwar Activities and Implications of His Thought

When Aikawa was drafted into the Guandong Army in Manchukuo toward the war's end in 1945, he quickly deserted and fled to the Soviet Union. Imprisoned in a labor camp in Birobidzhan in the Soviet Far East, he was used by the Soviets in what became known as the "POW Democratization Movement," an effort to reeducate imprisoned Japanese soldiers in socialism. Aikawa wholeheartedly threw himself into this "antifascist" education movement as editor of the *Khabarovsk Japan News* between 1945 and 1949 in the hopes of organizing a party vanguard for revolutionary struggle in Japan upon his return.[101] He was repatriated in 1949, and until his death in 1953 he was an organizer for the Japan Communist Party, a leader in efforts to reintegrate Japanese returnees from the Soviet Union and China, and a strong promoter of normalization with both socialist countries.[102] Aikawa wrote glowing accounts of the Soviet Union's rapid reconstruction and advancement toward communism after World War II. A new "Soviet humanism" was emerging, he wrote, in contrast to the empty desolation of abstract bourgeois humanism or the "fascist antihumanism" quickly reappearing in Japan and abroad. Labor was respected and people were rewarded according to their abilities and hard work; individuality, creativity, and responsibility were encouraged in the workplace; people had not only basic civil rights of free speech, religion, and organization but also social rights to food, housing, education, and health care; ethnic autonomy and equality was respected; people acquired a high level of political consciousness and an advanced knowledge of Marxism and Leninism; and everyone was able to participate in a vibrant cultural sphere of literature, film, art, and theater.[103] In this way, Aikawa asserted, Soviet humanism formed the basis of the Soviet Union's rapid recovery from wartime devastation and advancement toward communism.

Judging by his flight to the Soviet Union and firm dedication to their goal of socialist construction, it is clear that Aikawa never abandoned his earlier commitment to communism even at the war's height. Although he later denounced his wartime work as his "descent into a state-sponsored scholar of wartime production," he remained committed to his "systems" theory of technology, which viewed technology in terms of the interconnected processes that constituted society. He promised to revise his technology theory, but this work never materialized because of his untimely death in 1953 from exhaustion and illness from years as a POW.[104] The question, however, remains: why did he throw his support behind war and empire? Scholars of *tenkō*—ideological reversal among leftists during the prewar and wartime eras as a result of state pressure— have interpreted the phenomena in terms of "true" or "false" conversion.[105] Yet it is increasingly clear that the objectives of Japan's wartime fascists and communists like Aikawa were indeed reconcilable and at times mutually complicit, especially within the discourse of the technological imaginary. Similar to other Marxist intellectuals who witnessed increasing state repression and the cooptation of proletarian mass movements during the 1930s, Aikawa may have seen the war and the cultural mission of "constructing a New East Asia" as the only available path toward eventual socialism. In this way, the state's goal of rapidly modernizing Japan's economy through wartime mobilization and liberating East Asia from imperialism coincided with the *Kōza-ha* program of creating the appropriate conditions for socialist revolution—wartime modernization and national liberation would eliminate the "feudal remnants," which *Kōza-ha* Marxists believed had hindered Japan's advance toward communism.[106] Aikawa's postwar activities in "democratizing" and "modernizing" Japanese POWs in the Soviet Union and his organizational activities for the Japanese Communist Party and other groups to establish a "people's government" committed to peace, democracy, social equality, and solidarity with the Soviet Union and China after the American occupation attested to his continued commitment to placing Japan on the "proper" path toward communism—particularly the Stalinist version.

Aikawa's work demonstrates the extent to which technology formed an integral part of wartime fascist power. Whereas such scholars as Richard Samuels, John Dower, and Nakamura Takafusa have revealed how wartime ideologies, structures, institutions, and policies of technocratic rationalization and planning were integral to Japan's postwar economic miracle, they fail to examine how these constituted a distinct mode of power that continued well into the postwar era.[107] The state's wartime discourse on mobilizing technology proved quite compelling to many groups, such as leftist intellectuals, who proceeded to invest the term with their own meanings and objectives of social transformation. Yet in doing so, Marxists, such as Aikawa, articulated a totally systematized vision of Japan and its empire that proposed to make wartime fascism and imperialism much stronger and more efficient. People would be organized into "vocational organizations of science and technology" or worker assemblies that encouraged and thrived on creativity, sacrifice, and self-responsibility. Comprehensive zones of heavy industrialization centered on strategic natural resources and Asian labor power would be developed throughout Asia and firmly integrated with Japan and the rest of its empire. "Neorealist" films and other forms of contemporary media would instill a scientific attitude among the people as well as a passion to freely sacrifice everything for Japan's war effort. Far from simply forming the building blocks to postwar Japanese economic growth, Aikawa's version of the technological imaginary in many ways prefigured several key institutions and policies that once again mobilized the Japanese people after the war—a centralized science and technology bureaucracy, Japanese management ideology and factory quality control groups, Japanese overseas development policies that emphasized "comprehensive development" in Asia, and media campaigns that encouraged values of productivity for high-speed economic growth. Such postwar legacies of the technological imaginary and their implications for power and social control are the subject of this book's epilogue. In the next chapter, we turn to another important group besides intellectuals who helped institutionalize the technological imaginary: engineers and technology bureaucrats.

Chapter 2

Technologies of Asian Development

Engineers as Social Managers

Intellectuals were not the only ones who inquired deeply into the nature of technology in early twentieth-century Japan. As Japan rapidly transformed itself from a light to heavy industrial economy in the 1920s and 1930s, engineers became a prominent social force, and they shaped the intellectual discourse on technology. Terms they developed, such as "comprehensive technology" (*sōgō gijutsu*), the "standpoint of technology" (*gijutsu no tachiba*), and "technologies of Asian development" (*kōa gijutsu*), became part of the lexicon of Japan's technological imaginary as they struggled to elevate their social status, pushed for a unified national science and technology policy, and flocked to Japan's colonies to participate in research studies or infrastructure projects with the goal of developing East Asia. Whereas leftist intellectuals expanded technology's meaning to encompass the production of all areas of life, technology bureaucrats and engineers put forth a more explicitly technocratic definition of technology as the rational organization of Japan and its empire by visionary experts.[1] Government technical elites shaped Japan's technological imaginary and employed their conceptions in political programs and policy formulations during the Asia-Pacific War (1931–45). Miyamoto Takenosuke, a Home Ministry engineer, was the ideological leader of Japan's engineers' movement and later managed colonial technology policy as head of the Asia Development Board's (*kōain*) Technology Section and led efforts to create a wartime mobilization system for science and technology as the Cabinet Planning Board's (*kikakuin*) assistant director. His conception of technology as rational planning and comprehensive development were incorporated into the platforms of key national technology organizations, which were pushing for better treatment and more responsibilities for engineers in wartime policy making as well as into the policies of reform bureaucrats who managed wartime and colonial planning (see Chapter 5). In addition to the conceptions and policies of technology bureaucrats, this chapter examines the techniques of representation that they employed toward the colonies in order to propagate and naturalize their discourse of developing Asia through Japanese technology—a level that was just as essential as the ideologies and actual technologies used in the colonial context (discussed in more detail in Chapters 3 and 4). Empire was an essential component of Japan's technological imaginary, and for technology bureaucrats and engineers, an important element for advancing their social status and interests.

Engineers and Status Politics in Early Twentieth-Century Japan

Engineers began organizing themselves into sociopolitical interest groups in the 1910s as the Japanese government began systematizing scientific research and expanding science and engineering programs in its national universities. World War I's outbreak in Europe cut Japan off from essential industrial chemicals, pharmaceuticals, and precision technologies that had fueled its earlier economic development. The war also forced the state to invest more in constructing the social infrastructure of science and technology.[2] Civil engineers, who took the lead in organizing the engineering world, were mobilized by the Home Ministry for flood control projects and by other ministries to build railways, roads and bridges, ports and canals, dams, and water, sewage, and irrigation systems as Japan expanded its economic infrastructure. The increasing prominence of engineers and technical experts in state-sponsored projects contrasted sharply with their low rank and pay within ministries dominated by law degree bureaucrats, who were often younger and less experienced than engineers. Engineers were excluded from the regular promotion ladder reserved for law bureaucrats and given separate ranks of "technician" (*gishi*) or "technical officer" (*gikan*) instead of "section chief" (*kachō*) or "administrative officer" (*jikan*).[3] Such treatment reflected a wider tendency in society to view many engineers as equivalent to instruments or machines to be utilized—more associated with the natural world of things rather than with the loftier world of ideas, culture, or ethics.

Ōyodo Shōichi traces the emergence of the idea of the engineer as a social manager in Japan to the U.S. engineers' movement to increase their status in the late nineteenth and early twentieth centuries in the context of the development of the social infrastructure for science and technology and the rise of such high-tech, integrated electrical and chemical conglomerates as General Electric and Du Pont.[4] Frederick Winslow Taylor, whose ideas on the rational organization of workers and machines to exhibit maximum economy and efficiency later influenced other utopian visions throughout the world of organizing society into a rational, optimal social system, was a prominent example of an influential proponent of engineers as managers who emerged out of this movement. President Theodore Roosevelt provided activist engineers with a boost when he made natural resource management and "national efficiency" a top priority in 1907 with the establishment of the Inland Waterways Commission, which involved prominent engineering associations. With the increasing social visibility of engineers and state support of them as managers, many of the engineering societies began calling on engineers to educate themselves in economics and law so that they could take over more management roles in business and government. They argued that they were the ones most familiar with actual technical processes and therefore could best maximize efficiency and productivity. Along with Taylorism, this idea of the engineer as a social engineer who optimized society and resolved its various problems began taking root in Japan during the 1910s.[5]

Naoki Rintarō, the head of Tokyo's Rivers and Ports Bureau, was Japan's earliest proponent of this idea of technology as including social management. His 1918 compilation of articles, *From a Life of Technology*, served as a bible for young, idealist engineers at the time who wanted a more prominent role in society and who chafed at the law graduates' dominance of government administrative positions. In these early editorials for a prominent engineering journal, Naoki passionately argued that technology was a "living," human discipline dedicated to improving people's lives. Influenced by engineering professionalization movements in the United States and Great Britain, he criticized the view of technology as "means," whereby it was thought that the engineer's job was to merely select the best means of achieving a certain objective determined by bold, visionary leaders. Technology and engineering, Naoki argued, included "planning and implementation" as well as "construction and management." Miyamoto first heard Naoki's speeches when he was a disgruntled student in Tokyo Imperial University's School of Engineering, and as a result he began developing his idea of the engineer as an "engineer toward the details, but manager toward the whole" (*ichibu ni taisuru engineer, zembu ni taisuru manager*).[6]

Inspired by Naoki's arguments about the social role of engineers and angry at their low status in government and society, nine Tokyo Imperial University engineering graduates including Miyamoto and Kubota Yutaka (see Chapter 4) formed the Kōjin Club in 1920, whose main objective was to improve engineers' status and welfare as well as promote technology as the primary force for social development. In fact, they cited Thorstein Veblen and his 1919 call for the establishment of a "soviet of technicians"—a technological general staff who would rationally guide and manage the increasingly intricate and interconnected "social mechanism"—as one of the inspirations behind the group's formation.[7] Veblen's ideas were essential not only to the technocracy movement's rise in the United States but also to the formation of certain key concepts in wartime Japan, such as "comprehensive technology" and "technology for constructing East Asia."

Thorstein Veblen and the Intellectual Background to the Japanese Engineers Movement

Veblen followed debates in Germany on the spiritual dimensions of *Technik*, and according to Eric Schatzberg, he was the first to make "technology" a key term in the American social sciences. Technology for him was not just the physical system of machine processes but also the principles, knowledge, skills, and practices embodied in their operation.[8] In his seminal 1919 work, *The Engineers and the Price System*, he described modern society as an intricate "industrial system"—"an inclusive organization of many and diverse interlocking mechanical processes, interdependent and balanced among themselves in such a way that the due working of any part of it is conditioned on the due working of the rest." Technology was not just this system but also the "joint stock of knowledge and experience of the community" that accompanied it. Industrial experts, engineers, chemists, skilled workers, and technicians arose from this "joint stock of knowledge" and took up positions that were absolutely essential for society's welfare.[9]

For Veblen, technology (or what he described as "the state of the industrial arts") had its own ethics that was being subverted by nontechnical corporate financiers and absentee owners whose interests were tied to private profit rather than maximum output and "tangible performance." These "vested interests" had instead committed all sorts of "sabotage" ranging from inefficiently using natural resources, equipment, and manpower to purposefully slowing down production in order to maintain prices at a suitable level for the guarantee of profits. According to Veblen, the young technologists who constructed and operated the industrial system were "beginning to understand that engineering begins and ends in the domain of tangible performance, and that commercial expediency is another matter."[10] Efficiency, performance, optimality, and community welfare embodied the technologist's ethics in opposition to the ethics of the "captains of finance," who instead generated commodity shortages and mass unemployment through their decisions in the aftermath of World War I. As the industrial system progressed and became more intricate, Veblen argued, technologists would rise in prominence and a common class ethic of responsibility would develop and ultimately enable the formation of a "soviet of technicians." This technological general staff would then carry out a detailed survey of the country's energy resources, materials, and manpower and mobilize the support of skilled workers within the industrial system as well as the entire public. Ultimately, the soviet would allocate available resources in power, equipment, and materials to key nodes within the industrial system with the help of production engineers in each field. Regional subcenters and local councils would then ensure that resources were utilized in accordance with local needs and conditions.[11]

Thus, Veblen's notion of technology consisted of not just the industrial system but the knowledge, practices, and ethics that accompanied it. In contrast to capitalists who were perceived as purposefully "sabotaging" the economy for their self-interest, technology and technologists represented virtues of tangible performance, efficiency, and common material welfare. As society became more technologically integrated, more and more people would push for the organization of society based on such values because it was becoming increasingly apparent that the "vested interests" were responsible for the devastating imbalances in the economic system that created hardships for everyone. Veblen's ideas had a fertile reception in Taishō Japan as rapid, heavy industrialization and periodic recessions after World War I worsened social inequality and encouraged workers, farmers, minorities, and women to join popular protests. As Hiromi Mizuno shows, the Kōjin Club's founding was closely connected to worker struggles for better pay, working conditions, job security, and recognition as well as the growing socialist movement.[12] In the midst of these movements, the Kōjin Club quickly became the vanguard group for engineers who were seeking improved pay, more job opportunities, and better treatment in the public and private sectors, whose top positions were largely dominated by law graduates from elite national universities. Veblen's ideas on the engineers' ethics and values as well as their managerial role in overcoming the capitalist system's irrationalities were incorporated into the Japanese engineers' struggle to develop a class and occupational consciousness during the tumultuous Taishō years.[13]

Engineers and the Construction of a Technological Culture

Asserting a more immaterial notion of technology that was in tune with the emerging idea of the engineer as social manager (or "imagineer"), Miyamoto drafted the Kōjin Club's 1920 founding charter. It opened as follows:

Technology is cultural creation that fuses natural science and technique. Technology is creation, not means; therefore, it is absolute, not relative. Although cultural creation is not something achieved by technology alone, in some ways, human culture can be seen as technology. At the very least, since technology is weaved into an inseparable relationship with culture, technology should be developed in all areas of society. If one affirms culture, then by all means one should not reject technology.[14]

The charter immediately rejected the idea of technology as the physical means of production divorced from human agency and asserted that building advanced technical systems involved such cultural factors as ethics, creativity, and social practices, not simply the instrumental application of scientific principles. Therefore, as cultural creators involved in the building of all areas of society, it was the "responsibility of engineers to actively engage in governing [not only] one part of society . . . but . . . human life in its entirety."[15] As the engineer's work was immediately cultural, it involved not only resolving technical problems of design and construction but also wider questions of social utility, administration, and planning. In fact, engineers were more qualified than law bureaucrats to become skillful managers, they argued, because of their technical expertise.

The charter's next section most vividly evoked Veblen's ideas on the engineer's role as social manager. Describing capitalism as an "unhealthy social system" that subjugated workers to capitalists, engineers would become a "pivot" between capitalists and workers and lead them toward the common mission of establishing a "technological culture." Yet in a step back from Veblen's explicit radicalism, the charter added that this mediation of capital and labor was to be achieved through the "rational" means of electing engineers to parliament and political advocacy campaigns rather than by direct action or revolution. For much of the 1920s, as Mizuno shows, the Kōjin Club was torn between becoming a labor union of "brain workers" (zunō rōdōsha) firmly allied with the thriving proletarian parties and becoming an occupational association dedicated to improving engineers' status within existing public and private institutions.[16] In fact, in 1926 the Kōjin Club's more radical members steered the Kōjin Club into affiliation with the Social Masses Party, which advocated such positions as land reform, official labor union recognition, minimum wage and factory regulations, antidiscrimination laws for women, and equal access to education. Even Miyamoto envisioned the Kōjin Club to be somewhat similar to trade union organizations in England, such as the Fabian Society, the Labor Party, and the Federation of Professional, Technical and Administrative Workers, whose leaders he visited during his study tour to Europe and the United States. More conservative, high-level engineers, however—particularly those in the association's Osaka and Sapporo branches, where most worked for the government—resisted the trend toward proletarian party politics. By 1928, after much heated debate, these conservative members had gained the upper hand and dismissed the governing board's radical members, dropped the phrase "brain workers" from the club's bylaws, and ended all affiliation with political parties. After this, they focused on narrower occupational issues, such as expanding industrial education, publishing basic engineering texts, developing a standardized national examination system, establishing an employment office and school, and electing members to parliament.[17]

Thus, although engineers put forth a new sense of technology as cultural production and rational social management, similar to professional engineering societies elsewhere around the world, they were developed mainly for the purpose of improving their own social status within business or government and forging their own distinct identity rather than any kind of radical social change. As Mizuno shows, it took a more powerful symbol—the nation—to unify engineers from different class backgrounds into a powerful force that would ultimately shape the state's science and technology policy and colonial administration.[18] The "industrial rationalization movement," instituted by Prime Minister Hamaguchi Osachi in the aftermath of the 1927 financial panic, provided one key impetus to mobilize Kōjin Club engineers for state goals. This movement promoted the adoption of scientific production methods to increase efficiency, improve product quality, and reduce costs; the introduction of new management techniques to increase worker productivity and promote cooperation between labor and capital; and the integration of businesses to eliminate waste and competition.[19] By 1930, Kōjin Club members had more firmly aligned themselves with the more mainstream technology association, the Kōseikai, an organization of largely private-sector engineers. Thus, the goal of building a "technological culture" became more associated with state-sponsored ideas of social rationalization, management, and efficiency rather than the radical social reform of the Taishō and early Shōwa era of mass democratic politics.

Engineers for Empire: The Beginnings of "Comprehensive Technology"

The "Manchurian Incident" of September 1931, in which the Guandong Army invaded Manchuria and established a collaborationist regime, proved to be a boon for Kōjin Club engineers frustrated by the lack of improvement in their social status and the decline in employment opportunities. In a March 1932 editorial for their magazine, *Kōjin* (Engineer), Miyamoto proclaimed that the incident marked the beginning of the construction of a new state in Manchuria and Mongolia. Manchuria would become a "new world" for engineers, a "mecca" that could save Japan's engineering world, as nation building would require an expansion of technological and industrial facilities and therefore the employment of more engineers.[20] As a Tokyo Imperial University engineering student, Miyamoto was swept up by pan-Asianist discourse in the aftermath of Japan's victory in the Russo-Japanese War (1904–5). In his diary, he cheered Japanese expansion into Manchuria through the Twenty-One Demands (1915), writing that China was the place for "men . . . fighting for something substantial."[21] In the summer of 1916, he joined a study tour to Manchuria and enthusiastically proclaimed in his diary that China was primitive and undeveloped because of Western imperialism and therefore required Japanese administration and development.[22] These early attitudes burst forth again in 1932, when he expressed to a group of Japanese engineers in Manchuria his "earnest hope in [their] activities as pioneers of national development."[23] For him and many others, Manchuria was a "lifeline" that provided Japan with natural resources and a market for its manufactured goods. Japan should provide "technology" and "organization" to help "young China" develop those resources and "construct a paradise of Japanese-Chinese cooperation . . . based on the economic principle of complementarity," whereby Japan would in return gain access to China's cotton, coal, soybeans, and wool. This "paradise" would not be for the narrow interests of Japan's capitalists but would uplift the Chinese working classes, whose situation was closely tied to Japanese workers, from the warlord Zhang Xueliang.[24]

As Miyamoto and the Kōjin Club pushed the government to establish a long-term development plan for Manchuria as well as to expand and coordinate its technical institutions, they elaborated on their emerging notion of technology as "cultural creation" and social management in relation to colonial development. As an engineer in the Home Ministry's Civil Engineering Section, Miyamoto was sent to the Guandong Army's newly formed nation of Manchukuo in 1932 to explore possibilities for the expansion of public works projects. There he met leaders from Mantetsu, the Guandong Army's Special Affairs Division, and the Manchukuo State Council's General Affairs Agency to arrange for the employment of technical experts from Japan and to push for the establishment of a unified technology administration.[25] At the beginning of 1933, the state council established the National Roads Bureau (*kokudō kyoku*) not only to build a road network for national security and economic development but also to conduct flood control studies in the aftermath of several major floods in 1931 and 1932.[26] Naoki Rintarō, one of Japan's first proponents of incorporating ideas of social management and planning into the meaning of "technology," was appointed the new bureau's head. One month after his appointment, he announced in *Kōjin* that he was departing to Manchuria to engage in "creation." He proposed a notion of what engineers would later call "comprehensive technology" as a suitable framework for Manchurian development. Naoki wrote that road building, for instance, had to take into account not only narrow technical considerations but broader national defense, security, and administrative issues as well within their fundamental design. For example, hiring Chinese workers in surrounding areas to build roads would prevent them from joining "bandits" and contribute to local security and economic well-being. The National Roads Bureau would not only dig wells and reservoirs to provide people with water but also construct large mountain dams and reservoirs that provided them with drinking water, irrigation, and electricity. Increasing the number of canals, too, would improve China's inland commerce and improve popular will toward Japan.[27] In short, Naoki suggested that introducing technology to Manchuria meant more than simply bringing in Japanese technical expertise, organization, and infrastructure. It involved close coordination with military, political, economic, and cultural objectives. As Japan's empire expanded, technology was redefined to include comprehensive social management and planning to bring about East Asian development, security, cooperation, and prosperity.[28]

The notion of "comprehensive technology" had some of its roots in the government's flood control planning activities during the 1920s and 1930s. Floods began to recur frequently in Japan during the late Meiji era, after state engineers had dismantled most of the sophisticated Tokugawa "low dike" infrastructure that had earlier been intended to achieve an organic balance between accommodating and using water for multiple purposes of flood prevention, transport, irrigation, and forestry. In its place, Meiji engineers introduced Western "high dike" technologies of straightening, containing, and channeling rivers exclusively for flood prevention and control. Instead of viewing rivers and lakes organically as part of the wider local ecosystem, state engineers viewed them as a bundle of abstracted functions (flood control, water supply, electricity production) to be controlled and directed. Although these techniques were effective for most of the Meiji era, frequent and devastating floods returned in the early twentieth century as a result of deforestation and the loss of traditional mechanisms of slowing and absorbing river flow.[29] As budget problems, construction delays, and frequent natural disasters forced the government to focus on flood recovery more than prevention, Home Ministry bureaucrats and engineers began searching from the late 1920s for an alternative strategy toward flood control. Miyamoto spent much of these years conducting inspection tours of river improvement, rural development, and disaster reconstruction projects throughout Japan, and he was heavily involved in efforts to establish a more coordinated and comprehensive flood control approach. By 1936, the Flood Disaster Prevention Committee, which consisted of bureaucrats from the Home, Railways, Agriculture and Forestry, Communications and Transportation, and Commerce and Industry Ministries, drafted a proposal calling for the coordination of sixteen different areas related to flood control: road and bridge construction, forestry, drainage and irrigation, land reclamation, embankment construction, fish preservation, logging, and so on. Miyamoto had argued for the "comprehensive coordination of technology" toward flood control from 1930 onward as a way to create optimal and mutually reinforcing results among different engineering specializations. Unifying the scattered and overlapping technical jurisdictions in the Japanese and colonial bureaucracies became a key demand of the Kōjin Club and its successor organization, the Japan Technology Association, as part of their push for widespread acceptance of the idea of the engineer as a social manager rather than narrow specialist.[30]

"Comprehensive technology" also had its origin in the construction of multipurpose dams throughout the industrialized world. This meant building dams that simultaneously contributed to flood control, improved river transportation, land reclamation, irrigation, water supply, and electricity production. The concept of multipurpose dams was introduced to Japan in 1925 and was formally incorporated into Home Ministry policy in 1936.[31] In the United States, the Wilson Multipurpose Dam on the Tennessee River (completed in 1927) inspired similar projects in 1933 for the Tennessee Valley Authority (TVA), President Franklin Roosevelt's comprehensive program of flood control, electricity production, transportation improvement, and economic revitalization in America's southeast.[32] In the Soviet Union, inspired by Lenin's slogan that "Communism is the power of the soviets and the electrification of the whole country," the Dnieper Power Station in the Ukraine was completed in 1932. This was part of a larger plan to develop industry, improve transportation, and modernize Soviet agrarian life.[33] China's Nationalist government formed the Huai River Conservancy Commission in 1929. With the help of League of Nations financing and expertise, Western-educated Chinese engineers drew up extensive plans to build irrigation channels for land reclamation, a diversion channel to the sea, locks to improve river transportation, and dikes and levees to prevent floods. Plans for hydropower production were also proposed.[34] Throughout the world in the 1930s, national governments were developing comprehensive technical plans in order to transform and control the natural environment for agricultural and industrial development. In Japan, proposals toward the "comprehensive coordination of technology" and "river control" contributed to the formation of a new notion of flood control and development that "actively utilized the land" rather than simply defending it from natural disasters or passively relying on its natural benefits.[35] This notion of comprehensive technology, in which nature and society were transformed into an integrated and mutually reinforcing system through the coordinated construction of infrastructure projects that fulfilled political, economic, and cultural goals, however, only took on actual form in technical projects in 1930s Manchuria (see Chapter 3).

Toward Technocracy: Planning from the "Standpoint of Technology"

As engineers and bureaucrats were arriving at notions of "comprehensive technology" in Japan and Manchukuo during the early 1930s, the technocracy movement arose in the United States in the midst of the Great Depression. It created a stir in the Japanese press and among engineers, academics, and bureaucrats in particular, who were also searching for new ideas of social management during an economic crisis. A flurry of introductory texts written by journalists and translations of technocracy's key figures, such as Howard Scott, Graham Laing, and Frederick Ackermann, flooded the Japanese public sphere in 1933 and 1934. Taking cues from Veblen, proponents of technocracy argued that the price system, based on infinite wants and desires, was an imprecise and irrational way to organize economic production and distribution, especially with the rise of modern technology. Instead of improving people's livelihoods, technology's existence under the price system had merely led to more unemployment as factories tried to raise production and profits by replacing labor with machinery, which resulted in overproduction and a decline in consumption. The price system was also based on spiraling debt with a small minority of investors demanding ever-increasing returns on their capital, which in turn spurred businesses to constantly increase production and profits by cutting costs, thereby worsening the economic crisis. The price system's logic was alien to that of technology and production, which was instead based on optimal performance and precise measurements of energy conversion. As technicians rationally reorganized production and distribution on the basis of energy consumption and the optimal employment of modern technology, people would only have to work twice a week for eight hours a day to earn a yearly income of $20,000, they claimed. Precise energy credits would replace fluctuating money as the standard for exchange, distribution, and production. A cultural renaissance would result with people devoting more time to leisure, the arts, and personal development. Thus, following Veblen, proponents of technocracy believed that technology and engineers formed the very basis of what they called the social mechanism. Technology represented values that were in direct conflict with capitalism—precision, optimality, rationality, and social well-being as opposed to imprecision, waste, irrationality, and social dislocation.[36]

Technocracy's reception in Japan was enthusiastic but guarded. At a roundtable of academics, journalists, engineers, and bureaucrats on the movement, many participants welcomed the incorporation of technical data and values into social policy as well as the emphasis on technical expertise in solving social problems, such as unemployment and poverty. Yet several participants also criticized technocracy's commonalities with socialism, its frequently vague policies, and its imprecise use of statistics and examples.[37] Afterward, however, the participants announced the formation of the Technology Economy Association (*gijutsu keizai kyōkai*). In their prospectus, members of the association committed themselves to conducting detailed studies and proposing policies from the "standpoint of technology" in light of what they called the failure of liberal economics and Marxism. In the end, technocracy's chief influence in Japan was less its utopian vision of an equitable society based on rationalized energy distribution than its basic thesis that society formed a larger, complex mechanism based on principles of optimality, efficiency, and innovation, which required the expertise of a growing class of skilled managerial elites and workers. Although the technocracy boom quickly faded, the notion of a "standpoint of technology" as essential to socioeconomic policy and planning became central to the engineers movement's ideology.[38] Notably, in 1938, when Miyamoto began propounding his wartime notion of "technology for developing Asia," he continued to argue that industrial development in Manchukuo and China needed to be based on rational, precise technical planning as embodied by "Scott's Technocracy" and "Ōkochi Masatoshi's scientific industry" rather than on unbridled capitalism.[39]

Transforming Subjectivity Through Technology

Various factors, such as the engineers' increasing demands for higher positions in government, Japanese colonial expansion in Manchuria, the beginnings of comprehensive flood control planning, and the brief introduction of technocracy in the early 1930s, thus played important roles in forming a conception of society as a dynamic mechanism managed by visionary experts. As a Home Ministry engineer and one of the directors of the Japan Technology Association and the Society for Civil Engineering (*doboku gakkai*), Miyamoto was at the heart of establishing and popularizing such a conception. This notion of a managed social mechanism, however, also contained a crucial subjective component; it was not simply an object for detached technocratic manipulation. In an article on developing Tōhoku, Japan's poor northeastern region, he called for a "spiritual arousal" among its residents in order to establish a spirit of self-sufficiency and independence. He criticized the government's tendency to simply establish social programs and institutions or grant economic concessions to poorer regions.[40] In addition to a coordinated program of promoting river improvement, erosion control, land reclamation, and railway, road, and port construction for industrial development, Tōhoku required more agrarian centers and citizens schools to instill values of self-sufficiency and to correct "irrational" agricultural practices. Only by building industrial infrastructure and promoting rational practices and an independent spirit among the population could Tōhoku (and other economically depressed areas like Okinawa) wean itself off of state patronage and pay for its own development. "The goal of Tōhoku development is not to make it into a state kindergarten or elderly home," he argued. Development through technology therefore went hand in hand with the establishment of an ethic of self-sufficiency and rational self-discipline among the population.[41]

In 1930, as the state began to actively promote the heavy industrial and chemical industries, the Kōjin Club and other engineering associations petitioned the government to reform the education system. According to the club's Industrial Education Reform Committee, Japan's education system "paralyzed creativity" with its emphasis on memorization-based classical subjects over the natural sciences. Therefore, the education system churned out engineers and technicians who could merely copy foreign technologies rather than invent their own. Such an educational system was incapable of producing the human resources necessary for the establishment of an innovative, world-class industry that could break Japan's dependence on foreign patents and technologies. To alleviate this problem, the Kōjin Club proposed increasing basic science education at the junior and secondary levels, creating a parallel system of industrial middle and secondary schools alongside regular ones, increasing afternoon and evening schools for workers, and establishing a standardized national examination system. In the 1920s, the Kōjin Club published a very popular series of introductory textbooks in each engineering discipline, organized a series of standardized national exams with the promise of letters of recommendation for those who passed, and drew up plans to establish an evening school to supplement the lack of quality technical training in the education system. These efforts to create more opportunities for engineers and popularize science and engineering met with limited success; however, in 1930, as the state began mobilizing heavy industry, engineering groups renewed their efforts to increase technical skills and values among the wider population. Although they firmly believed in the engineers' leadership role as creators and social managers, throughout the wartime period they also campaigned for the expansion of quality technical training as well as the promotion of a spirit of creativity and responsibility among the populace.[42]

The China War and the Institutionalization of Comprehensive Technology

Miyamoto started using the term "comprehensive technology" regularly as he began to travel frequently to Manchuria and China to inspect and help plan civil engineering projects for the Home Ministry during the 1930s. He worked closely with former ministry engineers in the newly established puppet state of Manchukuo, such as Naoki Rintarō and Haraguchi Chūjirō, who were on the frontlines of conducting studies and drawing up blueprints for comprehensive flood control and multipurpose dam construction. Colonial engineers were the ones who really developed the conception of comprehensive technology through their hands-on investigations of unfamiliar environments and their negotiations with different interests in formulating what became known as the Liao River Improvement Project (see Chapter 3). Miyamoto and other Home Ministry engineers were instrumental in pushing the Manchukuo authorities to accept and institutionalize this emerging notion of comprehensive technology at the Manchukuo Flood Control Meeting attended by top military officials, engineers, and bureaucrats from concerned ministries in 1937. From this point onward, the Guandong Army threw its support behind comprehensive public works projects both as a pacification strategy and as part of its visions to transform Manchukuo into a heavy industrial base for an envisioned war against the Soviet Union.[43] Engineers followed suit by proposing further ambitious technical projects in parts of north China and vigorously expanding their efforts in Manchukuo (see Chapters 3 and 4). Emboldened by these events, Miyamoto began popularizing this conception of comprehensive technology at the state's highest levels and in the public sphere both as a guide to colonial policy and as an argument for unifying Japan's science and technology bureaucracies.

"Comprehensive technology" became truly established among engineers in Japan proper after full-fledged hostilities began with China in July 1937. In the previous year, Japanese reform bureaucrats in Manchukuo's Industrial Department in coordination with its major think tanks drew up a Five-Year Plan for Manchukuo Industry, which later became the basis for policies to establish a managed wartime economy at home. The Cabinet Planning Board, established in October 1937, became a powerful "comprehensive national policy institution" (*sōgō kokusaku kikan*) in charge of planning Japan's wartime mobilization system and managed economy, allocating resources to strategic industries, the military, and government institutions and guiding economic policy as the Imperial Army rapidly advanced into north China. Board bureaucrats drafted the State Total Mobilization Law (*kokka sōdoin hō*) in April 1938, which gave them widespread powers to allocate labor to key industries and manage labor disputes; control prices, production, and distribution of strategic materials and everyday commodities; enforce cartels and mergers in key industries; and restrict profits and financial speculation. Prime Minister Konoe Fumimarō announced Japan's wartime mission to construct a New Order for Greater East Asia in November 1938 based on the principles of building a "Japan-Manchukuo-China Economic Bloc," anticommunism, and the joint creation of an "anti-imperialist" culture. Together with this pan-Asianist campaign to overcome "particularistic" Chinese nationalism abroad, Konoe and the reform bureaucrats launched the New Order movement to overcome Western liberalism and capitalism at home through the establishment of a whole series of laws and institutions designed to mobilize the population spiritually for the goal of "constructing East Asia," encourage cooperation between labor and management, and maximize productivity and efficiency. The engineers' movement reached its peak in this climate of increased state planning and wartime mobilization and the official proclamation of "East Asian Construction" (*tōa kensetsu*) as the war's primary objective. They began to firmly ally themselves with reform bureaucrats (discussed more in Chapter 5) and their objective of designing a managed economy rooted in a corporatist national party.

As the China war intensified, the North China Expeditionary Army suddenly required more engineers to fix and build roads, operate and improve factories, and repair ports and commercial waterways. As Nakamura Takafusa describes, the Guandong Army had earlier sent teams of Mantetsu experts to north China to investigate the possibility of developing industry, natural resources, agriculture, finance, and transportation, as they proceeded to set up collaborative regimes in north China between 1935 and 1937.[44] The China Development Corporation (*kabushiki gaisha kōchū kōshi*) was established in Dalian in December 1935 to form and manage special companies, coordinate the entry of Japanese businesses, and supervise trade with north China. As full-fledged war broke out in 1937, the respective armies established the Special Government of the Republic of China in Beijing (December 1937) under Wang Kemin and the Reformed Government of the Republic of China in Nanjing under Liang Hongzhi (March 1938). In late 1937, while Miyamoto and other Home Ministry bureaucrats were designing the Liao River Improvement Project with engineers in Manchukuo, they were asked by the North China Expeditionary Army's Special Affairs Division in Beijing to inspect the Hai River region near the important port city of Tianjin. After their inspection, they discussed a plan with the military to develop flood control, port, river transportation, and road projects that were even more extensive than Manchukuo's Liao River plan.[45] This resulted in the drafting of a more detailed blueprint based on earlier Chinese investigations by the North China Water-Use Committee to build retention dams in the mountains as well as diversion canals into the Bohai Sea to alleviate excessive floodwaters that often built up in the plains around Tianjin. The dams would also be used for irrigation, electricity production, and improving river transportation inland. Projects were initiated to build a major diversion canal at an important river juncture near Tianjin, fortify dikes and improve drainage of the Yongding River, construct a canal to facilitate transportation inland to Shijiazhuang, and develop a major port at nearby Tanggu to expedite the flow of natural resources to Japan.[46]

Inspired by his experiences in China, Miyamoto immediately began mobilizing other engineers and bureaucrats in Japan to form the China Technology Federation (*taishi gijutsu renmei*) in February 1938. In its charter, they proposed the formation of a "comprehensive technological administration" to organize engineers to conduct studies, train specialists, and implement projects in China. This administration would lead the "rational comprehensive planning" of infrastructure projects within a proposed China Board that oversaw China's economic and development policy, which was then under the control of the respective invading army units' Special Affairs Divisions. During this time, Miyamoto wrote and spoke profusely about opportunities for engineers in China, whom he now described as "pioneers of North China development," and the need for an integrated technological approach toward China.[47]

In an April 1938 speech to members of the Hokkaidō branch of the Academic Association of Civil Engineers, a group he helped steer toward active involvement in Japanese colonial policy, Miyamoto laid out his clearest vision yet of introducing comprehensive technology to China based on his frequent technical planning trips to the continent. Japanese development in China, he began, was not simply a "technological invasion" to exploit China's vast resources for narrow commercial interests as the League of Nations powers and the United States did in the past under the guise of "technical cooperation." Rather than enter China with the "attitude of an aggressor" or worry about future competition or loss of technology, Japan should provide technology to develop China's resources and heavy industry for "mutual prosperity" and "eternal peace in East Asia." However, Japan should provide technology not simply haphazardly but comprehensively. To illustrate, Miyamoto gave an example from his Hai River inspection trip near Tianjin. The Hai River basin was the size of Japan's main island of Honshū and served a dense population of eighty million. Huge floods occurred around once every seven years, one as recently as 1937. During parts of the year, the high volume of sediment from faster streamflows obstructed river transportation and a planned port project in nearby Tanggu. A huge sandbar at the Hai River's mouth further slowed shipping, and a sludgy river bottom prevented land reclamation. Instead of tackling these problems one by one, Miyamoto noted, the authorities were proceeding with a comprehensive plan to build detention basins to store water upstream. The stored water in these reservoirs would irrigate Hebei Province for cotton cultivation, provide water to expand canal and river transportation around Tianjin, reduce sediment and enable the smooth functioning of a substantial port, and provide electricity for heavy industrial development. The mission of engineers, he urged the association's Hokkaidō members, was to provide such a standpoint of "comprehensive technology" in the planning of flood control projects or transportation networks. Unlike Japan, where the technology bureaucracy was scattered, China would become an arena where engineers could freely demonstrate their expertise and introduce integrated planning and construction for industrial development.[48] Miyamoto and the China Technology Association's efforts to promote the mobilization of engineers for "North China development" appeared to bear fruit when the North China Construction Agency (*kensetsu sōsho*; see Chapter 4) was established in Beijing under Wang Kemin's newly formed collaborationist government.

Fully pledging themselves to "East Asian Construction" and Prime Minister Konoe's New Order movement, Miyamoto and other technology bureaucrats simultaneously pushed their "comprehensive technology" conception domestically. The Japan Technology Association joined with such other technology groups as the Communications Heroes Association (*teishin giyūkai*), the Kōseikai, and the Six Ministry Engineers Friends Society (*rokushō gijutsukan yūshi kondankai*) to petition Konoe to promote technological independence from Western technology, correct the bureaucratic hiring system that discriminated against engineers, and increase the number of experts in policy planning and administration. "Technological patriotism" (*gijutsu hōkoku*) became a unifying slogan among engineers, and they seized on the state's increasing use of the term "comprehensive" or "unified" (*sōgō*) in its total war mobilization plans to argue for the establishment of an integrated industrial policy organ and, in particular, a centralized technology administration. Developing natural resources, constructing communications and transportation networks, promoting heavy industry, and expanding technical education and research facilities all required people with specialized expertise and technical management skills to achieve the most optimal, efficient results, they argued. In September 1938, the Industrial Technology Federation (*sangyō gijutsu renmei*) was formed as an umbrella organization for Japan's technology associations. Its charter declared that national policy for increasing productivity should be researched and planned from the standpoint of "comprehensive production technology spanning the natural and social sciences" and that the state should establish a policy centered on industry from the "standpoint of a clear and extensive comprehensive technology" in unison with the total mobilization of engineers and the cooperation of a "wider public that understands technology."[49] The notion of technology as comprehensive social management coordinated with popular mobilization toward the establishment of a rational, optimal, and efficient socioeconomic system was now firmly incorporated into the engineers' overall agenda. Wartime mobilization and rapid colonial expansion created a fertile environment for these engineers' conceptions to grow.

Shinohara Takeshi, founder and director of the Comprehensive Science Association (*sōgō kagaku kyōkai*) and editor of the Kōseikai's journal, *Kōgyō kokusaku* (Industrial Policy), wrote the Industrial Technology Federation's charter. Heavily influenced by the German physicist Ernst Mach and his "unified science" (*Einheitswissenschaft*) movement to establish a common empiricist and positivist framework among the different sciences as well as by American pragmatism, Shinohara explained his notion of "unified" or "comprehensive science and technology" in an article two months after the group's founding. Efforts to establish a planned wartime economy and a new order in East Asia made it all the more essential to overcome the rampant sectionalism, specialization, and lack of interest in larger socioeconomic issues among Japan's scientists and engineers, he argued. At the root of their narrowness and lack of social consciousness was a more widespread view of science and technology as merely being concerned with the minutiae of specific things or as the instrumental processing of the physical world's components. Instead, Shinohara argued, technology had to be properly grasped more broadly as "creating a form or design that most effectively achieves" a definite objective and then guiding all action toward the implementation of that form or design. As an active, dynamic process, then, technology included crafting forms and designs to guide action not only toward the material world but also toward human beings and society. In this way, one could talk about "social scientific technologies," such as administration and government. All natural and social scientific technologies needed to be rooted in a common empirical method that drew conclusions from objectively confirmed facts and experience and excluded bias, arbitrariness, and contingency, he added.[50]

Under the state's wartime goals of overcoming free-market liberalism and establishing a managed economy that maximized productivity (as opposed to profit), there were more opportunities for scientists and engineers to break out of their narrow specializations and help achieve such comprehensive goals, Shinohara continued. For example, in order to realize a particular technical design or form, the production engineer had to consider such various factors as resource availability and distribution, worker productivity and working conditions, work process organization, and the various interactions between them. Thus, engineers required an integrated knowledge beyond their own specializations that incorporated the knowledge of other natural and social sciences. "Comprehensive science and technology" involved the creative interaction of engineers and officials, natural scientists and social scientists, and managers and workers toward the common goal of increasing national productivity and the people's welfare. Elsewhere, Shinohara even envisioned the state as an "enormous energy conversion structure," a productivist machine whereby politicians and managers became creative engineers who coordinated production and the people were transformed into active scientist-engineers who contributed to raising productivity in their own specific occupations.[51]

Technology and China's Integration into the New Order in East Asia

How did Miyamoto and the technology bureaucrats represent China and the Chinese within their technological imaginary? Mizuno notes that for Miyamoto, China's role was to provide the natural resources to advance Japanese science and technology, which in turn would develop "backward" China.[52] But in addition to understanding how Japanese elites employed a universalistic language of science and technology to incorporate Chinese nationalism, we need to examine how they represented China and thereby created the image of it as ready for Japanese development. Such a discursive analysis sheds light on what Naoki Sakai calls "imperial nationalism," the process by which Japanese nationalism during the wartime era actively enunciated itself as a universal, multiethnic nationalism (or "genus"/*rui*) that incorporated and produced particular ethnic nationalisms (or "species"/*shu*). I argue that the concepts of "comprehensive technology" and "technologies of Asian development" among Japanese engineers constituted an imperial nationalist attempt to mobilize colonial subjects by transforming the empire into a rational, optimal, and efficient system. In various ways, anti-Japanese nationalisms and resistance were represented as irrational, antimodern, and obstructionist, thereby opening the way for their violent suppression. Thus, technocratic ideology operated not only at the level of ideas and concepts but also through the deployment of a powerful Orientalist structure of representation.[53]

As a high-level technology bureaucrat who inspected civil infrastructure projects at home and abroad and a key liaison with the Manchukuo authorities for the assignment of engineers to the colonies, Miyamoto had many occasions to travel there. We catch a glimpse of how he represented the Chinese within his developing "East Asian construction" framework in two travelogues written for engineering journals before and after the outbreak of full-scale war in July 1937. In October 1935, the Kōseikai organized the Oriental Engineering Congress, a three-week "cultural and educational exchange" tour made up of mostly technical representatives from Japanese government and business to meet with Chinese government, academic, and business leaders in many of north and central China's major cities, as well as in Manchuria. Miyamoto represented the Home Ministry and took the opportunity to travel in Manchuria on the side in order to view the progress of ongoing flood control planning, road and railway building, and urban construction projects. The congress convened when the North China Expeditionary Army was establishing collaborationist regimes and "demilitarized zones" in north China in order to limit the Chinese Nationalists' influence. Miyamoto and Japan's Oriental Engineering Congress delegation immediately witnessed the strength of Chinese nationalism upon their arrival in Shanghai—the site of fierce fighting between Chinese and Japanese forces in 1932, traces and tensions of which were still quite evident. Fearing that attending or sponsoring the conference might grant implicit recognition to Japan's Manchukuo puppet state, many Nationalist officials boycotted the event, and Chiang Kai-shek's Blue Shirts attempted to obstruct the meeting, which finally occurred through the intervention of Japan's consul-general, Shanghai's mayor, the head of Shanghai's Industrial Department, and the vice director of the Foreign Affairs Department. Miyamoto and the delegates continued to insist that the congress had no political agenda and that increasing academic and cultural exchange was the first step to overcoming what they called diplomatic "problems" and "misunderstandings" between the two nations.[54]

As a self-proclaimed liberal, Miyamoto believed that economic and cultural cooperation between Japan and a rising China was achievable because each nation had complementary needs. Quoting a speech at the congress meeting in Nanjing by Wang Jingwei—the premier of China's Executive Council who supported more cooperation with Japan against what he saw as Western imperialism's greater danger—he asserted that whereas Japanese heavy industry required China's resources and markets, China's developing industries required Japanese technical guidance.[55] "Young China's" engineers and bureaucrats impressed him greatly, reminding him of the visionary leaders who rapidly modernized Japan during the Meiji era. The construction of Greater Shanghai as a counter to the foreign concessions that controlled Shanghai's economic hub, the building of a new capital in Nanjing that preserved Chinese architecture, and the New Life movement to improve national morals and modernize feudal customs all gave him the impression of a strong national consciousness and eagerness for modernization. Miyamoto was moved by the speeches at the congress by Premier Wang and a Foreign Affairs Department official named Tang Youren, who both appealed for joint cultural and economic partnership to aid China's modernization. He admired their bravery in standing up to "impulsive anti-Japanese/anti-Manchukuo" nationalists by asserting that Japanese expertise and technology were essential to "national survival." Wang was shot and wounded several days later (allegedly by Chiang's Blue Shirts), and Tang was killed two months afterward, leading Miyamoto to lose hope in efforts to persuade the Nationalists that cooperation with Japan was the key to China's development.[56]

His second travelogue, written in February 1938 after the war's outbreak and a few months after his return from an inspection tour in Manchukuo and north China, had a different representational structure. The urgent mission of "East Asian Construction" replaced "Japan-China Friendship" as Miyamoto visited a wide array of technical projects undertaken by Japanese engineers, such as the Liao River Improvement Project, Fengman Multipurpose Dam, Tanggu Port, the Fuxun and Fuxin coalmines, and the Beijing-to-Tianjin road project. Instead of bold, visionary Chinese leaders and young, energetic engineers who were engaged in a wide array of technical projects and who eagerly sought Japanese advice, Miyamoto presented a picture of poor, helpless, and exploited peasants; a durable "national character" capable of enduring all sorts of hardship; and the Chinese "zest for life" and willingness to work hard under extreme conditions. For example, in arguing for such comprehensive flood control projects as the Liao River project in southern Manchuria, Miyamoto directly cited and borrowed imagery from Pearl Buck's classic representation of Chinese peasant life, *The Good Earth*, which was translated into Japanese in 1937. The Liao River caused a yearly average of thirty-five million yen in damages and had inflicted seventy-four million yen in damages during the recent August 1937 floods, Miyamoto noted. He then cited Buck's description of how each flood and subsequent drought caused peasants to flee the "good earth," exposing them to frequent bandit attacks and further poverty. Again applying Buck's representation of central and north China to southern Manchuria's conditions, he noted the "similar" lack of cultivable land, the decline in agricultural productivity, the existence of powerful landlords who monopolized scarce land, and the rapid growth of a subsistence-level "agricultural slave" class burdened by excessive debt and taxes who were forced by circumstance to sell their daughters into prostitution.[57] Thus, by objectifying the "Chinese peasant" as helpless and exploited, Miyamoto not only occluded an analysis of the specific histories of migration, class conflict, economic crises, and civil war that helped create such "helplessness" in Manchuria but also failed to mention Japan's role in appropriating lands for Japanese and Korean farmers and businesses. Similar to Buck, who depicted Chinese peasants as eternally tied to the earth's "restorative power" and their lives as an eternal struggle of humans against nature, Miyamoto also naturalized their condition by excluding any mention of the history of violence and conflict that contributed to their situation of "land scarcity" and poverty in the first place.[58]

In this way, he positioned the Chinese peasant as someone who inevitably required Japanese development. "From this perspective [of the Chinese peasant's situation], South Manchurian flood control constitutes a fundamental policy to 'preserve the borders and pacify the people' (*hokyō anmin*) and is necessarily a question that should be resolved first," he wrote. Comprehensive technology "rescues the peasant from his tragic plight," which if left unaddressed would make "the 'harmony of the five ethnicities' and the 'Kingly Way Paradise' impossible [to realize]."[59] Not only would projects like the Liao River Improvement Plan empower the

Chinese farmer, but it would also bring an additional 500,000 hectares of land into cultivation for Japanese and Korean immigrants to develop, thereby alleviating the problem of land scarcity and overcultivation, which hindered their advance into southern Manchuria. Thus, by tying the Chinese peasant's situation to such "natural" factors as floods, famine, land scarcity, and traditionally corrupt landlords and officials, Miyamoto opened the space for technology to correct these problems through the expertise of Japanese engineers. For him, the problem of rural poverty in the southern Manchurian countryside was not unequal class relations and resource distribution rooted in multiple layers of violence but rather a lack of technology and development that could overcome the "natural" obstacles to social improvement.

In a subsequent travelogue, Miyamoto made further observations about what he described as the Chinese "quality of national character" (*kokuminsei*). After hearing Guandong Army General Ueda Kenkichi assert the importance of flood control for "North China pacification" and joke about Chinese leisurely playing musical instruments on their rooftops as their houses were being surrounded by floodwaters, Miyamoto reflected on what he called the Chinese "zest for life." For him, such strange behavior represented their "latent power"—the same endurance and patience throughout history that led to the gradual decline and absorption of earlier invading Manchus and Mongols. Observing Chinese workers on the Tianjin-to-Tanggu road project, he described their "primitive life style" as one of living in "mat-rush beggar huts," which they "crawl into like crabs in order to sleep." Lower-class Chinese in Tianjin were apparently used to breathing in dust and living in dirt from constant car and truck traffic. The Han Chinese people's "vigorous life power" and "traditional life forms" made for sincere, diligent workers and transcended any concern for hygiene or discomfort. In the end, Miyamoto asserted, Japan should not ignore the Chinese people's "latent power" and must overcome their own narrow-minded "island mentality" to build a "continental national character" capable of leading and developing East Asia.[60]

For Miyamoto and other idealist engineers, technological modernization would form the basis of this new Japanese continental national character. When he visited Manchukuo in late 1935, he expressed great admiration for Russia's earlier comprehensive railway construction plans and their building of numerous weather and river observation stations. In sharp contrast to "scientific" Russia, Japanese pioneers in Manchukuo had introduced Shintō shrines and bar girls, he lamented.[61] However, a new generation of young engineers who "understood the true meaning of Manchukuo development" and were willing to lay down their lives for the cause were beginning to replace the self-interested "carpetbaggers" and narrow-minded careerists typical of earlier Japanese bureaucrats and engineers. Engineers he met on his trip, such as Watanabe Yasuzō, a civil engineer in Jilin who worked on a Songhua River flood control project, were representative of this new "continental" Japanese expert totally dedicated to Manchukuo's development. Watanabe complained to Miyamoto about a foreman he had to stop from beating workers, arrogant Japanese bureaucrats who talked down to influential local Manchukuo intellectuals, and Japanese "carpetbagger" officials who immediately wanted to return to Japan after quickly advancing their careers.[62] Such characteristics represented a particularistic, narrow-minded Japanese-ness, not the necessary "continental national character" that Japan required for East Asian development.

Miyamoto's representation of "the Chinese" as having a patient, enduring "national character" in the face of such seemingly timeless forces as natural disasters, geographical limitations, exploitative landlords, and corrupt officials placed them in the position of requiring Japanese expertise and technical development to overcome their "natural" constraints. Such representative strategies occluded the specific socioeconomic struggles and histories (including Japanese invasion and colonization) that produced their situation, rendering them into passive objects for Japanese development. The capacity to incorporate the Chinese into a multiethnic New Order in East Asia, however, required not only this type of technocratic representational strategy but also the formation of a "continental character" on the part of Japanese engineers capable of making Chinese workers, farmers, officials, and engineers participate in Japan's imperialist endeavor. In sum, Japanese engineers and experts developed an "imperial nationalism" (as opposed to a purely particularistic, selfish "ethnic nationalism") that manifested itself concretely in transportation networks, multipurpose river control projects, and urban planning to directly address the "natural" plight of the people. To borrow Sakai's language once more, large-scale projects of technological progress became the *universal* "substratum" or horizon to integrate *particular* Japanese, Chinese, Manchurian, and other peoples within the New Order in East Asia.[63]

Engineers and the Formation of the Asia Development Board

In the mid-1930s, Miyamoto and the newly invigorated technology associations began pushing for the establishment of a unified government administration for economic policy toward China, where comprehensive technology would become institutionalized. With the formation of the Cabinet Planning Board in October 1937 as the primary national policy administration for wartime mobilization, reform bureaucrats there immediately conceived of a similar organization to manage occupied China's economic affairs—what soon became the Asia Development Board. With the army's support, which required coordinated economic planning not only to fulfill immediate military requirements but also to govern the newly occupied areas, they overcame strong opposition from the Foreign Ministry and established it in December 1938.[64] The Asia Development Board had branch offices in Beijing (north China), Zhiangjiakou (Mongolia), Shanghai (central China), and Xiamen (south China). The board itself was in charge of planning economic development, overseeing the North and Central China Development Companies, establishing economic controls, formulating financial and currency policy, planning transportation and communications networks, and even conducting cultural activities to promote hygiene, medicine, education, religion, and scholarship among the Chinese people.[65] Most important for the technology associations, however, was the creation of a Technology Department, which coordinated all technical matters for the organization's political, economic, and cultural departments. Because a major aspect of the board's work was to produce research studies on China's existing industry, agriculture, infrastructure, geography, natural resources, and institutions, the board required a large number of technical experts and engineers to conduct them. Miyamoto was selected to head the department, and many engineers within the wider movement were also employed, an event the technology associations hailed as a pathbreaking achievement of their goal to involve engineers and experts in policy making. On paper, however, Miyamoto was still a "chief engineer" and the department was labeled "consultative" even though in reality he had powers equal to a departmental head and the department had an important function within the board.[66] This illustrated the discrimination that engineers continued to face within the state machinery.

Miyamoto immediately began planning the establishment of what would become the Asia Development Technology Committee (*kōa gijutsu iinkai*), a policy-making committee of technical experts from different government ministries. According to the committee's "Reasons for Establishment," its purpose was to plan and evaluate technical research studies to aid in the formulation of board policies as well as to help guide and manage business activity in China. With the objective of "total mobilization of the essence of [Japan's] modern technology," the scope of its activities encompassed a whole array of technical specializations: urban planning, hydropower, health and hygiene, mining, agriculture, roads and ports, and so on. The five subcommittees (transportation, flood control, agriculture-forestry-fisheries, mining and metallurgy, and hygiene and urban engineering) would integrate perspectives from different technical fields in order to achieve the "total employment of technology." For example, the document continued, flood control would simultaneously incorporate expertise in land and river transportation, irrigation, power production, port construction, and agriculture and forestry to generate the most optimal planning of flood control projects.[67] Miyamoto's draft also included a clause stating that the committee could directly make policy proposals to the board's head, although this was later removed after encountering resistance from other departmental and ministerial interests. Forty-three bureaucrats, businessmen, and intellectuals (including Miyamoto) were appointed to the committee, and eighteen board bureaucrats and engineers were assigned as coordinators (including Mōri Hideoto, head of the economic department, discussed in Chapter 5).[68] Most were members of technology associations, and they strongly shared their vision of the engineers' role as social manager and policy maker. Although the committee did not have the power to directly determine policy as Miyamoto had hoped, it nevertheless represented the incorporation of state engineers' standpoints of "comprehensive technology" and "East Asian construction" into the board's conceptual framework. These standpoints soon revealed themselves in the numerous studies and research missions either that the committee commissioned or that its members were actively involved in.[69]

"Technologies for Asian Development"

Miyamoto's selection as head of the Asia Development Board's Technology Department in 1938 was accompanied by the publication of several collected volumes of his earlier articles over the next several years, which reaffirmed his status as the engineering world's ideological leader. His work *Tasks of Continental Construction* was a collection of speeches and articles written during his nearly three-year tenure at the board, when he conducted "actual research" into China's natural resources and peoples in order to realize Japan's mission of constructing a new order in East Asia.[70] During this period, he formulated and publicized his notion of "technologies for Asian development" in numerous speeches around Japan and in the media. An analysis of this concept provides further insight into the ideological landscape behind numerous projects conducted by engineers in the colonies.

Miyamoto described three basic aspects to "technologies for Asian development": "rapid advancement" (*yakushinsei*), "comprehensiveness" (*sōgōsei*), and "local potential" (*ritchisei*). "Rapid advancement" meant establishing an independent, world-class "Japanese technology" capable of developing China's vast resources. Japanese technology needed to break its dependence on foreign technology and constantly develop innovative technology in order to permanently maintain its lead over China; otherwise the principle of complementarity (Japanese technology for Chinese resources) underlying the New Order in East Asia would be lost. In the same way that Germany had established its global dominance in science and technology by investing heavily in the electrochemical industries, Japan should focus primarily on inventing and exporting top-of-the-line technology in order to transform and improve China's resources.[71] Miyamoto frequently asserted that having an independent technology was the single most important factor for national development, even more essential than possessing abundant natural resources or capital. For example, in Germany, the invention of synthetic dyes eliminated the nation's reliance on India's natural salts, and the Haber-Bosch method of producing ammonia reduced Germany's dependence on Chile's saltpeter, thereby transforming Germany into a global leader in chemical dye and fertilizer production. Citing the Nazi propagandist Anton Zischka's 1936 best seller, *Science Breaks Monopoly*, Miyamoto asserted that developing advanced science and technology was essential to promote self-sufficiency at a time when the great powers were establishing monopolistic economic blocs in their respective colonial territories.[72] Thus, "rapid advancement" as a characteristic of "technologies for Asian development" meant the production of a continually innovating Japanese technology that was capable of transforming Chinese resources into new ones (e.g., Chinese oil shale into fuel, aluminum shale into aluminum, and rivers into hydropower) or improving production of essential resources, such as coal, cotton, wool, and salts.[73] Japan also needed to follow Germany's example of promoting applied science and advanced technology by expanding technical education in schools and factories.[74] In short, the total mobilization of science and technology was necessary to overcome Japan's longtime dependence on purchasing Western patents and technology, and therefore to create a truly independent "Japanese" technology for the development of Chinese resources. This culminated in the movement among engineers and bureaucrats for the establishment of a "New Order for Science and Technology" to improve wartime research and increase industrial productivity.[75]

"Comprehensiveness," the second principle of "technologies for developing Asia," was the notion of comprehensive technology discussed earlier that Miyamoto and other engineers had begun advocating from the early 1930s for both Japan and its colonies. Such a perspective of integrated, rational planning and coordination of expertise for comprehensive development projects was all the more necessary to achieve Japan's long-term objective of "East Asian construction," Miyamoto argued, particularly in the midst of wartime resource shortages. In order to avoid getting lost in a maze of bureaucratic details, a new type of engineer that was prominent in Germany, the "administrative engineer" (*Verwaltungsingenieur*), was necessary. The administrative engineer sought to achieve not only optimal technical results but political, economic, and cultural ones as well—in short, "comprehensive results." Japan should also learn from German "process engineering" (*Verfahrenstechnik*), which synthesized industrial chemistry, mechanical engineering, management, labor psychology, and industrial economics for the purpose of maximizing productivity and improving quality.[76] As Japan mobilized for war and expanded farther into China, technology bureaucrats increasingly turned to Germany as the model for technological mobilization, particularly after Germany's wartime *blitzkrieg* success in Europe.[77] Miyamoto viewed Germany's "scientific national character" as a model for Japan and argued that Japanese colonial engineers needed the same type of "effectiveness" (*Zweckmäßigkeit*) and "methodical planning" (*Planmäßigkeit*) that Nazi engineers demonstrated, for instance, in their completion of the Mittelland Canal, which linked most of Germany's canals into an integrated East-West waterway system. Because China required technological development in a wide array of fields, Japanese engineers had to plan projects carefully and scientifically and effectively incorporate geographical, political, military, economic, and technical objectives into their work.[78] For example, the development of coal resources in Shanxi and Chahar Provinces required a coordinated consideration of port locations, railway links, canal expansion, flood control, agricultural improvement, and electricity production. All of these were inextricably linked to north China's comprehensive development.[79] With this type of German comprehensive technology and scientificity, combined with an equally intense spiritual mobilization, anything was possible, Miyamoto argued.

Finally, "local potential" meant the ability to adapt Japanese technology to a specific climate, culture, and economic context rather than apply it uniformly.[80] Science possessed universality, and technology "organized" and "systematized" science toward a specific goal within a particular time and place. Although in many ways national borders did not apply to science and technology, at the same time they were molded by national interests and contexts, as exhibited by state patronage of national defense and heavy industry in the age of total war and global natural resource competition. An essential part of building a superior "national technology" was flexibility in adjusting to existing geographical, economic, natural, and human conditions.[81] In China, for example, Japanese engineers had to take account of the abundant cheap labor there as they introduced advanced technology. Rapid mechanization would simply cause more unemployment, and utilizing Chinese labor instead often made more economic sense. Recently, however, with the advent of labor shortages, engineers had to strike a balance between the two to achieve maximum effectiveness. By always keeping in mind the local socioeconomic conditions, engineers could maintain a "living technology" rather than a formulaic, instrumental one.[82]

Despite the war's outbreak and fierce Chinese resistance, Miyamoto continued to insist that the New Order in East Asia's construction would not be based on "force" and "leadership" but rather on friendship and cooperation. Similar to many Japanese leaders, he thought that the use of force was simply temporary and that Chiang would ultimately be convinced to ally his interests with Japan's New Order in East Asia rather than with the West. The New Order in East Asia, according to Konoe's 1938 pronouncement, represented not only the formation of a self-sufficient, anticommunist economic bloc in Manchukuo, China, and Japan and the establishment of a national defense state but also the creation of a new culture. This new culture, Miyamoto argued, would dialectically fuse "Eastern spirituality" with "Western materialism" and thereby correct the "unhealthy" tendency to overemphasize one over the other. Specifically, Japan would cooperate in elevating China's science and technology while helping to preserve the glory of its "Kingly Way" (*Ōdō, Wangdao*) culture.[83] Japanese people should eliminate their tendency to want to "lead" the Chinese based on their own standards of "cleanliness, scrupulousness, impatience, and narrowness" without showing any appreciation for China's own national character, history, customs, lifestyles, morality, and thought. In fact, Chinese leaders had already shown a "youthful vigor" in pursuing the "New Life Movement" under Chiang, for instance—a mass movement that promoted science, health, hygiene, and national consciousness. Instead of deprecating China's capabilities, Japanese should recognize their tremendous potential and eagerness for national development and show in practice how Japan's science and technology might help them achieve their goals.[84] Without grasping the "actuality of East Asian peoples" and rooting "East Asian construction" within their lives, the new order would remain nothing but an "empty ideology" that generated "needless antagonism" toward Japan.[85] In this sense, Japan needed to take a cue from the thousands of Western Christian missionaries in China who not only spread their religions but also engaged in small-scale technical projects to improve cotton quality, teach modern agricultural techniques, provide medical care, and assist in natural disaster relief in order to reach out to the Chinese people.

Asian Development and the Contingencies of War: The Five-Year Industrial Plan for North China

Miyamoto and other officials were caught off guard by the war's intensity, which they always insisted was an "incident," and the fierce resistance put up by the Nationalist Army, the communists, and other irregular forces. As the Asia Development Board and the military realized the enormity of the task of "developing north China," cracks began to appear within the rosy vision of a cooperative new order based on comprehensive technological development. Soon after it was established, the board drew up its outline in April 1938 to "assist" regional political regimes in the establishment of central governments and announced its broad cultural program and vision for economic development. Based on earlier economic investigations by Mantetsu researchers and the Guandong Army's Special Affairs Division, they drafted an ambitious economic plan that focused on stabilizing and unifying China's currencies into Japan's yen bloc, developing railway lines to strategic natural resource sites, building and improving international ports, and increasing coal, iron ore, and salt production for Japan's military industries. With strong army input, the board's three-year plan heavily prioritized the provisioning of natural resources to Japan to meet wartime productivity targets. By late 1939, however, it became increasingly clear to occupation authorities that this policy of prioritizing Japan's needs was failing miserably.

A series of events forced Japan's leaders to change their economic policy toward China, at least on the surface. Intense fighting devastated much of China's economic infrastructure and local sociopolitical institutions. Japan's attempts to introduce a unified currency in north China tied to the yen stimulated rapid inflation. Mines did not receive the necessary machinery, materials, and food because of clogged railways, partisan attacks on infrastructure, natural disasters, and Japan's export restrictions to meet its own domestic demand. Railway companies were ordered to prioritize the transport of natural resources and military provisions over foodstuffs, which worsened inflation, starvation, and popular unrest. Natural and manmade disasters (such as the 1939 Tianjin floods and the Nationalist Army's 1938 breaching of the Yellow River's dikes) created even more suffering, displacement, famine, inflation, and property loss. Factories and mines faced acute labor shortages because of the war, food shortages, and natural disasters. As a result, military and Asia Development Board officials realized that if they continued to prioritize the provisioning of natural resources to Japan without stabilizing the Chinese people's livelihoods, they would not only not meet Japan's wartime mobilization targets but also create even more social unrest. In the midst of war, natural disaster, and intense pressure to meet the increasing demands of Japan's wartime economy, occupation officials in China concluded that a more comprehensive economic policy was necessary to meet their other goals of "pacification" and winning Chinese cooperation.[86]

In March 1940, Japanese military authorities merged the Provisional and Reformed Governments of China into a unified regime in Nanjing led by Wang Jingwei under the slogan of "peace, anticommunism, and national construction." In April, Miyamoto flew to Beijing to attend the

Japan-China-Manchukuo Economic Conference, a meeting of bureaucrats from various Japanese ministries, Asia Development Board branch leaders, military officers, Manchukuo officials, and special company managers to rethink economic policy toward China. In addition to inspecting various mines, factories, ports, and rivers, Miyamoto and the reform bureaucrat Mōri Hideoto also met with Asia Development Board officials in Qingdao to discuss north China's industrial development. As Ōyodo notes, Miyamoto began to realize the "unscientific" character of "Asian development" in light of reports of food and resource shortages, rampant inflation, damages caused by natural disasters, and unreasonable demands and restrictions created by Japan's managed economy policies.[87] In June, he traveled to Beijing again at the request of the board's North China branch and military authorities to help draft a Five-Year Plan for North China Industrial Development. After he inspected major coal mines in north China and interviewed board bureaucrats, military officers, engineers, and special company officials with jurisdiction over coal mining, agriculture, rivers, canals, railways, salt production, electricity, communications, and ports for most of June, he returned to Beijing to draft the "Comprehensive Regulatory Outline Five-Year Plan for North China Industrial Development."

Drafting an industrial plan for north China provided Miyamoto with an opportunity to incorporate some of his ideas on comprehensive technology directly into occupation policy. It began by emphasizing the need for a short-term "comprehensive plan" given the war's constantly shifting nature and China's worsening economic conditions. Although it was "comprehensive," it also stressed "prioritization" by focusing primarily on increasing mineral resource (especially high-quality coal) production to meet Japan's goals of expanding military production, as well as rice and grain production to resolve north China's food shortages, restore social stability, and eliminate import dependence. Twenty development plans were drafted for coal, iron ore, and some steel production in different regions; canal, railway, and port construction; electricity generation for mines; grains, rice, vegetable, cotton, and salt production; and flood control. On the one hand, Miyamoto's overall industrial plan emphasized comprehensive coordination between the experts involved in each prioritized area to maximize effectiveness; on the other hand, it also called for them to set "rational production goals" and make concrete plans based on local context to reach those specific targets. He therefore incorporated the principles of integrated, coordinated planning and technical specialization, which were central to his notions of "comprehensive technology" and "technology for Asian development," into the military's plans.[88]

Recognizing the failure of Japan's policy of "mobilized excavation" of Chinese coal, Miyamoto's plan called for a gradual shift to "planned development" and "rational adjustments" to Japanese demand. It outlined a massive program to improve mining technology, secure essential building materials, increase the number of mining engineers, rationalize labor management, develop electrical power facilities, extend and improve railway links, and expand port facilities and river transportation. According to the production plans for coal, iron ore, and salts, most of these resources would be destined for Japan (around 25–30 percent for coal, 50 percent or more for iron ore and salt), and the rest was to be used domestically or exported to Manchukuo and other parts of China. With regard to agriculture, Miyamoto's plan prioritized rice and grains to solve China's food crisis and cotton to help Japan's declining textile industries. A "comprehensive production plan" that included land development, irrigation, river management, flood control, and canal construction or improvement was also included. According to the five-year plans for improving wheat, millet, sorghum, and corn production drawn up by experts in the board's North China office, the introduction of new agricultural technologies, pesticides, fertilizers, and irrigation channels would enable them to grow enough food on less land, thereby enabling any surplus land to be used for cotton. Infrastructure, such as port construction at Tanggu and port expansion at Qingdao, flood control and river improvement projects in the Hai River basin around Tianjin, the Shijiazhuang to Tianjin and the Baoding to Tianjin canals, and road improvement and urban planning projects, were also prioritized, and a whole influx of engineers and experts into China followed suit. In sum, Miyamoto introduced prioritization, comprehensiveness, and concreteness into the overall outline, which subsequently set the tone for the board's twenty other plans in agriculture, mining, and transportation.[89]

Miyamoto's plan for north China industrial development sought to transform the region into a more productive natural resource and labor extraction machine linked to Japan and Manchukuo's heavy industrial centers through a fast, efficient transportation network. Japan's wartime mobilization system provided the framework within which his visions of "comprehensive technology" and "technology for Asian development"

were being implemented. Although the industrial development plan resembled what Japanese critics described as the West's colonial "resource
exploitation" system and seemingly contradicted Miyamoto's notions of
winning Chinese cooperation through the promise of industrial development, he viewed this as a short-term necessity in the face of worsening
tensions with European powers and the United States, who were increasing economic sanctions and imposing their own industrial controls at
home. Within the technology bureaucrats' technological imaginary,
rationalizing the extractive mechanisms in the colonies for total war
constituted only one stage of their long-term dream of Asian development, which in 1939 Miyamoto estimated would take up to thirty years.[90]
Although the plan somewhat departed from his previous arguments for
Japan to help China develop a wider range of economic areas, similar to
his reform bureaucrat allies, he concluded that wartime mobilization was
the best way to overcome liberal capitalism and rapidly transform Japan
and East Asia into an advanced, heavy industrial region surpassing the
West. In this way, Miyamoto incorporated his grand visions of comprehensive development into the military's narrow, short-term agendas on the
battlefield, thereby rationalizing Japan's brutal exploitation of colonial land,
labor, and resources. Ultimately, the technological imaginary demanded
the complete mobilization and sacrifice of colonial life and resources for its
achievement.

Miyamoto became increasingly disillusioned at the possibilities of
achieving Chinese-Japanese cooperation over the course of several more
inspection trips to China in 1940. A few months before he drafted the
Industrial Plan for North China in July 1940, he witnessed the continued
strength of European capital in cities like Shanghai and the strong anti-
Japanese sentiment among the Chinese, and he heard reports about the
army's myriad logistical problems and material shortages, leading him to
conclude in his diary that "East Asian construction" was "quite distant,"
particularly in the case of central China.[91] In another diary entry that he
wrote in June 1940 as he conducted investigations for the north China
industrial plan, he wrote, "How do we make the Chinese appreciate Japan
and welcome Chinese-Japanese collaboration?"[92] "Many problems in the
prospects for China-Japan collaboration. Perhaps Japan does not have the
ability to embrace another nation," he lamented again in September 1940
after a discussion with Japanese embassy officials in Nanjing.[93] Miyamoto, however, continued to insist that industrial development was the

key to winning over the Chinese people's hearts. He told Itagaki Seishirō, the China Expeditionary Army's chief of staff, that Japan's development efforts would be like a "castle in the sand" if they continued to alienate the Chinese people by economically blockading Chongqing (Chiang's capital in Sichuan Province), and he sent out his opinions on resolving the war through technical development to top cabinet and military officials before his sudden death from overwork in 1941.[94] Thus, although he privately expressed doubt and frustration over the progress of East Asian construction, his utopian visions of "technologies for Asian development" always overrode any reservations that might have arisen in the face of fierce Chinese resistance and Japan's continued reliance on brute exploitation of Chinese labor and resources. Many young engineers were inspired by similar notions of building a new China and flocked there to pursue a whole array of technical projects, demonstrating the technological imaginary's tremendous power during the wartime era.

Conclusion: Technology as Imperial Nationalism

From the 1920s, engineers expanded technology's meaning to include creative vision, rational planning, and optimal social management as part of the struggle to improve their status in society and government. They adopted predominant representational techniques of colonized peoples as "unchanging" or "backward" in order to justify their roles as heroic developers of the Asian continent. By the mid-1930s, their conceptions of technology had entered into the highest levels of national policy, colonial planning, and state ideology. "Comprehensive technology" or the planning and coordination by experts of projects that achieved multiple political, economic, and cultural goals became institutionalized within such wartime "comprehensive national policy organs" as the Cabinet Planning Board, the Asia Development Board, and colonial administrations. Japan's avowed mission to establish the "New Order in East Asia" after 1937 also became infused with the conceptions of engineers and technology bureaucrats. Although scholars have largely emphasized the cultural messages of pan-Asian ideology, such as preserving Asian tradition and harmony (e.g., the "Kingly Way"), the Shintō conception of placing the "eight corners of the world under one roof" (*hakkō ichiu*), or a more assimilating conception of "Japan" that incorporated other ethnicities, few have focused on technology as an essential component of Japan's pan-Asianist ideology.[95] As a result of intense lobbying by technology associations combined with increasing patronage by the military and reform bureaucrats, such conceptions as "technologies of Asian development" took on concrete form in research studies, industrial plans, and colonial infrastructure projects. The technological imaginary's various promises of development and modernization served as a powerful horizon for the incorporation of various peoples into the Japanese empire. In doing so, the technological imaginary rationalized Japan's intensive and comprehensive exploitation of colonial land, labor, and life. Chapters 3 and 4 examine specific colonial infrastructure projects and the role they played in the formation of the technological imaginary as well as the effects these had on the colonized.

Chapter 3

Constructing the Continent

Technology in Wartime Manchukuo and China

In 1940, the film *Vow in the Desert* (*Nessa no chikai*), featuring the Manchurian-born, bilingual actress Yamaguchi Yoshiko (popularly known as Ri Kō Ran) and the star actor Hasegawa Kazuo, played in theaters across Japan. Part of a series of continental "goodwill films" produced under the "Greater East Asia Film Sphere" policy, it portrayed a love story between Li Fangmei (Ri), the daughter of a Chinese scholar of Japanese music, and Sugiyama Kenji (Hasegawa), the younger brother of Sugiyama Ichirō, a North China Construction Agency engineer supervising the building of a 1,000-kilometer "New Anti-Communist Road" between Beijing and Xian. After a long pan shot of an exotic camel caravan in the Xian desert, the film begins with Ichirō gazing over the Great Wall of China extending into the horizon and pondering aloud to Yang, his Chinese technician companion, about the "extraordinary power" of the men who built this immense monument. Impatient to return home, Yang disdainfully comments on the Great Wall's impracticality and its inability to accommodate truck traffic. Ichirō chuckles and replies, "Well then, why don't we build a Great Wall where you could drive a truck on?" and points to the empty space where the new highway was going to be built.[1]

Engineers from the North China Construction Agency—established by the North China Expeditionary Army in 1938 to supervise flood control, road construction, and urban planning—helped in the film's production at various Chinese project sites.[2] Within the larger narrative of a love story between Kenji and Fangmei, the film depicts scenes of a Chinese-speaking Ichirō trying to persuade a reluctant landlord of the road's benefits for promoting economic growth and alleviating rural poverty; communist agents obstructing road construction through sabotage and spreading rumors about the road's destruction of ancestral graves; Japanese officials discussing road building as an effective anti-insurgency tool; Yang expressing shame over his fellow countrymen's resistance to modernization; agency engineers struggling to rescue the road and nearby villages from a vicious flash flood; and several scenes of Chinese workers using the latest construction machinery. The Chinese heroine Fangmei wholeheartedly supports the highway project, and she helps realize its final implementation by successfully mediating between the Japanese engineers and Chinese resistance throughout the film. Although these scenes clearly exaggerated actual conditions, the film and Itō Hisao's hit song, "Song of Construction" (*kensetsu no uta*), "really made young people's blood stir," according to Ueno Mitsuo, an engineer who participated in canal construction in north China.[3] Perhaps the song's lyrics best capture their emotions upon setting off for China as engineers during the war—"To our joyous voices in song—Shine, oh golden clouds of the wasteland! It's morning, it's morning! Our song of construction is bursting forth across the continent!"[4]

This chapter continues the examination of state engineers or technology bureaucrats and their understandings of the term "technology." It analyzes how project engineers on the ground developed and employed the technological imaginary in Japan's wartime empire, particularly at three large-scale technology project sites in Manchuria and China. Engineers had been working in Japan's empire since Taiwan's colonization in 1895, building cities, roads, railways, ports, and communications infrastructure. But until the Manchurian Incident of 1931, Japan's public works projects were largely geared toward facilitating agriculture, mining, and commerce. As Kobayashi Hideo notes, Manchukuo's establishment signified a shift in the colonial economy from a railway and port-centered imperialism largely based on commercial agricultural and mining to one geared toward military industrialization.[5] With this shift toward investment in industrial infrastructure in Japan's rapidly expanding empire after 1931, Korea, Taiwan, Manchuria, and China became arenas for engineers to develop the technological imaginary on a much larger and more intensive scale through the design and construction of comprehensive engineering projects to systematize and integrate Japan's empire into a total war economy. Although these projects were sustained by colonial inequalities, racism, and violence and facilitated the increased exploitation of colonial labor and resources, they also constituted a utopian attempt to transform Japan's empire into a rational, optimal, and efficient system that mobilized and incorporated its various peoples.

As scholars have recently shown (and as discussed in Chapter 2) with regard to state engineers' conceptions of "technologies for developing Asia" and "comprehensive technology," idealistic notions of developing East Asia through technology served as a guide to policy and action for Japan's governing elites.[6] Although it has become increasingly clear that Japan's empire was a laboratory for a wide range of utopian-minded blueprints, policies, and projects that incorporated the very latest Western technologies and trends in technocratic planning, how these actually took shape and were implemented in the colonial context is less apparent. Largely absent from studies of Japan's "brave new empire"—as Louise Young aptly describes the utopian mindset of Manchukuo's technocrats—are the colonial relations, interests, and forces that went into making their ambitious blueprints.[7] The various modernist plans and projects of Japan's "brave new empire" seem to be largely formulated in the minds of skilled technocrats (and perhaps later challenged by other colonial stakeholders) rather than in dynamic relation to unfamiliar natural environments, clashing and shifting interests, uncooperative populations, technical limitations, and unexpected turns of events in the colonies. Without an analysis of these constitutive colonial dynamics, the Japanese engineers' projects (and their technological imaginary) often appear more durable, systematic, and coherent than was actually the case.

Until now, our analysis of "technology" has largely remained at the level of how intellectuals, engineers, and bureaucrats conceptualized and appropriated the term for their utopian agendas of "constructing East Asia." But these experts did not first formulate their conceptions intellectually among themselves only to later implement their modernizing designs upon a passive colonial landscape. Contrary to their confident conceptions and neat blueprints of rationalizing the colonies through the application of Japanese technology, an analysis of their plans and projects shows that these were fundamentally imbued with ambiguity, contradiction, incoherence, and contingency from the very beginning as a result of being continuously subject to conflicting interests and forces in the colo-

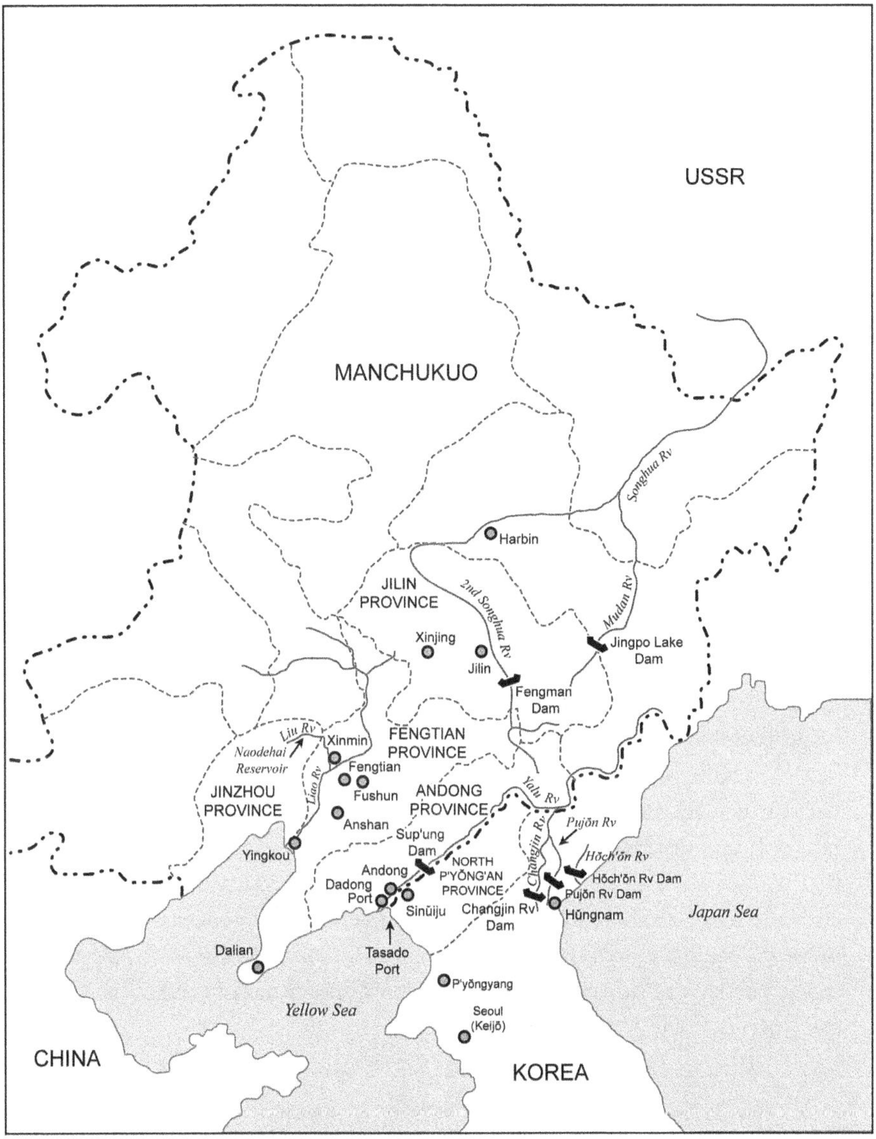

Figure 3 Key cities, rivers, and project sites in Manchukuo and northern Korea. Japan's major colonial dam projects are shown. The three dam projects near Hŭngnam in Korea were "conduit-style" (*suiro shiki*) dams, each consisting of a series of several steep drop-offs for the water to flow into the power stations at each drop-off point.

nies. The technological imaginary was much more than the body of ideological statements articulated by different Japanese elites in the public sphere. It emerged simultaneously in difficult negotiations with various people, institutions, and environments at different project and policy-making sites. This chapter therefore goes beyond a solely intellectual understanding of technology by examining the formation of the engineers' technological imaginary on the ground, specifically, their emerging notions of "comprehensive technology" as embodied in three separate projects. This conception meant the transformation of nature and society into a rational system in which each part contributed to other parts of the whole in efficient and mutually reinforcing ways.[8] Although such a conception was already gaining prominence among Western governments at the time, in Japan's case, this idea was formed in the context of imperialism and war in East Asia during the 1930s and 1940s and constituted an important ideological justification for Japanese imperial rule. Complex negotiations and processes were involved in planning and constructing a major flood control project in southern Manchuria, an urban planning project in Beijing, and a dam-powered coastal urban industrial zone on the border between Korea and Manchukuo (see Figure 3). This chapter denaturalizes the technological imaginary by revealing how one of its important manifestations, "comprehensive technology," was more an effect of specific practices in particular colonial contexts than a monolithic, well-formed ideological discourse designed and imposed from above by Japanese technocrats. Thus, this chapter seeks to avoid confirming "colonialist interpretations of their endeavors" that may result from an uncritical reading of the writings and studies produced by colonial experts in order to "support, strengthen, and maintain colonial rule . . . by appealing to modern notions of the developmental potential of colonialism," as the Korea scholar Andre Schmid notes.[9]

The Institutionalization of "Comprehensive Technology": The Liao River Improvement Project in Southern Manchuria

Japan's expansion into Manchuria in 1931 coincided with the development of what was called "comprehensive technology" among state engineers and engineering associations in Japan. As noted in Chapter 2, the term had some of its origins in the movement by engineers to increase their status in government through the promotion of themselves as managers of an increasingly complex society rather than as narrow specialists; the attempt to develop a more coordinated approach to domestic flood control among competing government ministries; and the global trend in engineering toward the construction of large, multipurpose dams that simultaneously addressed electricity production, flood control, transportation, water supply, and irrigation. Comprehensive technology, however, really developed in practice as prominent Kōjin Club leaders, such as Miyamoto Takenosuke and Naoki Rintarō, lobbied the government to send more engineers to develop the "new frontier" of Manchukuo and establish an integrated national technology administration there.[10] The establishment of such powerful institutions as the General Affairs Agency of Manchukuo's State Council, which combined policy planning and implementation in a wide array of areas under one roof, and the expansion of the Transportation Department (*kōtsūbu*) in 1937, which unified the National Roads Bureau (*kokudō kyoku*) and the Public Welfare Department's Civil Engineering Bureau into one centralized technology administration, enabled engineers to develop their notions of coordinated technical planning in Manchukuo relatively free from bureaucratic obstruction.

Haraguchi Chūjirō, a former Kōjin Club director, accompanied Naoki and the first group of Home Ministry engineers from Japan to head the National Roads Bureau's Xinjing Construction Office in 1933. In addition to planning and managing construction of the Xinjing-to-Jilin road, which formed part of a planned industrial corridor and larger Manchukuo road network, he established and ran the Xinjing Engineering Academy, a three-year vocational school that trained mostly Japanese but also some Korean, Mongolian, and Manchu students in basic civil engineering, mining, and construction.[11] Because the Roads Bureau also had jurisdiction over flood control, Haraguchi quickly turned his attention to the pressing problem of managing Manchuria's flood-prone rivers. Similar to what they commonly practiced in Japan, engineers first implemented a purely defensive policy against floods and focused largely on high dike construction and shore reinforcements in and around major cities.[12] Flood control of the Liao River region in the south—a major 234,700-square-kilometer river basin centering around the 1,345-kilometer-long Liao River that formed the nucleus of Manchuria's breadbasket and around which most of its population lived and its industrial cities were located—received particular attention from the Guandong Army and the State Council, especially

after enormous floods in 1932 and 1933 soon after Manchukuo's formation.[13] Angry at Japanese settlers' uncontrolled water use for paddy development and the Economic Department's prioritization of hydropower
development without any consideration of flood control, Haraguchi and
his team in the Roads Bureau's Second Division began formulating a
coordinated approach to managing the Liao River basin that combined
irrigation, hydropower production, flood control, and transportation
improvement.[14]

Their numerous studies noted the Liao River's particular problems,
such as the generally high silt content, the low yearly rainfall combined
with a concentrated summer rainy season, the winding nature and slow
streamflow in much of the central plains, the existence of major confluence points where several large tributaries flowed into one another, and the
high alkali level of adjacent lands. These factors combined to cause frequent
floods and droughts, unpredictable changes to river courses, frequent dis-
ruptions to transportation and communications, low harvests, and massive
property damage, population displacement, and poverty. Floods in southern Manchuria caused an average of 25 million yen in damages per year,
the worst ones easily exceeding 100 million yen. From these multiyear
studies, Haraguchi and other bureau engineers formulated a large-scale
and ambitious plan to redirect the Liao's tributary rivers by diverting
them and constructing drainage channels to relieve pressure at overbur-
dened confluence points and provide water for irrigating lands for future
Japanese settlement; building dams and water detention reservoirs to lessen
silt volume, reduce yearly flood waters, and supply water for increased river
transportation, urban and industrial use, and a new canal network; and
constructing levees, embankments, and weirs to prevent excessive sedimentation and better defend against flooding, thereby opening up surrounding lands for agriculture.[15]

By 1937, especially after the outbreak of hostilities with China in
July of that year, the engineers' emerging notion of "comprehensive technology" began to be incorporated into the policies of reform bureaucrat
state planners and the Guandong Army, who sought to transform Manchukuo into a total war economy. In August 1937, the Manchukuo government announced a national river investigation that emphasized the
integration of flood control, hydropower production, and land and transportation improvement in the interests of increasing Japanese immigration
and promoting Manchukuo's industrial development, thereby linking ar-

eas related to river management hitherto viewed as having entirely different purposes and being under separate ministerial jurisdictions. Also that
year, all engineering administrations were united under the Transportation Department with Haraguchi becoming the head of its Waterways
Division, which allowed for smoother coordination between different engineering specializations. After formalizing their plan for the Liao River
at the Manchukuo Flood Control Meeting in October, Haraguchi presented it to the newly created Liao River Flood Control Plan Commission
that December (see Figure 4). These important governmental gatherings
represented the growing institutionalization of "comprehensive technology," which had not yet been realized in Japan because of interministerial
conflict. They brought together Transportation Department engineers,
Guandong Army officers, Industrial Department planners, provincial directors, settlement agency officials, Mantetsu representatives, and such
top Home Ministry engineers from Japan as Miyamoto and Tatsuma
Shōzō. The December commission meeting was presided over by Hoshino
Naoki, leading reform bureaucrat and head of the General Affairs Agency,
and opened with a message from Guandong Army Chief of Staff Tōjō
Hideki, who emphasized the project's importance for "building an agricultural nation" (*nōgyō rikkoku*), promoting industrial development, improving public welfare, and preserving national land.[16] Launched in 1938,
the fifteen-year, 100 million–yen plan persuaded a reluctant Guandong
Army of the efficacy of large-scale, coordinated public works projects and
became a prototype for a whole range of comprehensive development initiatives in Manchukuo and China. It was a "silent pacification project worth
a thousand words," as Manchukuo's official propaganda organ described
it in 1938.[17]

Although Haraguchi and his team presented their project as the rationalization of Manchuria's environment and the bringing of "culture" to
its "wild" landscape, their studies and plans were in fact quite unstable
and originated in a diffuse array of sources.[18] Japanese engineers had little
experience with China's larger-scale rivers and more extensive and devastating floods. Nor were they very familiar with its climate and geography.
The Roads Bureau's Rivers Division expanded upon earlier studies on
rainfall, streamflow, water volume, and topography conducted by the
Russians and Mantetsu, as well as previous smaller-scale flood control
projects carried out by local governments, Mantetsu, and the warlord
Zhang Zuolin (in cooperation with Western treaty port powers near their

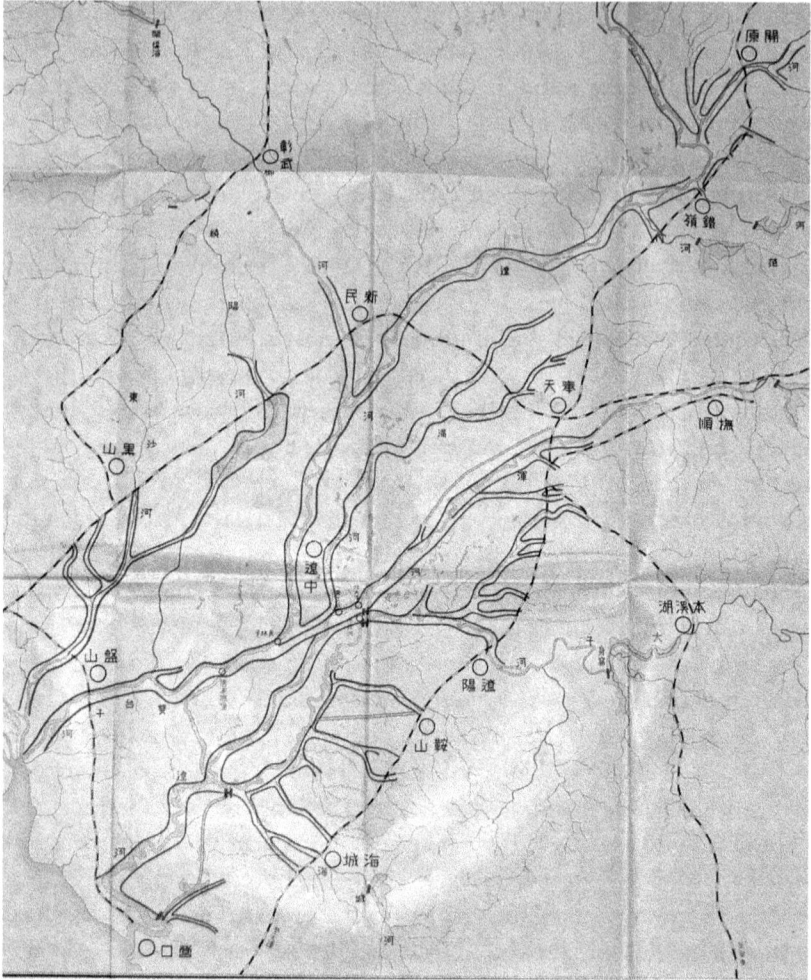

Figure 4 Abbreviated blueprint of the Liao River Improvement Project. The dark lines indicate planned dike construction along rivers or new river diversion routes. Short bold lines show reservoir points or water gates to control river flow (for example, Naodehai Reservoir is in the upper left corner). Mantetsu rail routes are shown as black and white barred lines. Circles are major cities such as Yingkou and Fengtian. Source: *Ryōga chisui keikaku shingikai gijiroku* (supplementary map).

zones of influence). One apparent source for the comprehensive plan was a Russian engineer named Unikowski, who formerly worked for Russia's Chinese Eastern Railway and later served as a technical advisor to Mantetsu after Russia's defeat in the Russo-Japanese War in 1905. In 1934, he delivered a widely published lecture titled "The Development of Manchukuo and the Problem of Flood Control," in which he described his experiences of a devastating flood around Harbin in 1903 and an overall plan he had developed since then based on his longtime observations of the Songhua River basin, which consisted of building multipurpose dams for flood control, irrigation, electricity production, and canal construction in northern Manchuria. Thus, earlier Russian expertise and scattered data provided by Chinese and Mantetsu sources partially laid the groundwork for the grand Liao River project.[19]

Japanese engineers actively appropriated existing Chinese knowledge and expertise, yet they continued to see themselves as introducing "modern" Japanese technology to "backward" China. As the Manchukuo government publicly disparaged Zhang Zuolin for letting the rivers "remain wild" and causing immense suffering among the Manchurian people as a result of his lack of a coherent flood control policy, the Waterways Division rapidly took over local Chinese "water-use committees" (*suiri iinkai*) in 1933 and incorporated their staff and experience, a practice Japan continued as they later expanded into north China in 1937.[20] Akigusa Isao, former Manchukuo engineer and head of the North China Construction Agency's Water-Use Division in 1938, in fact noted how Chinese engineers in the newly incorporated North China Water-Use Committee had already arrived at similar conclusions of a comprehensive water control policy along the lines of what the Japanese were proposing for the Liao River and other river basins.[21] Japanese engineers also mentioned adopting cheaper and more effective Chinese dike- and sluice-building techniques after the failure of their own, which were more suited to preventing Japan's smaller-scale and more frequent floods, and adopting a local technique of planting willow trees in arid portions of the river to effectively prevent river bank erosion.[22] In various articles and speeches, Miyamoto often praised the ancient Chinese notion of "water-use," which he said was fundamentally rooted in providing for public welfare as a whole, as opposed to the tendency in Japan to divide "water-use" (*risui*) into separate areas of flood control, irrigation, and hydropower, while other engineers claimed they were applying Ming Dynasty Chinese flood control principles of using a river's own natural qualities to manage floods.[23] Thus, the official discourse that proclaimed Japan as the developer of China and Manchuria through the introduction of "comprehensive technology" hid its hybrid origins and its actual basis in Japanese engineers appropriating different types of local and foreign knowledge.

The overall lack of knowledge among Haraguchi's early team of civil engineers was reflected in their frequent complaints about not having enough data on Manchuria's rivers to actually conduct any extensive, large-scale projects. For example, in 1938, Manchuria had only one precipitation measurement station per 1,600 square kilometers in comparison to an average of one every 160 square kilometers in many Western countries.[24] Upon Manchukuo's founding, engineers joined Guandong Army water provision teams and began setting up meteorological, hydrometric, and hydrological stations in key areas and employed local Chinese officials and schoolteachers (who "received constant guidance so that there would be no mistakes") to measure and gather data. At first, engineers used the Army Survey Division's rough 100,000:1 maps, which in turn were based on earlier warlord and Russian ones; however, this led to many errors, and therefore newer topographic surveys were commissioned. Hydrometers were set up at main river points to measure streamflows, but the frequent shifting of Manchuria's riverbeds caused by floods and damage from floating ice and wood made regular maintenance of the meters difficult, and the meters were checked only twice a year because of Manchuria's vast expanse, budget limitations, and the uncertain security situation.[25] By 1934, Mantetsu engineers had formulated a vague conception of flood control centering on a more active approach of constructing reservoirs and dams upstream to better utilize the water for multiple purposes rather than a purely defensive one of dike building, and even conducted their own detailed studies of several sites on the Songhua and Liao Rivers.[26] These Mantetsu studies, however, were limited, and when Haraguchi presented his more extensive Liao River plan at the Liao River Flood Control Plan Commission in 1938, he emphasized how it was still based on rather sparse data, and he pushed for a more comprehensive study of minimum, average, and maximum streamflows so that they could then more accurately plan the interrelated series of diversion canals, irrigation channels, dams and detention basins, and dikes, weirs, and levees they were proposing.[27] The entire project hinged on achieving a "delicate balance" between the various parts, he noted, and although accurate studies alleviated concerns about such thorny issues as the effects of raised water levels on downriver areas caused by dam building upriver or whether or not the proposed reservoirs could accommodate the enormous silt volumes, in the end even Haraguchi had to admit that managing the Liao River through this new technique of constructing detention dams rested on a large degree of speculation and hope rather than any kind of solid scientific and technical expertise.[28]

"Comprehensiveness" was not only rooted in foreign knowledge and grounded in difficult negotiations with an unknown environment. It was also shaped by conflicts between various colonial interests in Manchuria. At the 1938 Liao River Flood Control Plan Commission meeting, tensions between provincial and state officials, technology and economic bureaucrats, and the state and business interests, such as Mantetsu and the Manchurian Colonization Company (*Manshū takushoku kabushiki gaisha*; hereafter, Mantaku), quickly became apparent as Haraguchi presented a draft version of their Liao River Improvement Plan. The vice directors of Fengtian and Jinzhou Provinces in southern Manchukuo were the project's most enthusiastic supporters, and they even expressed frustration at the slow pace of the proposed implementation schedule. Takeuchi Tokui, Fengtian Province vice director, noted that one successful harvest every three years was considered a great achievement among residents used to natural disasters, and he discussed the difficulties provincial officials faced in making villagers feel part of a larger Manchukuo nation when they could not even properly pay schoolteachers or collect taxes because of the scale of frequent floods. He also expressed concern about the multipurpose dam idea, noting that agricultural demand for water might eventually exceed supply, and argued instead for the construction of separate dams for different purposes of flood control and irrigation. Whereas Haraguchi and other state engineers wanted to conduct comprehensive investigations before implementation to avoid costly construction mistakes resulting from bad data, local officials argued that delays would not only prolong the residents' suffering but also push up land prices as a result of speculation, thereby causing further hardship for Japanese immigrants. Criticizing the state's top-down policy, they urged central authorities to take a "total mobilization" approach instead, in which the people's "energy" and "strength" would be immediately utilized through dike building and reforestation campaigns conducted in lockstep with state goals of increasing immigration and agricultural productivity. Thus, although the provinces enthusiastically supported the engineers' Liao River project, they also did their best to alter their vision of "comprehensiveness" toward what they saw as the more urgent needs of flood control and to assert their own control over its implementation.[29]

Other institutions also continued to push for one aspect of the proposed project over another. Mantaku urged for the expansion of land reclamation programs to ultimately achieve their twenty-year goal of settling one million Japanese households, more research into effective fertilizers to combat declining agricultural productivity, and increased attention to reforestation to prevent soil erosion. The Guandong Army demanded the coordination of dike building with road construction to allow for military transport. Mantetsu urged more careful flood control investigations and planning to prevent costly changes to rail routes that might result from future shifts in river courses. Engineers also used the project to increase their own power and successfully won control over its administration through the newly expanded Transportation Department. Haraguchi had passionately argued for more comprehensive river studies and the resources to conduct them before the construction's beginning. Miyamoto, the influential technology bureaucrat, flew in from Tokyo to attend the Liao River Flood Control Commission meeting, where he strongly supported Haraguchi's pleas. He noted that although Japan had a long history of river improvement dating from the Meiji period, they still lacked sufficient data on many of Japan's rivers, causing enormous waste of state resources as engineers simply reacted to each natural disaster by conducting expensive local investigations after the fact and constructing dikes accordingly as opposed to pursuing projects based on comprehensive national data that would allow for more effective long-term planning, implementation, and damage control. At the meeting, Hoshino, who represented the economic planning orientation of reform bureaucrats (see Chapter 5), pushed instead for the project's immediate implementation in order to meet the economic goals set forth in the Manchukuo Five-Year Industrial Plan, and he received strong support from the provincial vice directors, the Guandong Army, and Mantetsu. Haraguchi reluctantly agreed to Hoshino's demands for quick implementation while continuing to express uncertainty because of insufficient data and cuts in their investigation budget, as other officials, such as the head of agricultural affairs in the Industrial Department, worried about a loss of state prestige among the people in the event of future failures caused by hasty investigations and bad data. Thus, from the very beginning, Japan's visions of "comprehensiveness" were infused with conflict and tension among different institutions asserting their interests, belying claims by engineers regarding their own expert coordination of flood control, electricity production, agricultural settlement, and transportation improvement for Manchuria's future prosperity.[30]

In the interests of accommodating the engineers who demanded the prioritization of comprehensive river studies and those groups who pushed for implementation without delay, the Transportation Department decided to pursue both directions at once. Haraguchi was placed in charge of the Liao River Flood Control Investigation Office in March 1938, which proceeded to set up branches in Liaoyang, Fengtian, and Tieling. They first emphasized conducting geological and topographical investigations for upriver dam and reservoir sites that formed the heart of the entire plan to use the rivers for multiple purposes. Basic statistics for all of the Liao River tributaries were compiled and published between 1938 and 1940, and the office was closed in 1941 after all investigations were completed.[31] As these studies began in 1938, the Transportation Department also started a major initiative in the western portion of the Liao River basin, the Liu River Flood and Erosion Control Project. The Liu tributary was a major source of the high silt volume in the Liao River, as it originated in the arid mountains near Mongolia, and therefore contributed to rapid desertification in the western Liao River regions. Mantetsu's important Fengtian-to-Shanhaiguan rail line, which connected Manchukuo to north China, was constantly threatened by rises and shifts in the riverbeds caused by high silt volume, leading to three costly bridge, rail, and station renovation projects since 1903. The silt also caused flash floods as a result of the elevated riverbeds and obstructed traffic at the important port of Yingkou at the Liao River's mouth. Thus, managing the Liu River was prioritized over other aspects of the Liao River flood control project because the Liu River threatened important transportation links that were essential to Japan's war efforts and its management was viewed as one of the keys to resolving southern Manchuria's chronic flood problems.[32]

Igarashi Shinsaku, one of the Liu subproject's main engineers who had significant experience investigating the entire Liao River region, revealed some of the ways that the plan emerged and evolved on the ground in a 1943 report he wrote for a colonial engineering journal. Overall, the Liu subproject had three interrelated parts: the construction of small anti-erosion dams and reforestation projects upriver near the Liu River's source; one large reservoir mid-river at Naodehai to trap sediment, manage yearly river flow, and provide irrigation; and dikes and levees downriver near the city of Xinmin to defend against future floods. Most of their efforts went toward building the Naodehai Reservoir because trapping the sediment was seen as the quickest and most effective way to resolve the area's severe flooding problems and put the water to good use. Because of the project's urgency, engineers simply made rough estimates of streamflow at the reservoir site based on scattered precipitation and streamflow measurements made earlier by the warlord regime and Mantetsu as well as their own investigations at four sites. Based on these statistics, they arrived at parameters for the dam and reservoir, which they predicted would decrease the Liu River's average streamflow from five thousand to three thousand cubic meters per second, as it had a storage capacity of 120 million cubic meters of silt and 50 million cubic meters of water. During the summer flood season, the silt would be sluiced at the dam, allowing clear water through, and during the dry seasons water was stored for controlled release whenever necessary for other purposes. Engineers estimated that the detention reservoir would reach near capacity in fifteen years, upon which the other anti-erosion projects upriver and flood control construction downriver would further combine to alleviate and control the river's sedimentation issues in the long term.[33]

In addition to being unfamiliar with the Liu's river environment, engineers continuously struggled with the area's geology and the river's large silt volumes, which challenged the principles of hydrology they had learned in Japan. In order to determine the optimal dimensions and locations for the Naodehai reservoir and dam as well as the various dikes, scientists at Manchukuo's Continental Science Board built a small-scale model in a laboratory at Xinjing that replicated a section of the Liu River near Mantetsu's rail line—complete with sand and soil from the area, which was used to re-create the banks and flowing silt, water provision mechanisms to maintain different streamflow rates and depths, and adjustable inclines to change river slopes to simulate various environments and climate situations. The region's geology was also deemed unsuitable for large dams because most of the rock was very brittle. They therefore invited several experts to conduct boring investigations to confirm whether their chosen site was indeed adequate for the dam despite the cracks in the geological foundation, which might lead to disastrous leaks in the future.

Much of the soil around the area also consisted of fine yellow sand, which was not ideal for supplying the necessary gravel for manufacturing-quality concrete. Engineers therefore had to additionally build a fifteen-kilometer road to several adequate gravel sites, send samples to the Continental Science Board to calculate the correct proportion of the area's gravel with other materials for mixing into durable concrete, and construct a materials testing facility onsite to ensure concrete and cement quality. Employing local officials to mobilize nearby Chinese labor, Naodehai Reservoir was completed in 1942, as were several anti-erosion and dike projects on different sections of the Liu River.[34]

Because of the continuously worsening war situation, much of the ambitious Liao River project was never implemented or completed. Aside from the Liu River project, engineers completed 85 percent of a large earthen irrigation and flood control dam on the eastern Liao River by mobilizing 1,740,000 mostly Chinese laborers of the Manchukuo Development Labor Service Corps between 1943 and 1945 and began a hydropower and irrigation dam project on the Taizi River.[35] Despite this lack of progress, the Liao River Improvement Project represented the first step toward institutionalizing a conception of "comprehensive technology" or the concept of technology projects closely coordinated with economic, political, military, and cultural objectives within the colonial regime. As engineers developed this conception in the actual project after receiving official blessing, ideas of comprehensiveness quickly spread to such other projects as multipurpose dam building (see Chapter 4) and urban and regional planning in different parts of Japan's expanding empire.

"Developing Asia": Japanese Engineers Advance into China

Immediately after their success in convincing the Guandong Army to commit to comprehensive technology at the 1937 Manchukuo Flood Control Meeting, Miyamoto and three other Home Ministry engineers visited north China to conduct an inspection tour.[36] Based on their subsequent report, the North China Expeditionary Army also became persuaded of the importance of coordinated large-scale public works projects and immediately requested Naoki Rintarō, head of Manchukuo's Civil Engineering Bureau, to send ten engineers and staff to plan road-building and river improvement projects at the beginning of 1938.[37] In March 1938, with the help of Home Ministry engineer Tatsuma Shōzō, the army established the North China Construction Agency in Beijing to unify all civil engineering activities under Wang Kemin's collaborationist government. The agency was placed in charge of immediate wartime reconstruction and development and the planning and building of roads, ports, waterworks, and urban areas (see Figure 5). Miura Shichirō, who had closely cooperated with Miyamoto earlier in recruiting young engineers to Manchukuo and China, was appointed as its head. Upon leaving Japan for Beijing, Miura announced that he and his initial group of forty-five engineers were off to "rescue" the Chinese people from years of civil war and exploitation. In an effort to rebuild the "cradle of Oriental civilization," they were going to conduct river improvement projects on the Yellow River, the Hai River near Tianjin, and the Yongding River in Beijing; construct a world-class port at Tanggu; and build an extensive network of roads and irrigation channels. It was the Japanese people's duty, he added, to guide China with modern technology and to "establish civilization" there.[38] The agency's establishment represented another achievement for the Japanese engineers' movement, to establish not only a powerful engineering administration rooted in "comprehensive technology" but also another base from which they could realize their visions of "constructing East Asia" free from bureaucratic interference.

The North China Construction Agency had branches in Tianjin, Jinan, Taiyuan, and Shijiazhuang as well as numerous project offices at road construction and river improvement sites. By 1939, there were 132 Japanese engineers in China, and by 1941 there were 314, which represented 25 percent of the agency's staff.[39] Japanese engineers were placed in technical positions or as advisors, and Chinese constituted the majority of administrative, technician, surveying, security, and secretarial positions. Yin Tong, formerly an economic bureaucrat for the Chinese Nationalist government and one of their top Japan experts and negotiators after he graduated from the Army Communications School in Japan, was selected to head the North China Construction Agency. According to a North China Expeditionary Army advisor to the collaborationist government, Yin was well known for his bold, straightforward criticisms of Japanese policy, and he was well respected by Japanese engineers for not being afraid to tell army officers his honest opinions.[40] He made efforts to invite the best Japanese civil engineers to China and continuously pushed for such large-scale projects as Yellow River flood control and urban and regional

Figure 5 Major cities, rivers, and project sites in north China.

planning.[41] Together with army "pacification officers" (senbukan) and Japanese and Chinese members of the New People's Association (shinminkai), a mass organization (later co-headed by Yin) that was the ideological wing of the North China Provisional Government, the agency presented its projects in terms of promoting long-term security, stability, and development as part of Japan's larger mission of restoring Asia's prosperity through its ongoing efforts to liberate China from Western imperialism, capitalism, and communism.[42]

Between 1938 and 1944, the North China Construction Agency actively pursued a three-pronged agenda of coordinating road construction, flood control, and urban planning in the midst of the North China Expeditionary Army's brutal military campaigns and intense attacks by Chinese communist guerrillas and other irregular forces. In order to create the "arteries of North Chinese development," agency engineers and technicians eventually improved, constructed, or paved about four thousand kilometers of roads.[43] Priority was placed on roads next to the main Chinese railways along which Japan's armies advanced—Beijing to Hankou, Tianjin to Pukou, and Taiyuan to Shijiazhuang. Another high-priority project was the Tianjin-to-Tanggu road, where the army was building a large port to quickly expedite the transport of natural resources to Japan.[44] In flood control, much of the agency's early efforts went toward defending Tianjin from huge floods in 1939 and partially containing one of the worst disasters in modern Chinese history, the 1938 breaching of the Yellow River dikes in Hebei Province by Chiang Kai-shek's forces to impede the Imperial Army's progress. The Yellow River floods eventually inundated seventy thousand square kilometers of land and killed and displaced hundreds of thousands of mostly poor farmers.[45] Citing the ancient Chinese proverb "Whoever governs China's rivers, governs the people," the agency took over the Nationalist government's North China Rivers Administration and implemented a range of flood control and water use projects.[46] Based on existing blueprints and statistics collected earlier by Chinese and League of Nations engineers, the agency drew up several plans similar to the ongoing Liao River project in Manchukuo and employed the same strategy of active, comprehensive flood control through damming and diversion. Projects included building a diversion canal where the Daqing, Ziya, and South Canal met near Tianjin in order to prevent water buildup upriver during the summer rainy season, a main factor contributing to the area's disastrous floods; a canal from Shijiazhuang to Tianjin to establish an "economic development corridor" through north China; irrigation canals from the Yellow River into the surrounding plains to promote agriculture; and an ambitious plan to build a multipurpose dam at Sanmenxia in western Henan Province on the Yellow River that would contribute to irrigation, electricity production, transportation, flood control, and industrialization.[47]

All of the agency's proposed projects were rooted in variations of the emerging discourse of "comprehensive technology," whether it was in the form of modernist cities strategically integrated with surrounding regions and national defense objectives or flood control projects combined with efforts to increase agricultural production, produce more electricity for industry, and improve transportation links between large cities. Similar to the Liao River Improvement Project in Manchuria, these comprehensive projects arose amid a web of interests, institutions, and forces in the colonies among which engineers constituted only one party. These difficult negotiations on the ground present a sharply different picture than the pristine conceptions of "comprehensive technology" put forth by utopian-minded engineers who confidently claimed that they were "rescuing" China with their expertise. A closer examination of another of the agency's top priorities, urban and regional planning, demonstrates the range and pervasiveness of comprehensive engineering projects during the war and illustrates the complicated dynamics in China at the foundation of the engineers' technological imaginary.

Urban Technological Imaginaries: The Case of "Pan-Asian" Beijing

In his October 1938 radio broadcast in Beijing—the new capital of the Provisional Republic of China—describing the North China Construction Agency's overall direction, Miura emphasized the importance of cities as national "military, political, economic, industrial, and cultural centers." As a result of years of civil war, the movement of China's capital to Nanjing in 1927, the outbreak of hostilities with Japan in 1937, and major floods in 1938, transportation infrastructure, such as railways, roads, waterways, and airstrips, and public health and hygiene facilities, such as waterworks, sewers, parks, and athletic facilities, in north China's major cities badly needed repair or were entirely lacking. Japanese engineers, he declared, would therefore establish the basic "functions of modern cities" by improving existing facilities and building new urban areas to "secure convenience for business" and "establish order in housing for the people."[48] The agency relied heavily on earlier experience in urban planning in Dalian, Xinjing, and Harbin in Manchuria. They invited Yamasaki Keiichi, chief of the Harbin Engineering Office's Urban Planning Division, to supervise its Cities Department and relied largely on the advice and studies of the Harbin Engineering Office's head, Satō Toshihisa.[49] Japan's modernist projects in Xinjing and Harbin were inspired by the Garden City movement's concept of self-contained, balanced cities neatly divided into industrial, commercial, residential, and agricultural zones surrounded by greenbelts, as well as the City Beautiful movement's ideas of beautification and monumentality to inspire civic virtue and patriotism.[50] These Manchurian cities became the models for urban planning in wartime China. But whereas earlier urban planning projects in the Guandong Leased Territories and Manchukuo focused primarily on designing modern, aesthetically pleasing cities to showcase to the world and local populations, urban planning in wartime China attempted to integrate cities more systematically with industrial, transportation, and national defense concerns in line with the rise of "national land planning" (*kokudo keikaku*) in the late 1930s. In Beijing's case, because of its history as the capital of the Ming and Qing Dynasty emperors, planners sought to tap into its imperial grandeur, whereas urban planning in Manchukuo focused much more on rejecting and erasing an image of "old corrupt China" in favor of building world-class modernist cities for the new nation-state.[51] Thus, Japanese urban planners in north China incorporated pan-Asian ideologies of "restoring Asian tradition" into the agency's founding conception of "comprehensiveness," which in this case meant understanding cities as integrated parts of larger regions and as organic syntheses of multiple functions and capabilities rather than as individual, self-contained units. As the Beijing Chamber of Commerce noted, "Beijing will symbolize the synthesis of Japan's politics, economics, culture, and science, the driving force of the New East Asia" and present a real achievement of "Japan's great endeavor of developing Asia."[52]

City planning and construction took place in the context of ongoing hostilities between Japan's North China Expeditionary Army and nationalist and communist forces. Whereas in Manchukuo, the Guandong Army often tried to scale back the grandiose ideas of urban planners or even delay their projects, in north China, the army immediately prioritized urban planning.[53] The agency's newly established Cities Department began investigations in May 1938 at six main cities in north China: Beijing, Tianjin, Jinan, Taiyuan, Shijiazhuang, and Xuzhou. Whereas Satō's earlier rough plans for the military had grander visions along the lines of Manchukuo's modernist urban projects, engineers in China had to adopt these to meet the immediate requirements of establishing military bases for regional security; constructing roads, health and hygiene facilities, and urban defenses; and resolving housing shortages in light of the sudden population influx, particularly of Japanese residents.[54] Without basic technical documents and precise survey charts, they finalized general outlines for each main city by the end of 1938 based on roughly drawn 50,000:1-scale maps, and began establishing basic laws on zoning, land purchasing and leasing, and construction.[55] Outlines for Baoding, Xinxiang, and the port cities of Tanggu, Qingdao, and Lianyungang were finalized by 1939. The engineers tried to balance the immediate needs of establishing basic security, transportation, and hygiene facilities with long-term planning of new neighborhoods and streets, distributing essential civic institutions, and forming zones for industry, agriculture, and housing for future urban development. They also considered the city's location, environment, history, politics, economy, culture, transport, and industry in an attempt to root "North China construction" in each city's particularities. Efforts to modernize urban facilities were balanced with plans to preserve Chinese cultural monuments and the traditional urban layout and architecture. Thus, urban planning in north China was conducted under the banner of pan-Asianist cooperativism, which claimed to respect the particularity and uniqueness of Chinese cities while asserting the necessity of modern, comprehensive urban development at the same time. As the army solidified control over north China's main cities in 1938, engineers shifted their focus from pursuing emergency measures designed to restore stability to planning urban and regional development, from constructing political cities to designing industrial cities, and from establishing central cities to envisioning "comprehensive regional development," according to Shiobara Saburō, head of the Technical Division of the agency's Cities Department.[56]

Beijing serves as a prime example of the engineers' urban planning ideals in north China and more generally the technological imaginary as embodied in wartime colonial cities. Beijing urban planning is important not only because it represented another way that Japanese engineers articulated their general concept of "comprehensive technology" but also because Beijing traditionally represented the center of Chinese civilization, which Japanese leaders were eager to exploit. Drawn up primarily by Yamasaki and Satō in close consultation with the North China Expeditionary Army's Special Operations Division, the Asia Development Board's North China Office, and the North China Political Council in 1938, the first Beijing Urban Planning Outline envisioned the city as a political, military, and cultural center of north China with a forecasted population of 2,500,000 in twenty years.[57] By constructing a new city that included modern facilities for defense, transportation, industry, commerce, and housing immediately to the west of Beijing's central city walls, they would alleviate overpopulation and the lack of hygiene facilities within the old walled city. At the same time, in line with the pan-Asianist ideology of developing China while maintaining its cultural particularity, they would make Beijing into a more prominent tourist city by initiating preservation projects for its many historical sites.[58]

Satō was fiercely critical of haphazard, "unharmonious" building by Japanese within Beijing's walled city, which he argued created a sense of "veritable pandemonium." Not only did he insist on preserving Beijing's cultural "flavor" in the walled center, but he also incorporated a plan to fully restore the Old Summer Palace (burned down during the Anglo-French invasion of 1860) by transforming it into a park that preserved the existing Chinese-style structures and landscape.[59] The new city to the west, Western Suburban New Town (hereafter, "New Town"), was supposed to "develop the elements that symbolized the new construction of East Asia in contrast to the classical characteristics of Beijing's walled city," Shiobara wrote in a 1939 article for the *East Asian News*.[60] It was to be a "harmoniously planned" city with a "modern Oriental flavor," he added elsewhere.[61]

After encountering difficulties over choosing a site that was both close to the walled central city and geometrically aligned with it, they settled on a sixty-square-kilometer area around four kilometers immediately to the west of Beijing's center. To the north was the Western Suburban airfield, to the west Babaoshan Mountain, and to the south the Beijing-Hankou railway. About half of the area was set aside as "green zones" for fields, forests, agricultural land, and reserved areas for future expansion. The New Town's northern section was reserved for the military, and the central and southern areas were for housing, government and company offices, shops, schools, hospitals, and businesses to accommodate a population of 150,000 to 200,000. Parks would regularly dot the city, and wide squares were positioned at or near intersections to the wide, tree-lined boulevards crisscrossing the city in the interests of health, defense, and urban beautification. A new central railway station would be built slightly south of the city center near a central market and light industrial area, and a 100-meter-wide "Asia Development Road" was planned to run through the town's center leading up to a 100-hectare Asia Development Square (later changed to Yamato Square) in the north. An octagonal "Yamato Altar," symbolizing the slogan of "The Eight Corners of the World Under One Roof," would be placed at the square's center. This altar was supposed to serve as the counterpart to the famous Altar of Heaven, built by Ming Dynasty emperors immediately to the south of the Forbidden City for praying for good weather and abundant harvests.[62] (See Figure 6.)

As a whole, the New Town's designs embodied Japan's pan-Asianist ideology of guiding China toward modernization while preserving its great culture and traditions. Satō and others respected the clear north-south, east-west grid layout of Beijing's inner and outer cities. As the imperial capital, Beijing represented the model of Chinese urban planning for centuries. The old city's north-south axis—formed by the numerous palace buildings, squares, and roads—began at the Bell Tower, passed through the Forbidden City's center, through Tiananmen and Zhengyanmen Gates into the outer city, and ended at Yongding Gate at its southern end. The buildings along this line formed the governmental hub of the Ming and Qing Dynasties. Chang'an Street was the main east-west thoroughfare and intersected with the north-south axis at Tiananmen Gate. It was decided that the New Town's street pattern would mirror the grid layout of Beijing's inner and outer cities, while at the same time incorporating the latest trends in urban planning.[63] Chang'an Street would be extended westward and eastward outside the city walls to connect to a new industrial area planned to the east and the New Town to the west. In the New Town itself, the road would be widened and made to intersect with the north-south Asia Development Road. Purposefully mimicking the Beijing inner city's north-south axis, this road constituted the central artery onto which governmental buildings and businesses would face. The road's north-south axis was surveyed so as to line up precisely with the Temple of Buddhist Incense at the Summer Palace's center to the north,

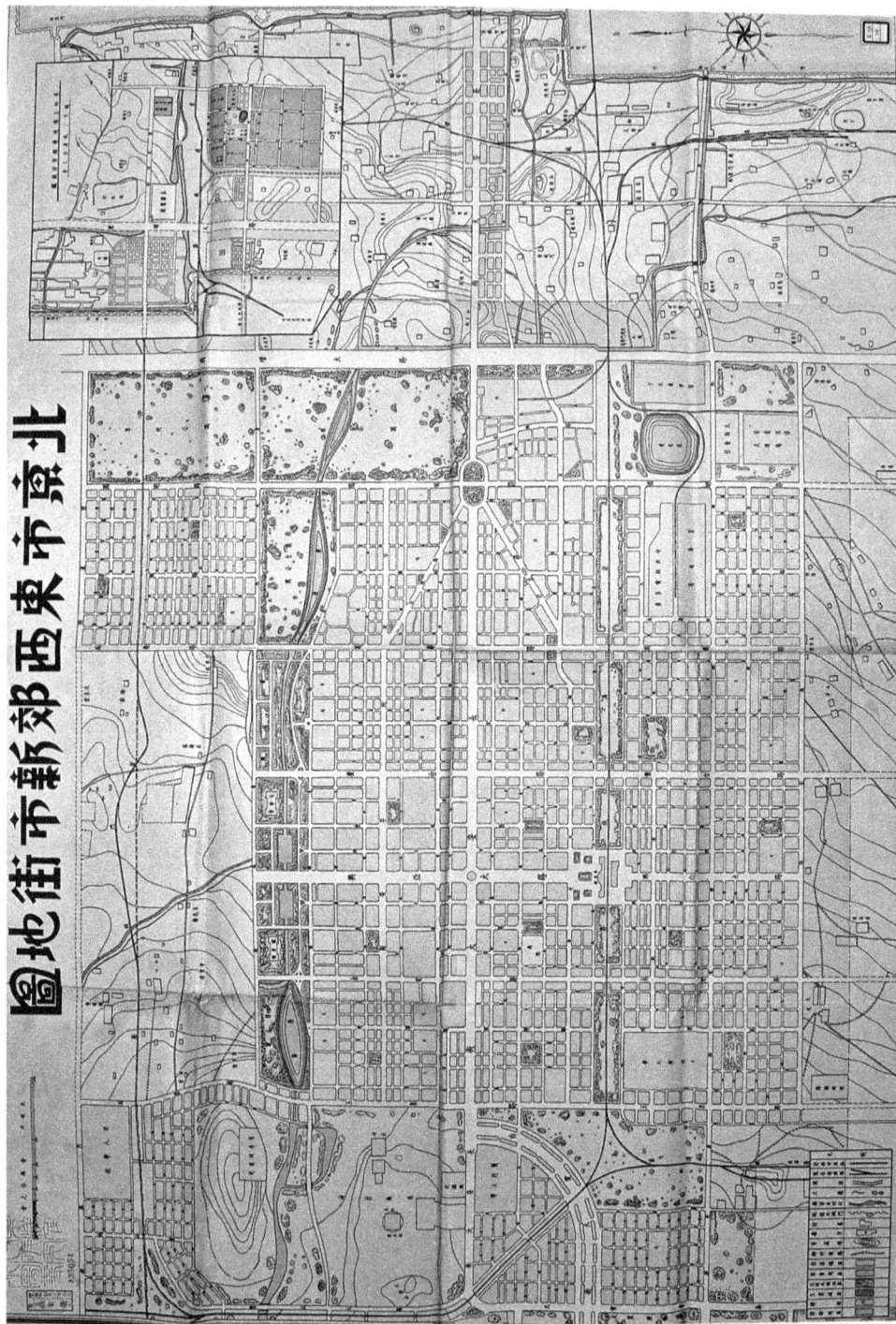

北京市東西郊新市街地圖

which the Qing Dynasty ruler, Empress Dowager Cixi, spent large sums of money to reconstruct and enlarge.[64] Chinese craftsmen were also put to work on restoration and preservation projects at the Temple of Heaven, the Five Pagoda Temple, the Summer Palace, the Imperial Academy, the National Palace Museum, and the Beijing wall gates, among many other places.[65] In sum, by incorporating the earlier layout of Beijing's imperial city into contemporary ideas of urban planning, creating precise links to the "traditional" inner city and summer palace, and designating Beijing as a tourist center of China's heritage, Japanese planners reinforced the North China Provisional government's pan-Asianist, "renewing the people" (*shinmin*) ideology of rescuing an old, timeless China while at the same time connecting it to a trajectory of Japanese-led modernization and development.

The New Town merged earlier Chinese city design with concepts borrowed from the Garden City and City Beautiful movements in international urban planning that were implemented earlier in Manchukuo. Both the New Town and the walled city were to be surrounded by green belts to both limit expansion beyond Beijing's walls and create space for future suburban "hygiene towns" (*eisei toshi*). While the New Town would become Beijing's residential, military, and governmental center, a one-and-a-half- to three-kilometer section immediately to the east of the old walled city would be reserved for light industry and a new railway station. Another industrial area was planned for Tongzhou to the south. The entire outline was governed by the Garden City movement's idea of balance between residential, commercial, and industrial areas utilized earlier in Manchukuo. Zones were designated as special housing, housing, commercial, mixed, and industrial, and another layer of green, scenic, and

Figure 6   Blueprint of Western Suburban New Town for the Beijing Urban Plan, 1940. This map largely details the New Town (planned as a strict grid) and illustrates how the New Town connects with the western walls of Beijing's old city on the map's far right edge. In the upper right-hand corner box, there is a small blueprint of the planned eastern industrial town to the old city's east. Chang'an Street is the wide main east-west boulevard running through the map's center that connects the old city with the New Town. Intersecting Chang'an Street is the New Town's main north-south artery, Asia Development Road, which runs into the main public park area in the north, Asia Development Square. The rail station is located in the large square in the New Town's center where both main arteries meet. Green belts and parks surround much of the New Town, and there is an abundance of squares and rotaries. Source: Kensetsu sōsho Pekin-shi kensetsu kōteikyoku, *Pekin-shi tōseikō shin shigaichi chizu* (1940).

beautification zones was added, accompanied by strict rules on building, lot size, frontage, and height. The New Town itself represented a kind of garden city–type, self-sufficient community with residential areas near numerous parks and athletic fields for people's health; such basic civil services as schools, banks, and post offices; shopping and amusement districts near the railway station and residences; fields for growing vegetables on the southern outskirts; a small factory zone around the railway; and even a golf course at Babaoshan Park to the northwest and vacation house spots near Longevity and Beijing West Hills to the north.[66]

Monumentality was combined with beautification to instill a type of civic virtue in the populace, along the lines of ideas from the City Beautiful movement. Around 45 percent of the total urban planning area was reserved for roads and parks, compared to 35 percent in Manchukuo's capital of Xinjing, which would firmly place Beijing among the world's grand capitals. The 100-hectare, well-wooded Yamato Square faced the New Town from the north, and its "Altar to the Eight Corners of the World Under One Roof" was the center of an arena for public festivals, military parades, and events. Government and commercial buildings would face out onto the main north-south Asia Development Road, which was designated a scenery and beautification zone and facilitated rapid, convenient transportation. Behind the main roads were residential areas. The government and public facilities district on the New Town's eastern side was located within a green zone with parks, squares, and athletic fields near wide, tree-lined boulevards. Thus, in addition to incorporating the precise geometry and grandeur of the Chinese imperial city, urban planning in Beijing reflected ideas on systematizing space and life, encouraging economic self-sufficiency, producing strong and healthy citizens, and creating a rationalized civic layout that could capture the people's imaginations.[67]

The North China Expeditionary Army exerted a great degree of control over the urban planning process, much more than was the case in Manchukuo, where architects and urban planners had more say because of a lack of expertise and experience within the Guandong Army.[68] A large part of the outline was therefore dedicated to road improvement and construction—for example, building the main east-west and north-south arteries that connected the New Town, Beijing's central inner and outer cities, the military base, and the planned industrial towns.[69] A top-secret telegram from a vice minister in the Army Ministry to the North China

Expeditionary Army chief of staff dated February 13, 1939, gave detailed instructions on road widths, anniversary gates, and rotaries necessary for military parades and rapid transportation and ordered parks and pools to be built on every block (every one square kilometer) for air defense.[70] Railways were to be diverted to the New Town and the industrial zones, more storage yards and switchbacks built, and smoother connections to the long-distance railways to Tianjin and Hankou facilitated. Two more airfields were planned as well as canal projects to allow for boats larger than five hundred tons to ultimately travel between Beijing and Tianjin.[71] Thus, Beijing urban planning incorporated the military's strategic priorities of transporting goods and troops rapidly to meet the requirements of the widely expanding war. In this way, the Beijing project reflected emerging conceptions of regional and national land planning among agency engineers, who from 1941 began conducting regional studies in north China and planned comprehensive regions that most optimally integrated urban planning with industrial development, population distribution, transportation efficiency, and river management in order to establish regional self-sufficiency for national defense.[72] The meaning of "comprehensiveness" shifted as the military, urban planners, and preservationists tried to work out how to balance multiple functions.

Population pressures on cities in the aftermath of open hostilities also shaped urban plans in north China. As poor Chinese refugees and rural propertied classes moved to the more secure cities and Japanese officials and company employees streamed in, authorities were presented with such problems as overcrowding and lack of adequate housing and health facilities for the increasing numbers of sick and injured.[73] Between 1937 and 1940, Beijing's Chinese population increased by around 240,000, while its Japanese population shot up from around 4,000 in 1937 to 85,000 in 1941.[74] Japanese authorities were caught by surprise at the sudden influx of Japanese into north China's cities, and as war refugees and people seeking work streamed in, demands for housing space and prices increased. In Beijing's case, the overcrowding problem was particularly acute in the central walled city area. Housing seizures by the North China Expeditionary Army for barracks only made the problems worse.[75] Of most concern to military authorities was the fact that many Japanese were living among the Chinese, often in tiny rooms within hastily modified tenement houses, leading to many "misunderstandings and conflicts" caused by cultural differences and insensitivity among the new Japanese residents.[76] Japanese were also fiercely competing with each other for quality housing, leading to an influx of criminals and land brokers as well as a rise in landlord-tenant disputes.[77] According to Satō, they decided to build the New Town in order to maintain "Japanese-Chinese Friendship" by adhering to the strict principle of "non-Japanification" of China and "non-Sinification" of Japan.[78] They also did not want to drastically alter the walled city area, as this would hinder efforts at turning Beijing into a tourist center. Political considerations brought about by population pressures and a pan-Asianist ideology of "preserving China" thus played a large role in the decision to build the New Town primarily for Japanese residents. The green belt immediately around the traditional inner and outer cities was designated for future Chinese expansion.[79] Clearly there was some resistance to total separation, however. One article noted that Japanese special companies and businesses would face difficulties with a hasty separation policy because they largely depended on nearby Chinese employees.[80] The authorities in the end decided to allow Chinese bureaucrats, employees, and businesses closely associated with the Japanese to apply for New Town residency.[81] Similar to Manchukuo's capital of Xinjing, the New Town was presented as an example of "East Asian harmony" and had a target of accommodating one hundred thousand Japanese and one hundred thousand Chinese who were closely connected to them.[82] But in the end, building "modern" facilities for the Japanese population was the plan's main priority and driving force.

Designing separate but linked Chinese and Japanese neighborhoods dated back earlier to 1906, when Mantetsu engineers planned Dalian (Dairen) in the Guandong Leased Territories as Japan's colonial showcase.[83] Based on earlier plans by the Chinese Eastern Railway's chief engineer, Vladimir Sakharov, Dalian was designed as an airy "garden city" with connected districts based on function. However, it also reflected colonial ideas on hygiene that designated Chinese neighborhoods and day laborers as sources of disease and epidemics that periodically ravaged Manchuria, thereby resulting in policies of strict separation and seclusion. Such ideas were reflected in later urban planning in Manchukuo and Beijing, although there was never any official ethnic cordon sanitaire as in other colonial contexts.[84] Urban planners in north China often referred to such city projects as the New Town as "hygiene towns" or "modern culture cities" in contrast to the crowded walled sections of occupied Chinese cities that lacked many urban facilities.[85] These new towns primarily for

Japanese residents were to be close to work places, and housing would be clustered around a central neighborhood with shrines, temples, government offices, schools, hospitals, public halls, movie theaters, parks, squares, and athletic fields. Markets or department stores would be placed every five hundred meters.[86] New towns were rooted in not only colonial notions of hygiene as reflected in ethnic separation policies but also cultural ideas of the Japanese as more civilized than the Chinese, whereby Japanese neighborhoods more urgently required all of the trappings of modern urban life.[87]

The North China Construction Agency immediately began surveying, formulating zoning laws, purchasing land, and setting up offices and guard posts at the New Town site from June 1938. Because of the war, they decided to focus primarily on meeting immediate road and housing construction requirements at the New Town and eastern industrial zone—in particular, the road connecting the central walled city to the New Town residential areas. At first, a joint Chinese-Japanese special company was selected to manage the construction, with the Beijing city government providing the land and the Japanese authorities providing funds, materials, and technology; however, this failed as a result of unspecified "difficulties" surrounding land procurement, and the agency proceeded to take control of the entire project, setting up the Beijing City Construction Division in 1940.[88] Historical preservation and the construction of roads, sewers, drainage, and flood control structures were paid for from the agency's regular budget, while housing, industrial, and commercial areas would be largely self-funded under the category of the agency's "special budget."[89] Following earlier precedent in Xinjing whereby the Capital Construction Bureau also operated as a development company, the idea behind this special budget was to construct fixed lots and then sign thirty-year "land-use" leases with companies or individuals who promised to construct housing, shops, or factories within two years.[90] Using the lease money, the agency compensated former owners and built the necessary public facilities. Future lease revenue would be used to pay back initial loans from the North China Reserve Bank.[91] Advertisements and pamphlets extolling the business advantages (such as transportation links and central location) and urban conveniences were distributed and tours conducted to attract lessees.[92] The agency even made a film in 1942 titled *Paradise West Suburban New Town* and showed it at Beijing's movie theaters, which apparently led to a rise in the number of Chinese signing lease agreements.[93] Rental prices were determined by location, proximity to services and main roads, and purpose of use.

Engineers and bureaucrats rarely discussed how the agency actually purchased lands from Chinese landowners and relocated residents in Beijing, choosing to focus instead on describing the overall plan and its implementation. Chinese city employees conducted the purchasing negotiations, and landlords apparently cooperated fully, allowing for construction to begin in 1940, one engineer's account noted.[94] According to a 1943 study by Mantetsu's Economic Research Association on procedures used in Jinan's urban planning project, after issuing a land sale ban in the planning area, the agency and the city government conducted investigations and determined compensation prices for different land categories, which was then approved by the Jinan Planning Committee, a newly created supervisory organ presided over by the head of the North China Expeditionary Army's Special Affairs Division and staffed largely by Japanese officers, bureaucrats, and engineers.[95] The army and agency created an elaborate process called *kashisage*, in which the city would lease agency-developed land to individuals or businesses and use this revenue in turn to "rent" the lands from the owners. This contrasted with the process of *haraisage*, or forced purchasing of lands, commonly used in Manchukuo urban planning. According to an employee at the agency's Property Division, they switched to this system of "renting" the land from the original property owners (thereby returning land titles to them) because of strong resistance over loss of livelihoods and the communists' use of the issue to foment anti-Japanese sentiment.[96] Lands were officially purchased or "leased" by the city, and Chinese officials performed the actual serving of notice and implementation. Individuals and businesses in turn applied to the city for leases and paid an application fee, a nonrefundable guarantee, and a yearly rent for different lots in the urban planning zone. The process of appropriating lands for north China urban planning obviously caused significant discontent among some of Beijing's Chinese residents, and the army and agency had to quickly create an elaborate legal procedure centered on the largely Chinese-staffed city administration to legitimize the land purchases, which were ultimately enforced by army-dominated planning committees.[97] Legal maneuvers, pressure from Chinese middlemen negotiators, and autocratic colonial power created the possibility for implementing the engineers' technological imaginary on north China's cities.

Quite contrary to their representations of urban planning as a process of modernizing China's "chaotic" cities, the construction itself was haphazard and plagued by conflict and difficulties. For the required labor, the agency worked through Chinese foremen (*batou*), who received a lump sum for the group of workers they supervised.[98] Wages for Chinese foremen and workers were extremely low, apparently creating tremendous hardship for them during a time of rapid wartime inflation.[99] The North China Expeditionary Army's Special Affairs Division took charge of labor mobilization in cooperation with the Asia Development Board's North China Coordination Department until July 1941, when the North China Labor Association was established to centralize all labor recruitment and procurement inside and outside the region. The New People's Association—co-directed by Yin Tong, who was also the North China Construction Agency's head—worked with Special Affairs Divisions in each city to recruit and mobilize labor in line with quotas drawn up by the Asia Development Board.[100] Construction began in 1941 but did not go very smoothly, as they suffered from chronic labor shortages caused by workers fleeing to partisan areas or being sent off to Manchukuo, materials shortages arising from rampant inflation and transportation delays, low productivity caused by communication problems and differences in technical customs, and budget shortfalls.[101] When the Asia Development Board recommended postponing urban planning projects because of the worsening war situation, the North China Expeditionary Army's chief of staff had to telegraph Japan's Army Ministry to insist that this would have negative consequences for the Chinese and Japanese residents and emphasized that alleviating the housing shortage in China's cities was an absolute necessity.[102] A Japanese engineer for the Beijing city administration's Construction Division recounted how they hired Chinese craftsmen to make key parts for mechanical crushers and trammels needed to pave roads; utilized informal connections with army officers and bureaucrats to secure necessary materials of poor quality; invented makeshift contraptions needed for construction; and used local stone quarries that were frequent targets of partisan attacks.[103] Thus, the story of urban planning was not one of visionary engineers unilaterally transforming China's urban landscape through their expertise but involved them in negotiations with a complex array of actors and forces beyond their control, which continuously shaped the entire project and their larger visions.

By February 1942, eight hundred houses were built in the New Town for around two thousand one hundred residents, who were mostly Japanese employees for such special companies as North China Transport, North China Development, and North China Electricity (some Chinese employees lived here as well). The main connecting road with the central walled city was partially paved and basic water and sewage facilities built. An elementary school, residents' association office, medical clinic, market, and other public facilities were all in operation by this time. In the planned industrial area to the old city center's east, roads, industrial water facilities, and factory sites were prepared. Eight light industrial factories signed leases, and four were in operation by 1942. In 1943, however, the Beijing Residents Association rented 100,000 *tsubo* (a little over 300,000 square meters) of planning area land to utilize as a public farm for food provision as the war situation worsened, and all construction ceased.[104]

Although the experienced agency engineers presented their work within a larger narrative of comprehensively rationalizing and systematizing space while preserving the traditional forms of China's cities, the actual process of urban planning contrasted sharply with the coherence and systematicity of their visions grounded in a confidence in technology's power. Satō's grand, modernist designs for Beijing based on his Manchukuo experience clashed with the more immediate needs of reconstruction, security, basic infrastructure, and housing particular to north China's wartime context. Pan-Asianism and preservationist strains to limit urban development merged uneasily with plans to expand transportation, industry, and civic infrastructure, resulting in the New Town–versus–Old Beijing conception. The military's immediate defense needs prioritized road design, park and rotary placement, transportation routes, industrial zones, and security areas over the residents' long-term needs and comfort. Population pressures and worries about the "mixing" of Chinese and Japanese shaped the idea of separate towns and neighborhoods, yet clashed with the practical requirements of Japanese businesses that relied on Chinese employees and services nearby. Modernist designs emphasizing systematicity, balance, civic virtue, and monumentality were designed to appeal to the Chinese, some of whom were allowed to live in the New Town; however, at the same time, the plan was rooted in a conception of the Chinese as unhygienic and disorderly, resulting in the plan for separate "hygiene towns" away from the "traditional" walled city area. Finally, the actual construction process required pressuring reluctant Chinese landlords and quickly devising an elaborate legal process for eminent domain and leasing, obtaining scarce materials and funds, hiring cheap Chinese workers through labor brokers, and advertising to reluctant Chinese and Japanese residents and businesses that were emotionally attached to the old city center. These difficulties combined to somewhat scale down the original plans to the level of simply meeting the basic housing and infrastructure needs for the Japanese population. In sum, urban planning was not so much a matter of Japanese experts systematizing "chaotic" Chinese cities through the introduction of "comprehensive technology" but rather a continuous attempt to reconcile conflicting, uneven agendas—Manchukuo's grand urban planning conceptions with immediate infrastructure needs arising from the war, Beijing's existing classical designs with the needs of a modern "hygiene city," the military's defense and transportation needs with conceptions of Beijing as a tourist city and political center, and a policy of separating Chinese and Japanese with the reality of mutual dependence and interaction. An analysis of Japanese technologies of urban planning (and more broadly, the technological imaginary) that leaves out any consideration of these constitutive tensions and conflicts may unwittingly reproduce the modernizing perspective of engineers and confirm their own interpretations of themselves as Asia's heroic developers and preservers.

Dams and the Advancement of Comprehensive Regional Planning

Among the various versions of comprehensive technology being developed in the colonies, those centering on large-scale dam construction proved to be the most compelling to Japanese engineers as they learned more about the technologies of multipurpose dams during the 1930s. The construction of Sup'ung Dam on the Yalu River between Manchukuo and Korea, the world's second largest dam at the time, became the occasion for one of the first instances of what generally became known as "national land planning" (kokudo keikaku) during the war's height in the 1940s (and well into Japan's postwar high-speed economic growth era).[105] The frontier border region around Sup'ung and six other envisioned dams along the Yalu River—encompassing Dongbiandao (Tōhendō) in Manchuria and North P'yŏng'an Province in Korea—was to be transformed into a "comprehensive industrial region" (sōgō kōgyō chitei). National land planning was Japan's response to global trends in regional planning to alleviate urban overcrowding since the 1920s and particularly German concepts of *Raumordnung* ("spatial order") during the 1930s that emphasized curbing urban growth, evenly distributing industrial zones, and preserving rural areas in line with national defense objectives.[106] Along with ongoing efforts at total war mobilization, engineers in Japan's Home Ministry began conducting research into planning efforts abroad from 1935 and quickly formulated the administrative and legal apparatus for national land planning in 1940, culminating in the passing of the "Outline for the Establishment of National Land Planning" in September that same year.[107] This idea of the state efficiently distributing industry, agriculture, population, sociocultural institutions, and infrastructure in order to promote comprehensive development, however, was not simply formulated in the minds of Tokyo experts learning the latest in Western planning techniques but also emerged out of specific projects in Japan's empire, where engineers and bureaucrats were devising different notions of "comprehensiveness" (*sōgōsei*) from the mid-1930s in the form of river control, urban and regional planning, and multipurpose dam projects.

Suzuki Takeo, professor of economics at Keijō Imperial University and one of the Korea Government-General's top advisors, discussed the colonial origins of regional and national land planning in an article for *Chōsen* (Korea) in February 1940. Whereas Japan had established the basic institutions and laws of modern urban planning in 1919 and acquired significant experience in urban reconstruction after the 1923 Great Kantō Earthquake and in designing modernist cities in Manchuria and elsewhere, Suzuki argued that urban planning needed to evolve in the 1930s for several reasons. First, the shift from coal to hydropower now allowed planners to establish industrial areas far from urban centers. Second, the rapid advance of transportation enabled planning on a regional and national scale. Third, Japan was shifting from a liberal capitalist to a managed economy, in which the state rather than the free market planned the rational distribution of industry in relation to energy, natural resources, labor, and transportation. Finally, national defense concerns necessitated industrial deconcentration and the regional planning of food provisions for larger cities. Internationally, the Tennessee Valley Authority (TVA) in the United States, Stalin's Five-Year Plans in the Soviet Union, and *Reichsplanung* (national planning) in Nazi Germany represented new trends in regional and national land planning that combined state efforts to distribute industry, increase production, establish productive links between city and country, promote national defense, and improve self-sufficiency. In this way, conventional planning limited to the confines of one urban area was becoming increasingly outdated and anachronistic during the 1930s, according to Suzuki.[108]

While Home Ministry engineers in Japan began institutionalizing national land planning only from 1938, urban planning in Korea had already been shifting toward a vision of comprehensive land development from its beginnings in the Korea Town Area Planning Order of 1934, Suzuki noted. With Manchukuo's founding in 1932 and the Guandong Army's efforts at total war industrialization, Korea also began pursuing the "parallel development of agriculture and industry" (*nōkō heishin*) as it transformed itself into a "supply base" for the Asian continent. Hydropower's rapid development in the 1930s enabled Government-General engineers to plan the establishment of regional economic blocs as opposed to focusing solely on individual cities. In this manner, unlike Japan's case, Korean urbanization in the 1930s occurred in the context of colonial industrialization and the formation of a total war system, which in turn gave urban planning in Korea a distinctive "national land planning type of character" before such an approach was articulated and officially incorporated into Japanese national policy, Suzuki argued.[109]

In line with ongoing Home Ministry research into international regional and national planning trends from the mid-1930s, colonial engineers began to incorporate what they learned into urban planning projects, such as in Beijing, where planners attempted to integrate the capital into the larger region, as well as overall national defense initiatives rather than only focus on the city as a self-contained unit. With Manchukuo's commitment to a "hydropower first, thermal power second" policy in the 1937 Five-Year Industrial Plan and subsequent promotion of large-scale dam projects (see Chapter 4), comprehensive regional planning was further articulated in the form of designing heavy industrial zones near dams, ports, and natural resources to fulfill the national defense economy's requirements. The Dadong port coastal industrial urban zone project became a prime example of what the emerging discourse of national land planning looked like on the ground and in fact prefigured the type of comprehensive development projects pursued during Japan's postwar high-speed growth era. In much the same manner as colonial officials and engineers working on various projects throughout the colonies continuously worked to naturalize their technology and expertise as forces of progress and development, colonial officials like Suzuki (as well as postwar Japanese bureaucrats) sought to present such comprehensive regional and national planning as inevitable trends. Their grand claims of balancing industrial-agricultural growth, rationalizing urban-rural economic linkages, controlling industrial and urban expansion, and improving urban health and hygiene through the introduction of the latest planning techniques, however, hid the specific practices, conflicts, and negotiations among different actors, institutions, and forces in the colonies as well as some of the power dynamics that went into making their comprehensive plans. Whereas at the ideological level Japanese planners and officials emphasized the increased coherence, rationality, and systematicity they were bringing through the introduction of Japanese technology and expertise, an analysis of comprehensive regional planning around Sup'ung Dam reveals some of the ways their plans and blueprints were actually produced through difficult negotiations at colonial project sites rather than in the minds of experts.

Larger-scale regional planning was not seen as particularly natural to lower-level Japanese engineers working in colonial provincial government offices, and they immediately questioned the emerging conceptions of "comprehensiveness" developed by state engineers and reform bureaucrats in Manchukuo's government as part of the 1937 Five-Year Industrial Plan. Since Andong's establishment as a province in 1934, local engineers had conducted more limited urban planning projects in its capital, Andong City, along the lines of similar efforts in Manchukuo's other major cities, which generally aimed at promoting local industry, improving flood control and transportation, introducing strict zoning laws, and providing for ample green belts and parks for urban hygiene and health. Although the old trading port and lumber city began receiving attention in 1935 for its heavy industrial potential as a result of nearby discoveries of abundant iron ore and coal resources, the decision to build a Dongbiandao railway, and plans to develop Yalu River hydropower, its port had become increasingly dysfunctional as a result of increased siltation caused by deforestation from lumbering activity upriver, a factor that greatly hindered its future industrial development. Kuroda Shigeharu, head of Andong Province's Civil Engineering Division, noted that top Guandong Army planning officers and Mantetsu officials began visiting in 1937 and proclaiming Andong's "golden age" of industrialization in light of the beginning of Sup'ung Dam's construction nearby. When ordered to investigate sites for a huge industrial port, Andong engineers concluded that the area was only suitable for small fishing ports and pushed instead for a more limited plan to renovate Andong's own port through localized river improvement projects. In a subsequent visit to Xinjing in 1938 to win state approval, Andong engineers met with Kishi Nobusuke, vice director of Manchukuo's Industrial Department, who pushed them to conduct wider investigations for a larger port site adjacent to large tracts of land suitable for a huge industrial zone in light of the state's plans to locate heavy industrial factories there to develop the region's rich resources. After further discussions with reform bureaucrats and government engineers, Kuroda and his team were gradually won over to their "larger vision" and even began criticizing their own earlier planning "narrowness." Once they began thinking in terms of comprehensive links between natural resources, industry, ports, hydropower, and urban planning in the region, they expanded their investigations and discovered a wide plain where a tributary flowed into the Yalu River some thirty kilometers southwest of Andong City as a promising site for a large port and industrial zone.[110] In this way, comprehensive land planning emerged out of clashes and negotiations between provincial officials thinking about local development and reform bureaucrats seeking ways to incorporate the Yalu River dams into their national visions of a heavy industrialized total war economy. Such conflicts between province and state as well as between Manchukuo and Korea would continue as the project took on actual shape.

In close consultation with top reform bureaucrats and leading state engineers, Andong Province engineers formulated a massive plan around a small existing town that the Manchukuo Planning Committee renamed Dadong Port (*Daitōkō*), which state authorities later described as the "port of the century."[111] Under their new framework of "comprehensiveness," they delineated a 215-square-kilometer "comprehensive coastal industrial region" (*sōgō rinkai kōgyō chitei*) around the port for urban planning (fifty square kilometers designated for immediate city construction), noting how its ideal proximity to abundant natural resources, convenient transportation links, cheap land and labor, and ample hydropower and industrial water would help reduce business costs by 10 percent, thereby increasing national productivity and prosperity.[112] The new port would accommodate four-thousand-ton ships and have the facilities to accommodate two million tons of goods per year; the modern city would house four hundred thousand residents; the industrial zones would be provided with two hundred thousand tons of water per day; and a new railway and road network would connect the port to Andong City.[113]

But despite the systematicity of their visions, their blueprint was not really grounded in any kind of firm technical data when presented to Manchukuo authorities in 1938. For example, a joint investigation of Manchukuo, Mantetsu, and Andong experts—led by Naoki Rintarō and Haraguchi Chūjirō, two of Manchukuo's top engineers—had to be conducted in February 1939 to determine if the port site froze during the winter months as others in Manchuria often did. Although Naoki and his team confidently concluded that the site could indeed facilitate a year-round warm water port, the actual investigation was based on only two weeks of temperature and weather measurements, a few pictures of the area discovered from earlier years, and a visual confirmation that the tides flushed out and broke up ice from upriver, which revealed a large degree of ambiguity and guesswork behind the expert investigation.[114] As it turned out, future studies proved their conclusions to be correct but for unforeseen reasons, as Sup'ung Dam's construction upriver warmed the river's water temperature more than expected, thereby reducing overall ice flow volume.[115] Their report and subsequent investigations also revealed some of the difficulties the project would later encounter, such as the abundance of shallow zones in the river's proposed shipping lanes to the sea that would require massive dredging and the building of diversion levees to maintain sufficient depths for large ships; erratic wind and weather patterns farther out to sea, which might seriously affect shipping traffic; and the soil's weak qualities, which required engineers to build raised foundations before constructing any type of building or facility.[116] As was the case with dams and other infrastructure projects, nature proved somewhat resistant to expert conceptions.

Japanese heavy industrial interests forced changes to the Dadong port plan, which was adopted in April 1939 after several coordinating meetings between the Manchukuo government, Guandong Army, Korea Government-General, Mantetsu, and Andong Province the previous year.[117] Ayukawa Yoshisuke, who had recently transferred his Nissan industrial conglomerate to Manchuria and founded the Manchurian Heavy Industries Company (*Manshū jūkōgyō kaihatsu kabushiki gaisha*; hereafter, Mangyō) at the instigation of Manchukuo's government, had immense hopes for the Andong project. Upon visiting in 1938, he said: "In many ways, Andong is the world's most ideal location, and I believe I will realize my objectives of coming to Manchuria here. Andong is where I will leave my future business. I will live forever in Andong."[118] Two of Mangyō's companies, Manchuria Light Metals and Manchuria Automobile Production, were set to build large factories there to take advantage of the port and cheap electricity from hydropower as well as process the coal and iron ore from one of Mangyō's other regional companies, Dongbiandao Development. The project began in earnest with the establishment of the Dadong Port Construction Agency in June 1939. Kondō Kenzaburō, who previously headed Manchukuo's Urban Planning Department and helped establish the basic laws and technical standards for city planning, became the agency's deputy head (and soon head) of daily operations.[119] Yamada Hiroyoshi, a senior figure in Japan's urban planning world and agency advisor, defined the project's basic philosophy of regional planning as one of rationalizing the relationship between the land and such other factors as transportation, the economy, national defense, and people's livelihoods in order to generate an "organic" relationship between cities and regions.[120] Yet an unexpected rush of applications by Mangyō factories led to an oversubscription of zoned land and therefore an expansion of the original plan to reharmonize various aspects of urban planning, such as industry, housing, green belts, water provisioning, and transportation.[121] Demand from heavy industry increased the urban planning area from 215 to 323 square kilometers (125 square kilometers for immediate city construction) in 1940 and again to 375 square kilometers (145 square kilometers for city construction) in 1942, nearly triple in size and double the budget of the original 1939 plan.[122] In this manner, agency engineers never possessed a hard-and-set plan but had to continuously adjust their existing frameworks of urban planning toward emerging conceptions of "balanced" regional planning as well as accommodate the needs of heavy industry under Manchukuo's managed total war economy.

Japanese urban planners built on the garden city–inspired techniques they had employed earlier in other Manchukuo projects while at the same time attempting to incorporate a broader and more systematic regional perspective. The basic plan created three city centers along a rectangular urban planning zone stretching twenty-five kilometers along the Yalu River's banks extending from the proposed Dadong port to Andong City, which constituted a fourth city center and separate urban planning area of 116 square kilometers. The new coastal industrial urban region was expected to house one million residents—two million if Andong City and surrounding environs were included.[123] The large factory lots were to be built immediately adjacent to the river and equipped with their own docking facilities, connecting roads, and electricity and water sources. Ayukawa's Mangyō subsidiaries alone planned to invest one billion yen to produce an estimated five hundred million yen in manufactured goods per year, and at one point, the Ford Motor Company even sent an inspection team and applied for a factory site.[124] The Greater Andong region was to be transformed into a heavy and chemical industrial production and export center producing one billion yen in goods per year based on an estimated eight hundred thousand kilowatts of electricity to be provided by the seven Yalu River dams.[125] In the interests of health and hygiene, a one-kilometer green belt was placed between each of the three city units as well as a two-kilometer green belt around the planning region's entire perimeter to evenly balance out the heavy industrial areas. The amount of land allotted to parks and fields amounted to 25 percent of the entire planning area, 13 percent of the actual city construction zone, which amounted to twenty-five square meters per person.[126] The suburban green belts were designed to control urban sprawl, allow for sufficient agriculture, and enable managed expansion, and a variety of pedestrian parkways, sports fields, child play areas, and large recreational parks were included to alleviate the effects of industrial air pollution and provide for the population's health and leisure (as well as evacuation sites in the event of air attacks).[127] A seventeen-kilometer-long, twenty-five-meter-wide superhighway—the first in Asia—modeled after Hitler's *Autobahn* was planned along the outskirts to connect the four urban areas, and when completed, cars could travel at 120 km/hour and make the trip from the Dadong port to Andong City in fifteen minutes.[128] Mantetsu was also constructing thirty-five kilometers of railway to connect all four city centers to allow travel between each center in ten minutes or less and ultimately to link up with its international network at Andong City.[129] (See Figure 7.)

In the urban areas, a grid road network was formed primarily around one sixty-meter-wide central artery running through each city's industrial zone, and green belts were planned on either side of all major roads over forty meters wide to set them apart from residential areas, control growth, and mark boundaries between industrial, residential, commercial, and administrative areas.[130] Public squares and rotaries at major intersections and stations were included, and all roads more than twelve

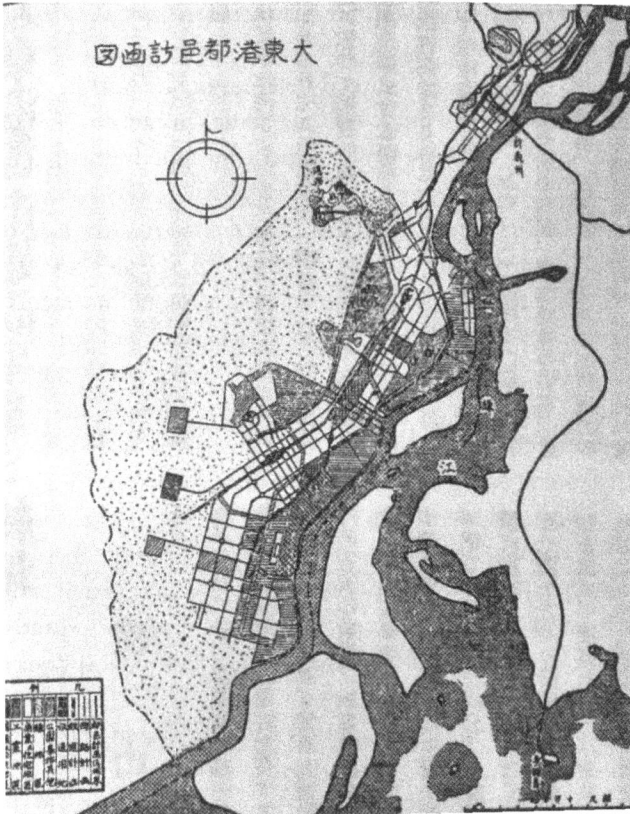

Figure 7   Blueprint of the Dadong Port Coastal Industrial Urban Zone near the mouth of the Yalu River. The grayish areas filled with darkened parallel lines indicate industrial zones, which are largely along the river and the planned port facilities. The white areas are commercial and housing zones. The thick black lines running through the urban planning areas and connecting to Andong City in the north and then branching into Korea are Mantetsu rail lines. Gray lines indicate large roads. The superhighway runs above the railway line and connects the urban areas. Tasado port in Korea is on the map's bottom right and is connected by rail. Source: Yokoyama, "Daitōkō toshi kensetsu jigyō to enchi," 112.

meters wide had sidewalks.[131] Kondō even included a plan to alleviate the effects of heavy industrialization through the development of a "special entertainment area" at Wenchizi Hotsprings, six kilometers northwest of the Dadong port city center, which would include hotels, amusement parks, athletic fields, sanatoriums, gardens, pastures, vacation homes, shopping centers, and restaurants.[132] Thus, departing from earlier practices focusing on one city, Japanese urban planners attempted to design a comprehensive regional unit that systematically integrated heavy industry, agriculture, transportation, national defense, public health, and people's livelihoods. Their plans represented one actual instance of national land planning among the flurry of largely paper conceptions drawn up in Japan and its colonies in the late 1930s and early 1940s.

As was the case with Sup'ung Dam's construction and other instances of large-scale infrastructure building, balance and comprehensiveness looked far better on paper and in official pronouncements than at actual project sites. Manchukuo engineers presented their enormous multipurpose endeavor as the transformation of "virgin land" and celebrated their ability to freely plan and implement their ideas based on "modern technology's essence" without any major obstructions.[133] Yet such assertions of technology's progressive, transformative nature hid the messy, uncertain, conflict-ridden, and haphazard reality of the entire project conducted at the height of Japan's war in Asia. Tensions between Manchukuo and Korea as well as between national and provincial governments plagued the planners throughout the construction, and the war forced them to make several compromises to their conceptions of evenness and systematicity.

Before the Dadong port site was discovered and the plan formulated, Manchukuo authorities had already settled on an existing port at Tasado immediately across the river in Korea as a primary export base for Dongbiandao's natural resources. The Korea Government-General responded with a six-year port improvement plan begun in 1938 to facilitate one million tons of freight per year and accommodate three-thousand-ton ships.[134] However, Andong Province engineers pushing for port improvement and industrial development in their own capital city criticized Manchukuo's plan for leaving Andong behind and ignoring their own area's attractive qualities, such as abundant industrial land and cheap labor. At around the same time, the Government-General purposely delayed approval of a Manchukuo project to improve Andong's port that involved a partial closure of the Yalu River ostensibly out of worry over the project's possible effects on flooding.[135] As the Dadong port plan later took shape and Korea and Manchukuo officials arrived at an agreement to simultaneously develop large ports on both sides of the Yalu River, they then had to manage concerns among their respective populations that the ports would unnecessarily compete and overlap, thereby adversely affecting each other's economic interests. For example, in order to address the business community's concerns about the project's benefits, Naoki Rintarō spoke to Andong residents in 1939. He invoked his past experiences in planning the Osaka and Tokyo ports and pointed out how the older ports of Kobe and Yokohama were not hindered by their construction and growth near Osaka and Tokyo, respectively.[136] Similar arguments of "co-prosperity" were made in Korea as officials emphasized the Sinŭiju-Tasado region's great potential for heavy industrial development and the fact that Dongbiandao's vast natural resources would first arrive at the Tasado port, as it would be completed several years earlier than the Dadong port.[137] Thus, despite confident promises by technocratic planners to incorporate various interests within a common vision of comprehensive regional development, officials had to continuously manage local concerns about the two projects' uncertain economic benefits.

Part of the reason why Japanese colonial officials often appealed to technologies of planning in their statements and invested significant resources into such comprehensive projects was that technocratic planning promised benefits that could transcend social, political, and economic divisions. Yet this ideological power of technocratic planning also rested on many of the same colonial institutions and techniques employed at other large-scale infrastructure projects in Japan's empire. Such mass organizations as the Concordia Association were relied upon to ensure that plans would bring about the promised results of balanced development and prosperity. In a speech to Andong's Concordia Association, Naoki Rintarō criticized the commonly held view that merely building ports and other infrastructure would automatically bring about economic benefits, again citing his earlier experiences in constructing Osaka's port, which was underutilized for several years because of nearby Kobe's popularity. He urged his audience to engage in what the British called "selling the port," or mobilizing residents to open shops, provide services, build amusement facilities, and actively market the port's benefits to Japan and the wider world rather than assume that the mere existence of technical infrastructure would automatically attract shippers and other business.[138] In short, Naoki argued that technology did not merely exist in a sociopolitical vacuum but required coordinated ideological campaigns and concerted social mobilization as well. Such past efforts as the residents' establishment of a committee in 1935 to campaign for the construction of a trans-Dongbiandao railway and their success in attracting more factories and businesses than there was available land for attested to the degree of mobilization among the local business community around technology.

As with other cases, struggles over land posed one of the greatest challenges to the regional urban planning project and Japan's overall ideology of impartial technocratic management. New strategies were required to convince thousands of resistant landlords within the planning zone to part with their lands. In 1939, the Dadong Port Construction Agency formed a Land Value Investigation Committee to determine a base price range for different types of land and ensure their smooth purchase.[139] Yet Andong authorities only managed to buy 27 percent of the required land by 1941, the year they were supposed to have completed all purchasing, at the "fair" prices they had set in 1939 in order to prevent speculation. Instead of continuing to negotiate with landlords individually, the province decided to issue bonds to the remaining landlords, which would pay 5 percent interest until the city managed to lease or sell the zoned industrial, commercial, or residential land to new tenants. By 1943, 83 percent of the land was purchased through this unilateral financial ploy in an attempt to co-opt local dissent.[140] Thus, comprehensive land planning was not just about creating rational and harmonious relations between cities and regions or industry and natural resources but also involved devising new techniques of appropriating land for state goals in the face of widespread resistance.

In the end, the intensifying war undermined the planners' technocratic visions more than anything else, causing major changes to their plans and recurring doubts about the project's feasibility. Materials shortages, delayed deliveries, and sudden budget restrictions plagued the agency so that only 15 percent of the roughly 300-million-yen budget was expended two years into the eight-year project.[141] Despite the plan's expansion in 1940 as a result of the rush of factory site applications, that year was described as the low point, and many people doubted if the project would ever truly take off. Under the new wartime restrictions, engineers began prioritizing the port facilities and shipping lanes, major roads and railways, and industrial zoning aspects of the project over urban planning, betraying their earlier visions of balance and systematicity. Their goal now was to provide the necessities for heavy industries to immediately begin wartime production and exports in the hope that this would then generate enough revenue to finish the project's remaining urban planning portions.[142] The beginning of electricity production at Sup'ung Dam in August 1941 and its transmission to Andong from May 1942 as well as the increasing development of coal and iron ore resources by Ayukawa's Dongbiandao Development Corporation pushed planners to focus more on establishing heavy industrial factories to consume the electricity and immediately process the area's natural resources. Another reason for the prioritization of heavy industry over urban planning was that although there was a rush of applications for factory sites of all sizes, not all of them immediately concluded contracts with the agency, which suggested that some were having doubts because of the war. Thus, scarce resources were directed instead toward preparing sites for larger special companies, such as Manchuria Automobiles, Manchuria Light Metals, Manchuria Mining Development, Manchuria Soybean Chemical Industries, and Manchuria Carbon, all of which were either in the process of building factories or beginning production by 1943. The war also affected the area's population growth, which rose by only 10 percent between 1939 and 1943, thereby casting further doubt about the viability of the one-million-person Greater Andong plan and providing another reason to not use scarce resources immediately for urban planning. Statistics show that a little less than half of the industrial, commercial, housing, and mixed zones had finalized contracts, which were an essential revenue source for the project.[143]

In addition to the contracted factory sites, by the war's end in August 1945, the agency managed to build the high-speed freeway (unpaved) and thirteen kilometers of railway, plant trees for the green belts, complete the dock foundations and one operational dock, and begin dredging the uneven river floor and building diversion levees for the shipping lanes.[144] Although technocratic visions of regional and national planning, such as the Dadong port project, sought to naturalize and solidify Japan's conception of a Greater East Asia Co-Prosperity Sphere through orderly technical blueprints and the latest planning expertise, these plans were never stable to begin with, as they were continuously formed by clashing political and economic interests, the natural environment, and the war's exigencies. Fluctuating demand for zoned land, increased wartime need for strategic resources, coordination with hydropower dam construction and natural resource development, materials shortages and budget shortfalls, competition between colonial regimes, difficulties in purchasing land from landlords and managing local economic concerns, and unexpected population changes all shaped the comprehensive regional plan. Technical blueprints representing Japan's empire as a rational and efficient system were formed within this nexus of conflicts and contingencies, not solely in the minds of Japanese engineers equipped with the latest technical knowledge from the West.

Conclusion: The Specter of Comprehensive Technology

Technology was an essential component of Japan's pan-Asianist wartime discourse of "constructing East Asia." State engineers, such as Miyamoto Takenosuke, defined technology as more than simply machinery and specialized knowledge; in their view, technology also involved social qualities of planning, management, and creative vision. In the struggle to improve their status, Manchukuo and China became "meccas" and "new frontiers" for them to realize what they saw as technology's true essence in the form of ambitious schemes to develop and modernize Asia rooted in a firm belief in their roles as social engineers. As Mizuno, Mimura, and Ōyodo have argued elsewhere, the empire was the space for them to realize their technocratic dreams of social engineering. Although these scholars have contributed greatly to shifting focus away from the military's spiritualist, emperor-centered ideologies for war and empire, the story of how the colonial context itself shaped technocratic ideologies is largely absent from their narratives.

The technological imaginary not only was formed in Japan among engineers and bureaucrats or simply applied whole-scale by them in the colonies but took shape through the colonization process itself. Comprehensive visions of coordinating flood control with electricity production, irrigation, and transportation improvement did not simply exist first on paper but were formulated in Manchuria through negotiating little-known landscapes and climates, acquiring new knowledge and expertise through exchanges with Chinese technicians, harmonizing competing Japanese agendas of industrialization, agricultural settlement, and flood control, and dealing with Chinese resistance and shifting war events. Building modernist cities was not merely a story of visionary Japanese planners bringing their experience and the latest Western design theories to China but a tension-filled process of reconciling a desire to preserve tradition with demands to modernize urban infrastructure; harmonizing Garden

City–influenced urban planning with the layouts of the imperial Chinese city; balancing the military's immediate requirements for national defense with long-term plans to build organic, self-sufficient cities that integrated economic, cultural, and social objectives; and addressing rapid, uncontrolled population growth through policies of ethnic separation that clashed with the everyday realities of interdependence. Comprehensive regional planning was not simply an adaptation of existing Western planning expertise to the colonial context but was simultaneously shaped through conflicts between local and state engineers as well as between colonial governments, adjusting visions of balanced regional and urban planning with the substantial requirements of wartime heavy industry, selling the project's economic benefits to skeptical locals, and responding to the shifting exigencies of the war. By shifting attention away from the engineers' own technological imaginary, we see that they were less the active subject implementing their rational visions of comprehensive technology and more the product or effect of a confluence of conflicting forces, interests, and institutions in the colonies. Chapter 4 focuses more closely on the wide range of colonial institutions and techniques of power that were employed to produce this ideological effect of Japanese expertise rationally and systematically "constructing East Asia" that was an essential part of the technological imaginary. It also analyzes some of the socioeconomic effects of large-scale technology projects by examining two of the most prominent examples of comprehensive technology in Japan's empire—Fengman and Sup'ung Dams.

Chapter 4

Damming the Empire

Hydropower and Comprehensive Technology

As visions of "comprehensiveness" were crystallizing in Manchukuo and north China from 1937 in the form of large-scale river control projects and urban and regional planning efforts, reform bureaucrats (see Chapter 5) joined forces with engineers to plan and construct what the General Affairs Agency head Hoshino Naoki called "a great monument to Manchukuo"—the Fengman Dam on the 849-kilometer-long Second Songhua River, twenty-four kilometers upriver from Jilin.[1] Measuring 91 meters tall, 1,110 meters wide, and 2,220,000 cubic meters in mass, the gravity concrete dam was equivalent in scale to some of the world's largest dams, such as the Boulder and Grand Coulee Dams in the United States. The reservoir itself was 610 square kilometers in area and 170 kilometers in length—around 80 percent of the size of Lake Biwa, Japan's largest lake.[2] Constructing the dam required nearly two million cubic meters of concrete, which was more than enough to pave a fifteen-meter-wide Tōkaidō Road (the traditional route from Tokyo to Kyoto) and equaled half of Manchukuo's annual concrete production.[3] Fengman was expected to produce seven hundred thousand kilowatts of electricity for the promotion of heavy and chemical industries, eliminate floods in a 170,000-hectare area and transform it into settler paddy land yielding an increased rice harvest of three million *koku*, raise and level off the Songhua's water level during the fall and spring to allow for shipping downriver, secure drinking and industrial water for Manchukuo's growing cities, allow for fishery development upriver, and even promote tourism around the huge reservoir.[4] Thus, Fengman represented a key focal point for the engineers' continuing pursuit of "comprehensive technology." Because Fengman was Asia's first multipurpose dam, Japanese authorities promoted it as an ideal representation of "East Asian construction" and arranged visits by Manchukuo's puppet emperor, Pu Yi; members of Japan's imperial family, such as Princes Chichibu and Takamatsu; and collaborating Asian prime ministers, such as Phibun of Thailand, Ba Maw of Burma, and José Laurel of the Philippines.[5] The dam also received a steady stream of visitors from industrial associations, newspaper reporters, university students, local governments, and chambers of commerce in Manchukuo, Korea, and Japan.[6]

Similarly in August 1941, government bureaucrats and engineers for the Yalu River Hydropower Company in Korea celebrated the beginning of electricity transmission from Sup'ung Dam, which they called the "project of the century" and a "victory for Scientific Japan."[7] The dam was part of a plan to build seven dams along the Yalu River and transform the frontier region near the river's mouth between Manchukuo and Korea into a coastal urban industrial zone—complete with year-round ports, improved transportation links to natural resources, and modern cities with large factory and residential areas that would herald the "Yalu River Era of Developing Asia."[8] Capable of producing seven hundred thousand kilowatts and measuring 106.4 meters tall, 899.5 meters wide, and 3,110,000 cubic meters in mass, the gravity concrete dam was the world's second largest after Washington's Grand Coulee Dam. Built between 1937 and 1945, it required 850,000 tons of cement and an estimated twenty-five million laborers in manpower to finish.[9] Both world-class dams not only produced cheap, abundant electricity for wartime industrialization but also helped form and naturalize a regime of techno-scientific expertise for large-scale comprehensive development in tandem with the other wartime infrastructure projects in China and Manchukuo. This regime served as an important pillar of Japanese imperial power, survived the war's end, and later reemerged in different forms of "developmentalism" throughout Asia.

Scholars have recently examined the role of science and technology in planning and justifying Japan's imperial rule across Asia and thereby challenged earlier views of war and empire as the product of spiritual fanaticism and a runaway military.[10] Although they have shown that Japanese imperialism, particularly in Manchuria, was infused with a modernist utopianism and a spirit of experimentation that resulted in the construction of planned cities, advanced heavy and chemical industries, thousands of kilometers of roads and railways, and vast telecommunications networks, few of these scholars have examined how these construction projects operated as a form of imperial power and mobilization. They have noted that these schemes rested on the "iron fist of a despotic alien power," for example, or were otherwise related to utopian ideologies and imperial agendas, but they do not adequately address the techniques of power involved in their very construction, particularly the processes behind the formation of the technical expertise that enabled and legitimized their projects.[11]

In justifying itself, Japan's technological imaginary often invoked the aura of stability, order, and progress which was in turn associated with the materiality, rationality, and grandeur of large technical systems and physical structures. But expert technical plans and material infrastructure were neither preexisting wholes in the minds of Japanese planners nor monolithic entities brought by engineers to the colonies but rather messy *effects* produced by a range of complex practices and negotiations that involved many actors, institutions, and forces—and often resulted in conflict. Analyzing the contentious histories behind the formation of technical expertise in the colonies reveals how Japan's technological imaginary tried to legitimate and naturalize itself among the colonized as well as the types of power utilized to create this external image of stability, rationality, and progress. Two of Japan's most prominent civil engineering projects during the war—the construction of Fengman and Sup'ung Dams—helped constitute powerful regimes of colonial development expertise that were used to justify and entrench Japanese imperial rule. Similar to the previous chapter, this chapter denaturalizes Japan's technological imaginary by examining the conflicts inherent within the various processes and negotiations involved in the formation of technical expertise at the project sites. However, this chapter focuses on the techniques of power employed in the process of constructing large-scale infrastructure projects as well as the actual power effects of Japan's technological imaginary among colonial populations, thus complicating and deepening our understanding of the colonies as a technocrat's paradise and space for modernist experimentation.

The dam projects were largely completed, and therefore an analysis of them contributes greatly to our understanding of the relationship between technology and colonial power. Large-scale dam projects best embodied the "comprehensive technology" conception that was gaining popularity among colonial engineers because they could effectively contribute to multiple purposes of electricity production, transportation improvement, agricultural reclamation, and industrial planning. As such, colonial bureaucrats and engineers expended significant resources into their completion and success and in the process formulated the techniques of power that entrenched their technical expertise and justified Japanese imperial rule. Examples from both cases demonstrate the power dynamics surrounding colonial dam construction in general.

Dams as the Foundation for Nitchitsu's Empire and Korean Industrialization

Japan's colonial dam construction attained its greatest extent in Korea and was closely tied to the history of Noguchi Jun's Japan Nitrogenous Fertilizer Corporation (*Nihon chisso hiryō kabushiki gaisha*; hereafter, Nitchitsu). By 1945, Nitchitsu had built a total electricity capacity of four million kilowatts in Korea (largely from hydropower), which nearly matched the Japanese mainland's total capacity of 4,500,000 kilowatts.[12] Historical narratives of dams and other colonial technology projects usually center the story on individual engineers and businessmen, praising their vision and expertise. In Korea's case, accounts of Sup'ung and other dams often emphasize Noguchi's "bold determination," leadership, and planning ability or the technical skills of Kubota Yutaka, his chief dam engineer, or simply focus on how company officials and experts designed and implemented the dam while encountering various difficulties or obstacles along the road to its successful completion.[13] It is important to expand the narrative framework beyond the level of individual capitalists and experts, as figures like Noguchi and Kubota also personified larger historical processes of the ongoing expansion and circulation of Japanese capital in Korea. The Korea Government-General's economic policies were essential to laying the foundations for Nitchitsu's rise into an industrial conglomerate. After annexation in 1910, the Government-General began transforming Korea into a provider of foodstuffs for Japan and a market for Japanese manufacturers. They rearranged land ownership into the hands of the *yangban* landlord class, thereby increasing tenancy and creating a vast pool of surplus labor. From 1920, they launched the "Program to Increase Rice Production," which reclaimed land for paddy development, improved irrigation, and encouraged farmers to use fertilizer-sensitive seeds, thereby solidifying Korea's status as a supplier of cheap rice to Japan in the aftermath of the 1918 Rice Riots. Noguchi's expansion of Nitchitsu's fertilizer business into Korea in the 1920s responded not only to increasing Korean demand but also to the successful commercialization of synthetic ammonia after World War I in Europe—and the mass production of ammonium sulfates requiring cheap access to abundant electrical power in order to compete with inexpensive Western fertilizers and other Japanese conglomerates.[14]

Industrialization began in earnest with Manchukuo's establishment in 1932, as Korea was transformed into a "supply base" (*heitan kichi*) for Japan's advance into China. By 1930 the effects of Japan's "starvation export" policy were being felt, creating an increasing surplus labor population in Korea made even worse by the Great Depression. In this context, the Government-General took a more active policy of promoting industrialization as a way to diffuse social crisis. Although Korea's Corporation Law was first liberalized in 1920, between 1931 and 1936 Korea's gates were opened further to Japanese industrial capital with land price controls, subsidies, and most important, a policy guaranteeing cheap, abundant electrical power. The Government-General began awarding hydropower concessions in exchange for mandating a percentage of power for general use, and fierce competition ensued between Nitchitsu and other conglomerates like Mitsubishi and Mitsui for water-use rights over Korea's rivers. Beginning with its completion of Pujŏn River Dam in 1929, which provided 200,000 kilowatts of cheap electricity for its chemical complex at Hŭngnam (the world's third largest), Nitchitsu pursued a "see-saw game" of parallel hydropower dam and chemical factory construction. After winning water-use rights from Mitsubishi in 1933, the Changjin River Dam was completed in 1938, supplying 330,000 kilowatts for Hŭngnam's increasingly diversified chemical products and powering industrial growth in P'yŏngyang and Keijō (Seoul). Plans began in 1936 to build the Hŏch'ŏn River Dam scheduled to produce an additional 335,000 kilowatts, two-thirds of which was allotted for general use throughout the Korean peninsula. Thus, the unprecedented 700,000-kilowatt Sup'ung Dam project was not only part of Nitchitsu's long-term strategy of rapid expansion in competition with other Japanese and international conglomerates but also important for the Government-General's promotion of Korean industrialization and incorporation into Japan's East Asian wartime economic system. Nitchitsu benefited enormously from the Government-General's active encouragement of capitalist investment, such that by 1942, its subsidiaries owned 36 percent of Korea's industrial fixed capital and were establishing factories throughout East Asia.[15]

Thus, although the vision and determination of Noguchi and his team of experts were certainly a major part of the narrative of colonial dam construction, we also have to give credit to the circulating force of capital and the laws, institutions, and technical processes that made that circulation and hence their visions possible—land rationalization and agricultural improvement schemes to increase rice production, the invention and global proliferation of synthetic ammonium sulfate fertilizers requiring abundant electricity, intense competition with other Japanese and Western chemical firms, colonial laws to attract investment, and the creation of a cheap industrial labor force. All of these capitalist practices and forces did not stem from Noguchi the capitalist alone but formed a complex circuit that ultimately enabled him and his engineers to propose and build the world's second largest dam, among other things. Whereas Sup'ung Dam's story often took the form of Noguchi and his engineers' devising ambitious plans followed by their successful implementation in the empire, such a focus privileges their agency and fails to explore the broader colonial power arrangements and forces that made them possible. Such an approach in fact may come close to reproducing the overall standpoint behind this 1943 statement by one of Noguchi's admirers: "The American continent existed before Columbus was born, but modern America was created by Columbus's discovery; modern Korea was created by Noguchi's discovery."[16]

Rationalizing Rivers

Dam construction consisted of more than implementing engineers' plans and blueprints. It also involved transforming rivers from free-flowing waterways into regulated hydropower-producing systems, a task that required an entirely new way of understanding, negotiating, and shaping the environment. The acquisition of new knowledge that ultimately enabled large-scale dam construction in Japan's empire in the late 1930s, however, was not simply a progressive process of acquiring new expertise or scientific data about an unknown environment. In conducting river studies, engineers were also engaged in the dynamic constitution of a new object of calculability in the form of comprehensively managed, high-electricity-yielding rivers. Although Korea and Manchuria's rivers were presented by engineers as quantifiable and manageable in the form of officially compiled statistics, the entire process of "rationalizing the river" in this manner was in fact quite unstable and imprecise and involved various extrascientific practices.[17] An examination of Korea's Second Hydropower Study (1921–29) provides insight into the types of engineering practices used on the ground to create this new regime of calculability, which made hitherto unfamiliar environments visible and controllable in the form of authoritative graphs, maps, and charts for hydropower development.

Government-General engineers only realized Korea's hydropower potential during the second study, when they began collecting data on the maximum and normal streamflows of the country's major rivers to determine if they were sufficient for building the high dams necessary to meet the electricity needs of Korean industrialization and urbanization.[18] First, government engineers decided on the most promising dam sites based on earlier hydropower and flood control studies as well as the Army Survey Division's topographical maps. They took into account a variety of factors, such as weather conditions, topography, and geology and estimated construction and transmission costs. After forecasting electricity output for each location, the engineers conducted meteorological and hydrometric tests as they discovered new sites and established precipitation and streamflow measurement stations at the most promising ones. Their budget, however, limited their building as many stations as they required or conducting as many tests as they would have liked. For example, there was only one hydrometric station per 1,600 square kilometers, which forced engineers to make uniform generalizations based on sparse data about streamflow across vast river basins. The final published study disguised such generalizations through what appeared to be organized statistical charts, giving the sense of a relatively uniform environment subject to expert manipulation and control.[19]

Obtaining the measurements was difficult, because it required negotiations with different institutions and people. The Communications Division (*teishin kyoku*) supervising the study had to conclude formal agreements with other government divisions to share their stations. Koreans, who were "somewhat illiterate" with a poor command of Japanese, had to be trained to take daily precipitation and hydrometric measurements as well as submit monthly reports (see Figure 8). Local post offices and police stations had to be enlisted to oversee these Korean observers. Poor weather was a frequent problem, with floods that washed away gauges or prevented teams from reaching remote stations. Korea's rivers were wide and sometimes doubled or tripled in width during the summer rainy season. State engineers were therefore often forced to conduct makeshift streamflow measurements or rely on local hearsay to calculate maximum water levels. Finally, the need to determine a uniform streamflow coefficient that could calibrate varying water velocity gauges resulted in the hasty construction of a gauge inspection facility in 1924. Using a sedimentation tank at a water treatment facility outside Keijō, they built a 48.5-meter-long track for a special rail car equipped with a velocity gauge that reached into the tank. By moving the car at different speeds and taking readings, they arrived at a device coefficient that would then uniformly measure streamflow across

Figure 8   Policeman overseeing Korean man taking precipitation readings during the Second Korea Hydropower Study, 1921–29. Source: Chōsen sōtokufu teishinkyoku, *Chōsen suiryoku chōsasho*, 76.

Korea. Thus, seemingly scientific devices designed to accurately measure the environment disguised the uncertainty and ambiguity within their very structure.[20]

The environment did not simply lie outside the above practices, ready to be transformed into precise data and shaped by Japanese technical expertise. It was actively constituted through many unstable, extrascientific processes, such as training Koreans to take measurements, attempting to make erratic rivers conform to new knowledge and devices, relying on generalization and local hearsay for data, and building makeshift facilities to anchor the new forms of measurement. Rivers were never fully "rationalized" during dam construction and continued to defy scientific expectations through their frequent fierce floods, which betrayed the images of control and transparency presented by both colonial governments' statistics.

Planning and Designing Dams

Dam construction involved not only engineers struggling to understand and frame unfamiliar environments but also difficulties in acquiring the necessary expertise, doubts among many about their ability to successfully complete such unprecedented projects, and chronic institutional conflict which undercut public proclamations of a unified "Scientific Japan" boldly "constructing East Asia." In sharp contrast to the tremendous idealism of official proclamations on Fengman and Sup'ung Dams, when Manchukuo was founded in 1932, state engineers were actually very pessimistic about utilizing Manchuria's rivers for hydropower for similar reasons, as discussed earlier with regard to Korea. Japan's predominant technique of building "conduit-type" (*suiroshiki*) dams—damming and diverting river flow through large pipes over long distances toward steep drop-offs to low-lying power stations—seemed unsuitable to Manchuria's climate and geography.[21] Manchuria had low yearly rainfall but condensed summer rainy seasons, slow and gently sloping rivers that froze for much of the year, and vast plains, which seemed to make it unfavorable for dam development.[22] But as National Roads Bureau engineers began conducting basic investigations of Manchuria's rivers and devising the new comprehensive flood control strategy in response to devastating floods in 1932 and 1933, they also began learning more about the technology of large-scale gravity dams, which enabled more control over rivers, and soon became persuaded of their feasibility. The top Manchukuo reform bureaucrat Hoshino Naoki noted that soon after he arrived in Manchuria, an elderly Russian engineer named Shelkovsky, who for thirty years had measured water levels on the Songhua River, visited him in 1933 to sell an ambitious plan to build two dams that would not only control its frequent floods but produce abundant hydropower and aid in irrigation as well. Intrigued, Hoshino purchased the plans and pushed the Guandong Army and relevant government departments to approve a national river investigation, resulting in the first Roads Bureau hydropower studies. In 1934, Naitō Kumaki, vice president of Japan Electricity (*Nihon denryoku*), visited Hoshino to argue that Manchuria had abundant hydropower resources and asked permission to develop a site that exactly matched Shelkovsky's plan, leading Hoshino to push for more expansive state investigations.[23] The Temporary Industrial Research Bureau, led by the reform bureaucrat Shiina Etsusaburō, was established in December 1934 and began conducting its own studies into hydropower production on the Songhua, Liao, Taizi, and Yalu Rivers as part of a larger investigation on planning Manchukuo's future industrial development.[24] Despite the growing enthusiasm among engineers and bureaucrats for hydropower development, Mantetsu—the institution with the most experience in civil engineering in Manchuria and wary of the rising power of reform bureaucrats—continued to be skeptical toward the idea, and the Guandong Army was rather unwilling to devote resources to uncertain comprehensive infrastructure projects during the new nation's early years.[25]

Whereas the Government-General in Korea had historically adopted policies friendlier to capital, radical officers in the Guandong Army and reform bureaucrats were designing a managed economy in Manchukuo from its establishment in 1932. Hydropower development was officially incorporated into state policy from 1936 onward, a period commonly referred to as Manchukuo's "second period of economic construction," when three major policies were launched—the Five-Year Industrial Plan, the "millions for Manchuria" Japanese emigration plan, and development of the northern frontier.[26] Envisioning a future war with the Soviet Union, then pursuing its second Five-Year Economic Plan under Joseph Stalin and increasing its military forces along Manchukuo's border, radical officers aligned with Guandong Army Chief of Operations Ishihara Kanji and reform bureaucrats initiated plans to rapidly transform Manchukuo into a total war economy based primarily on heavy and chemical industries. As Manchukuo's Industrial Department in cooperation with army officers and Mantetsu's Economic Research Association drew up the Five-Year Industrial Plan, National Roads Bureau engineers and employees of the Manchuria Electricity Company (*Manshū dengyō kabushiki gaisha*; hereafter, Manden), a semiprivate special company founded in 1934 charged with unifying Manchukuo's power production, pushed for hydropower projects under their respective jurisdictions. By 1936, such Roads Bureau engineers as Naoki Rintarō, Haraguchi Chūjirō, and Honma Norio (head of the water-use division) had sufficient data to make the case for comprehensive river management to the Guandong Army as represented by the Liao River Improvement Project, which included the construction of multipurpose dams. A similar comprehensive plan was developed for northern Manchuria's Songhua River, which along with the Yalu River had the most potential for hydropower development.[27] In January 1936, Manchukuo's government approved plans for dam construction along the Songhua, and the Roads Bureau immediately began conducting site investigations and designing what would eventually become Fengman Dam.[28] Manden engineers cooperated throughout, hoping to ultimately win control over dam construction, but in August 1936 the Guandong Army decided to maintain state management over hydropower development out of concern that the private sector would not control costs.[29] The Hydropower Construction Bureau (*suiryoku hatsuden kensetsu kyoku*) was subsequently established in January 1937 under Naoki's leadership, demonstrating the growing influence of engineers at the Manchukuo state's highest levels. As Naoki noted in a 1938 article, the bureau was founded upon the motto of "comprehensive river-use planning," and state management ensured that the multiple objectives of flood control, transportation improvement, and hydropower production would be adequately and efficiently met in contrast to earlier wasteful projects that pursued similar goals in a haphazard, uncoordinated manner.[30] Unlike in Korea, where private-sector engineers took the lead in dam development, utopian-minded state engineers in alliance with statist reform bureaucrats and military officers defeated private interests, such as Manden and Mantetsu, in asserting their control over large-scale infrastructure projects in Manchukuo.

Passed in January 1937, the 2.58-billion-yen Manchukuo Five-Year Industrial Plan sought to increase electricity production over five years by 946,000 kilowatts to a total of 1,405,000 kilowatts, of which 590,000 kilowatts would be from hydropower, thereby saving on coal consumption and providing cheaper, more abundant electricity to meet the military's ambitious industrial targets.[31] In February 1938, the total electricity objective was raised to 2,570,550 kilowatts by 1941 (1,240,000 kilowatts or nearly half from hydropower) as a result of the outbreak of full-fledged hostilities with China in July 1937.[32] Three large hydropower projects were prioritized in the first plan—Sup'ung for southern Manchuria to power the industrial zones of Dalian, Anshan, Fengtian, and Andong; Fengman for northern Manchuria to power the large cities of Xinjing, Harbin, and Jilin; and Jingpo Lake (Kyōhakuko) for eastern Manchuria.[33] Emboldened by the state's shift toward a "hydropower first, thermal power second" policy, the bureau began a four-year comprehensive Manchukuo hydropower survey, largely conducted by air because of continuing security issues, which led to the selection of forty-six promising dam sites that would not only produce a total of 7,500,000 kilowatts a year but alleviate the region's chronic flooding as well.[34] Boring tests at the beginning of 1937 revealed that the Fengman site on the Second Songhua River was the most promising, and bureau engineers quickly focused most of their efforts there while planning for five other dams on the river.[35] But whereas state and private engineers were "theoretically" confident about their ability to build a large-scale gravity concrete dam, neither Mantetsu nor Manden had sufficient technical experience for the ambitious project, leading some in the government to briefly propose transferring the project to Noguchi Jun's Nitchitsu Corporation in Korea. Nitchitsu had already demonstrated their ability to build large hydropower dams on the Changjin and Pujŏn Rivers, and Manchukuo authorities were beginning to cooperate with them and the Korea Government-General on starting dam projects along the Yalu River.[36] But bureau engineers insisted on carrying out the project themselves and decided to send Kuga Tokuhei, Fengman's head of construction and one of Japan's most experienced dam builders, on a three-month North American tour to study dams and construction machinery technology in June 1937. A public tender to subcontract the construction ended in failure in October 1937, as the few construction companies that applied submitted extremely high bids, which demonstrated the lack of confidence among private-sector engineers at the time to undertake such a large and dangerous project.[37] In the end, Hoshino relied more on a general faith in the abilities of state engineers than on any proven technical expertise or experience when he helped win final approval from the Guandong Army and Japanese government for the 100-million-yen project.

During his three-month whirlwind tour of North America, Kuga visited the Grand Coulee and Boulder Dams (then considered the world's most advanced) as well as several of the Tennessee Valley Authority (TVA) dams and reputable machinery factories to study firsthand the latest in high dam technology. Japan's capability in dam construction developed primarily during the 1920s and 1930s as American expertise flowed into Japanese civil engineering circles.[38] Kuga himself had conducted an earlier study tour in 1933 of major dams in Europe and the United States (including Boulder Dam) and subsequently became the head engineer for Japan's highest dam at the time—the Tsukahara dam (completed in 1937), which utilized some of the latest construction techniques employed at the Boulder Dam as well as in the TVA's dam projects.[39] His 1937 travelogue to North America is instructive because it reveals the conditions of Japan's dam construction industry at the time and some of the problems they would have to overcome in order to successfully build the Fengman multipurpose dam.

Kuga began by noting that dam construction methods in the United States were no different than those in Japan in terms of such basic techniques as ways to use cement mixers for different environmental conditions and methods to determine concrete rigidity, and the ease with which he debated with American engineers demonstrates the narrow technical gap between the two countries. However, Kuga was still very impressed at the degree and precision of mechanization involved in the projects he visited, and most of his photographs were of the cranes, mixers, batch plants, buckets, and conveyor systems. At the recently completed Boulder Dam, Kuga was surprised at how the builders managed to set 2,400,000 cubic meters of concrete in only two years without its resulting in any major leaks. This was a result of the use of special low-temperature cement and an expensive "pipe cooling method" to remove the internal heat inside concrete during construction, which Japanese dam builders could not yet afford to replicate. At Grand Coulee Dam, the main objective of his trip, he was impressed at the skilled use of hammerhead cranes, dragline excavators, bulldozers, tractors, and diggers to effortlessly remove and smooth out geological obstacles with great precision. Concrete setting was done with amazing speed, twenty-four hours a day and thirty days a month without interruption, and the cement mixed with very little water resulted in the most durable concrete Kuga had ever seen. He was also very impressed with the techniques of scientific management and the efficient division of labor at the site where the crane and railcar operators, signalmen, and bucket controllers worked "completely as one body" and each worker only did the specialized job assigned to him. In a criticism of Japanese work culture, he noted that American workers were much more productive than their Japanese counterparts because they were better paid, worked fewer hours, and had an environment where they could freely interact with and express their opinions to upper management, unlike in Japan, where managers largely stayed in their offices and foremen treated workers like "bulls or cows." Thus, as Manchukuo state engineers confidently declared the power of Japanese technology to shape and control Manchuria's environment for the people's welfare, they were also keenly aware of their own technical and organizational limitations as they closely followed the latest American dam-building techniques.[40]

From the early 1930s and especially after the various wartime embargoes imposed by Western nations after 1937, engineers and government officials in Japan began to decry Japan's dependence on foreign technology and emphasize "technological autonomy" instead, which culminated in the passage of the Outline for the Establishment of a New Order for Science and Technology in May 1941.[41] Such civil engineering projects as Fengman and Sup'ung Dams were promoted as examples of Japanese technology's superiority—for example, engineers took pride in Fengman's "world-class" mixing facilities that produced seven thousand cubic meters of concrete per day; its fully mechanized transport, sorting, and delivery systems; the enormous manmade reservoir; and the unique construction techniques that were employed.[42] Kuga described the entire project as Japan's demonstration of the benefits of comprehensive river management to the Manchurian people, while the project's head, Honma Norio, viewed it as a shining example of "Manchukuo's power" to the world.[43] Indeed, such Japanese machinery and heavy industrial companies as Hitachi, Fuji Electricity, Ōji Steel, Ishikawajima Shipyards, and Daidō Concrete were closely involved in providing machinery, equipment, and materials for the project.[44] At the same time, however, the Hydropower Construction Bureau still depended quite heavily on foreign technology. Because Japanese manufacturers did not have the capability to quickly produce the ten record-breaking generators (65,000 kilowatts) and water turbines (85,000 kilowatts), the bureau decided in 1937 to order most of them from overseas—five generators and turbines from Westinghouse in the United States (including two smaller ones for internal power), three generators and turbines from Germany's

AEG, and five generators and turbines from Switzerland's Escher Wyss (including two smaller ones for internal power).[45] According to Honma, both AEG and Westinghouse were quite surprised at the order and the dam's overall scale, as they had never produced generators and turbines of that magnitude before, and AEG even had to expand its factory facilities in response.[46] The bureau ordered two additional generators and turbines from Hitachi in 1941, which copied the recently delivered foreign models and completed them by 1942.[47] The construction machinery also relied heavily on some of the same foreign technology that Kuga had admired at the Grand Coulee and Boulder Dams during his 1937 trip—electric shovels from Bucyrus-Erie, Kinyon pumps from Fuller-Kinyon, bulldozers from Caterpillar and Hanomag, and drifters from Ingersoll-Rand acquired through Japanese trading companies.[48] Thus, whereas Fengman Dam was publicly celebrated as a prime example of Japanese technology advancing onto the world stage that was gaining attention from international experts, in reality it relied rather heavily on foreign products and expertise.[49] In the end, forging the necessary institutional will to implement the project and negotiating the competing demands of private versus public management and Japanese versus foreign technology proved just as difficult as managing unfamiliar river environments.[50]

"A Sumo-Match with Nature": Constructing Dams and Managing Rivers

The Fengman project's supervisor was Honma Norio, the Hydropower Construction Bureau's head of construction affairs and bureau chief from August 1937. Honma had previously worked in Korea for eighteen years, quickly rising to become head of the Keijō Civil Engineering Office, where he managed a variety of river control and bridge-building projects before being invited by Naoki to join Manchukuo's National Roads Bureau in 1933. Frustrated by the lack of advancement opportunities in Korea's state engineering circles and excited by opportunities in newly established Manchukuo, he briefly headed the bureau's roads division before taking over its water-use division.[51] In a 1939 speech to Japan's leading industrial engineering organization, Honma explained that "total development of national land, total utilization of rivers through comprehensive technology" was one of Manchukuo's top policies.[52] Preparatory construction of a supply railway from Jilin, construction roads, and housing and office facilities began immediately in January 1937, as engineers hurriedly designed the multipurpose dam in order to begin construction as soon as possible. Determining the design, dimensions, and materials for the multipurpose dam depended on a whole host of factors: the forecasted maximum water volume and streamflow rates generated by yearly floods during the summer rainy season, minimum water volume and streamflow rates during the early spring and fall when rainfall was very low, calculations of water volume needed to sustain planned levels of electricity production, minimization of building costs, speed and ease of construction, and the river's winter icing period. The dam's multiple objectives did not seamlessly flow together either but required different ways of utilizing the dam, all of which had to be factored into the design and planning. For example, for flood control, the dam needed to be kept fairly empty to accommodate floodwaters, whereas sustained hydropower production required a relatively full reservoir. Dams had to hold back water to utilize later to maintain sufficient depths for river transportation, yet at the same time, they needed to release water to meet irrigation, water supply, and electricity needs.

Cobbling together the earlier Zhang Zuolin regime's water level records at Jilin from the great flood of 1923, Russian Chinese Eastern Railway streamflow data from 1923 onward, Mantetsu's precipitation records at main points from Changchun southward, and the Korea Government-General's precipitation measurements at Chunggangjin on the Yalu River border from 1915, engineers were able to compile basic hydraulic data on the Songhua River.[53] The National Roads Bureau and later the Hydropower Construction Bureau also collected their own information on streamflow, water levels, precipitation, and sediment by constructing eight hydraulic stations along the river from 1933 and compiled graphs that attempted to capture its seasonally erratic hydraulic patterns.[54] Based on these various sources, engineers estimated average yearly streamflow to be 525 cubic meters per second and designed the dam so that it could accommodate streamflow rates of up to fifteen thousand cubic meters per second and therefore safely withstand the yearly summer flood season. In order to ensure that the streamflow downriver would never exceed three thousand cubic meters per second, they designed an overflow drainage system that could release water at three thousand cubic meters per second when the reservoir was low and at 7,700 cubic meters per second when high.[55] Based on their scattered hydraulic data, they also estimated that a hydropower station could produce a minimum of 270,000 kilowatts and determined that the reservoir would have to be designed so as to maintain a minimum water level of 248.5 meters and a maximum water level of 263.5 meters during the year to sustain that level of hydropower production.[56] Thus, although comprehensive river management was the project's principal objective, the dam's design was fundamentally based on rather incomprehensive data and incorporated some relatively rough estimates into its overall design.

In November 1937, dam construction began with the temporary closure of the river's right half (see Figure 9). The Hydropower Construction Bureau decided against starting construction by closing off the entire river because of its sheer size and rapidity. Instead, they first partially blocked the river by building a temporary earthen dam (cofferdam) on one side and then began the concrete setting process behind that structure, removing it once the concrete dam was of sufficient height over the river's surface and the dam's internal drainage channels were prepared. The process was repeated on the other side until both sides of the dam were connected and the river flowed through the channels in preparation for the river's permanent closure and the reservoir's filling. The power station was built on the right bank and set to eventually accommodate ten generators capable of producing a total of 700,000 kilowatts of electricity. Over the course of its construction, the bureau continuously underestimated the river's force, causing delays and design changes. Concrete setting was largely done in the winter before streamflow dramatically increased in the spring and summer as winter ice melted and the rainy season began. Every year, engineers estimated an adequate target height for the dam structure based on past water level statistics so that the dam structure and construction areas could withstand the increased streamflows. In the spring of 1940, after they began building the left cofferdam from the previous winter, which diverted river flow toward the partially completed dam structure on the right side, a large flood caused major damage to an excavated portion on the right side and narrowly avoided flooding the power station area, causing a one-year construction delay.[57] In the following spring of 1941, water levels rose two meters higher than expected, again almost damaging the power station. Construction was once again delayed because they had to wait longer than expected for water levels to recede. Unforeseen water level rises again in October 1941 prompted engineers to design temporary flaps for the power station's water intakes and

to build higher-than-planned temporary drainage pipes in the dam structure in preparation for the following summer's flooding season.[58]

Although it is not clear when the decision was made, Honma noted
that construction plans were changed at some point to build the dam two
meters higher, resulting in a cost increase of ten million yen as a result of
the larger size, higher land compensation payments resulting from an expansion of the submersion area, and the need for extra flood control reinforcements farther upriver.[59] The dam's flood gates were closed in October

Figure 9   Fengman Dam under construction. Source: Cover of *Dengyō* 10, no. 109 (1944).

1942 in preparation for electricity production the following spring, and engineers had their first real test of the dam's flood control capabilities in the summer of 1943. The floods that year in the upriver regions were the highest in twenty to thirty years, as streamflows reached nine thousand cubic meters per second in August. At 60 percent of its planned height and with the reservoir filled to around 20 percent of its capacity, the dam managed the summer floods quite well and kept streamflow downriver to 2,500 cubic meters per second, although reservoir levels were already un-usually low before the floods because of less-than-average water volumes earlier in the year. Despite these abnormal conditions, engineers neverthe-less expressed confidence at the dam's ability to control future floods after construction was completed.[60] In the postwar era, however, when the Nationalist Chinese government asked an American technical advisor to its Natural Resources Committee named John Cotton to inspect the dam, which was 89 percent complete upon Japan's surrender in August 1945, he recommended removing 7.5 meters off of the overflow spillway to improve its drainage capacity and lessen the water pressure on its structure while at the same time resetting the concrete at greater volume and strength— much to the anger of Japanese engineers forced to stay behind as advisors who worried about the effects on hydropower production. In the spring of 1950, after the People's Republic of China was established under Mao Ze-dong and the Chinese Communist Party, Soviet dam experts examined the available hydraulic data on the river and concluded that the chances for a huge flood that summer were high. They therefore recommended immediately setting another forty thousand cubic meters of concrete, and Chinese engineers responded quickly by setting 57,360 cubic meters be-fore summer. After 1950, the Chinese followed the advice of Russian hy-dropower engineers in Moscow, who devised a renovation plan to improve hydropower and flood control capacities, strengthen the dam structure and spillway, and automate the power station facilities.[61] Whereas Japa-nese engineers represented and later remembered their work within a larger narrative of technology overcoming nature, as expertise transforming the earth's surface and even as an ultimate expression of their own "masculin-ity," in reality it was more of an unpredictable "sumo-match with nature," as Honma described it in a 1944 interview, in which the entire process of planning and construction was fundamentally imbued with a complex of unknowns, contingencies, and guesswork.[62] The unpredictable forces of nature and the engineers' general lack of knowledge of Manchuria's environment consistently revealed the weak basis of Japan's ideology of rationalizing and systematizing the colonial landscape.

In their articles for academic journals, presentations to professional associations, and accounts to the media, top-level engineers often presented difficulties within a progressive narrative of Japanese science and technology accompanied by bold resolutions to each and every problem. But their articles and accounts also revealed that their superior expertise and preexisting intelligence did not direct their work as much as the actual implementation of the project itself. As shown above, solutions to unforeseen problems were worked out on the ground, and scientific expertise took shape through difficult, unpredictable negotiations with the environment. Colonial civil engineering expertise was always a hybrid assemblage of metropolitan scientific learning and know-how acquired in the colonies rather than a well-defined body of knowledge confidently adapted to the environment, which was how civil engineering publicly and professionally presented, perpetuated, and justified itself.[63]

Contesting "East Asian Construction": The Socioeconomic Effects of Dam Construction

It was not only nature that compromised and challenged Japanese expertise but also the business interests, landowners, and residents whose livelihoods were affected by the dams. From the beginning of each dam's construction, company officials, bureaucrats, and business leaders promoted their benefits for regional industrial development to local residents and businesses. When the Fengman project was announced in 1936, Jilin's city government, the Manchukuo Concordia Association, and the Chamber of Commerce began enthusiastically promoting the dam's benefits for the city's growth. Until then, Jilin was a declining city with a population of around 140,000 along the Songhua River in northern Manchuria and was being increasingly eclipsed by the growth of the capital Xinjing to the west and Harbin to the north. The city's only real industry largely consisted of one of Manchukuo's largest cement factories, and its economy on the whole imported more than it exported, relying primarily on the timber trade for income. Jilin's Chamber of Commerce argued that in addition to helping eliminate floods and raise rice production, the dam's cheap and abundant electricity would ignite a "great industrial revolution" by transforming Jilin into a "chemical industry city" that rivaled Fengtian in southern Manchuria.[64] The Chamber of Commerce vigorously promoted the city's ideal conditions for the establishment of chemical industry, such as access to cheap and abundant electricity, land, labor, coal, and industrial water, as well as its prime location in Manchukuo's center and the convenient rail connections.[65] In October 1938, Manchuria Electro-Chemical Industries (capitalized at thirty million yen) was established and began building a "comprehensive electro-chemical conglomerate" in the city's northern section. Its objective was to create a "tree-like management structure" centered on carbide fertilizer production, whereby interrelated branch factories producing synthetic materials, dyes, fuels, and other chemical products would spin off from the main company, which guided overall investment and innovation.[66] Noguchi's Nitchitsu Corporation in Korea helped establish Jilin Artificial Gas in October 1939 (capitalized at 100 million yen), another chemical conglomerate that liquefied coal to produce fuel, and Tōyō Bōseki was scheduled to complete a large textile factory in the city by August 1940. In response to this investment wave, the Jilin city government expanded its urban planning project from 140 square kilometers in 1937 to around 400 square kilometers in 1939 with an envisioned population of 1,200,000 in thirty years.[67] An electrified high-speed railway was planned to Xinjing via Fengman Dam, reducing travel time to one and a half hours and laying the basis for a future industrial corridor between the two cities.[68] In July 1939, Jilin's mayor and the Jilin Tourism Association issued a petition to the national government to designate the dam's reservoir area as a national park. They also submitted a ten-year, 2,800,000-yen plan to transform the lakes, inlets, and islands created by the newly formed reservoir into an internationally recognized tourism center that would include a scenic highway, ski resort, golf course, boat houses, fishing areas, summer villas, hotels, relocated and restored historical sites, and a dam tourist center. In sum, Japanese business interests and city officials hitched their hopes to the dam's promises of regional development and formulated their own grand visions of the city as a center for the chemical industry, "Manchukuo's sanitarium" and even "East Asia's tourist playground."[69]

Although the Jilin Chamber of Commerce emphasized the dam's benefits to the region and downplayed criticisms of the project, it is clear that Fengman's construction had considerable negative effects on the local economy. The Manchukuo government also later backed away from some of its glowing promises of comprehensive economic development and abandoned the dam's irrigation and paddy development scheme because of resistance from bureaucrats in Japan's Agriculture and Forestry Ministry, who apparently treated Honma and other engineers as "traitors" because they argued that increased rice production in Manchukuo would adversely affect rice farmers in Japan.[70] The exact extent of the economic damage caused by the dam after it was completed is unclear, but there is evidence of major effects on fishing, logging, mining, and commerce, and authorities in turn promised that newly constructed fish farms, land transportation facilities, and improved waterways would help alleviate any negative effects on the residents and businesses.[71] The effects on Jilin's important timber industry, which depended on unobstructed river traffic for its livelihood, seemed particularly significant. Jilin processed around 20 percent of Manchukuo's timber in 1938, and prospects for the industry were rising as the war boosted demand.[72] A government publication briefly mentioned the damage that the river's closure would cause to the industry; however, it noted that new land and rail facilities as well as improved shipping on much of the river caused by the dam's raising of water levels would eventually alleviate the problem.[73] The Chamber of Commerce argued that Jilin's rapid transformation into a "chemical industrial city" would ultimately provide employment to the area's affected residents.[74] Aside from brief mentions of timber industry interests organizing themselves to compromise with the government and of residents cooperating with land purchasing and relocation efforts, the documentation of negotiation and resistance at the dam is sparse.[75] However, because conflict with regional economic interests and local residents had always been an issue that plagued earlier dam building in Japan and Korea—and had even led to an earlier migration of Japanese timber transporters to Manchuria and Korea—there was probably more to the story than the relatively smooth process described in Manchukuo's official publications.[76] An analysis of the more substantial documentary record on how Korea's authorities dealt with the Yalu River timber industry around Sup'ung Dam and how resident Koreans thought about the dam's construction better illustrates how local interests and the colonized also attempted to shape the technological imaginary put forth by colonial governments and experts. It also uncovers some of the actual socioeconomic effects of colonial dam construction.

The Yalu River region's powerful timber industry immediately began to deploy the language of development to contest the state's own discourse once it became clear that Sup'ung Dam would adversely affect its interests. Every winter people cut down the timber and floated it downriver in rafts during the spring to Andong and Sinŭiju at the Yalu River's mouth for processing. In Korea alone, thirty thousand people were engaged in cutting and transporting trees, and thousands of Chinese and Japanese also depended on the trade. In the peak summer season, around 960,000 cubic meters floated downriver, mostly to Sinŭiju, where primarily Mitsui's Ōji Paper Company and Korea's Forestry Management Bureau processed them. From the beginning, industry representatives demanded nonobstruction guarantees, which were incorporated into the Yalu Hydropower Company's construction permit. Against the company's claims of hydropower's benefits for industrial development, they asserted timber's strategic importance to Japan's war economy. They also insisted on the company and the state's duty to protect the people's livelihoods.[77]

In response, Yalu Hydropower put faith in its own technical expertise and began designing six special logging channels despite serious internal doubts about its ability to calculate the proper inclines to facilitate the river's varying and unpredictable streamflows.[78] The company's plan was to unbundle the rafts before they reached the dam, rebundle them into smaller sizes, transport them by tugboats to the dam's floodgates, direct them through the special channels, and then rebundle them on the other side.[79] Company and state engineers conducted one week of tests at a timber yard in November 1938 to determine the average time to guide logging rafts of different sizes through a sluice gate and the average time to travel over one hundred meters using eight workers to guide the rafts. However, the engineers had difficulty simulating the proper glide effect present in an actual channel.[80] Additionally, two days of tests were conducted using two tugboats along a four-kilometer river section to determine average travel times and the numbers and proper boat sizes required to facilitate peak trade.[81] In the midst of pressures from the timber industry, the power company, and the Government-General, company and state engineers hastened to shape the environment according to their wishes.

Yalu Hydropower, however, delayed channel construction, and by the time the river was closed in November 1939 and logging season began the following spring, only two temporary logging channels had been completed. Engineers had not resolved issues related to adjusting the water levels within the channels and determining their proper incline in relation to the river's varying streamflow. Thus, when the rafts arrived, the bundles broke up and the logs were damaged from the water force within the channels. In response to this failure, engineers proposed a more gradual slope and more water-level-adjustment mechanisms along the channels.[82] Yet Yalu Hydropower was still unable to facilitate the following year's logging season, because the delays associated with the dam's construction forced engineers to build three logging channels at much lower levels than originally planned.[83] Although the company insisted on its technical ability to resolve the issues and guarantee people's livelihoods, dealing with such demanding factors as high timber volume, uncertain technology, construction delays, and the unpredictable river ultimately overwhelmed its expertise (see Figure 10).

Figure 10 Sup'ung Dam during summer flooding season (1941). Source: "Sekai dai ni no Suihō damu kankō," *Shashin shūhō*, no. 194 (November 1941): 6–7.

These technical failures devastated the region's timber industry and exacerbated an already tense relationship between the involved parties. Andong saw a 53 percent decline in timber volume, and its main processing factory closed in 1941.[84] In Sinŭiju, industry representatives continuously held Yalu Hydropower accountable to its earlier promises. Operators reacted cautiously to the company's initial logging channel plans at a meeting in April 1939, and they asked the company to invest in roads and railways in case the channel failed.[85] When technical problems became clear in November 1939 and rumors circulated that the river would become unnavigable, Japanese and Korean timber interests and city councilors immediately formed a countermeasures committee. They traveled to Keijō and petitioned high-level bureaucrats and military officers in the Government-General to pressure the company to keep its promises of ensuring the smooth transport of lumber and other goods from upriver upon which the livelihoods of sixty thousand people depended.[86] An agreement was subsequently concluded between Yalu Hydropower and the Yalu River Timber Industry Union in April 1940 to make more concerted efforts to facilitate the trade.[87] But when logging season subsequently peaked and the volume arriving at Andong and Sinŭiju reached only 5.7 percent of that of the previous year, the Sinŭiju Chamber of Commerce and Korean councilors fended off calls for a mass rally and demanded Kubota Yutaka, Yalu Hydropower's managing director, to address their grievances directly instead.[88] At an angry meeting with Kubota, industry representatives pointed to the discrepancy between the company's promises to facilitate forty rafts a day and the reality of accommodating only four; the inadequate facilities for breaking up and rebundling rafts; and the economic effects of factory closures and unemployment. Kubota apologized and promised to tackle the technical issues while arguing that his engineers were already making progress in resolving them.[89] However, as the difficulties of fully accommodating the timber industry became clear, company officials asserted that the industry's sacrifice was inevitable in the wake of hydropower's adoption into national policy and then more aggressively pushed the benefits of future industrial and urban development.[90]

According to the Sinŭiju Timber Association's director, despite increased company efforts to facilitate their trade, 1940 still ended disastrously with only 70 percent of the timber eventually transported around the dam. Because large factory timber was prioritized, small operators suffered, which resulted in the closure of processing factories, the slowing of production, small shops' eventual inability to fill orders, and the collapse of financing. In February 1942, the North P'yŏng'an Province governor finally brokered a deal whereby Yalu Hydropower purchased lands in Manp'o, which had better railway facilities, and paid the relocation expenses for some private companies.[91] Additionally, the company agreed to provide subsidized electricity for the relocated factories. By year's end, Yalu Hydropower purchased Sinŭiju's largest remaining factory, and the timber industry reorganized itself into a unified association under wartime controls. In sum, whereas the company and state officials emphasized the dam's benefits for promoting heavy industry, regional timber interests expanded that discourse to include discussions of balanced development and rights to livelihood. Although the company responded by employing the same faith in technology's power that had already been embodied by the entire project in designing the special channels, the effort failed, leading to a hastily drawn-up and vigorously contested settlement. Japanese officials presented these issues as difficulties to be overcome through technology or future industrial development; however, their expertise and discourse of development were challenged throughout as the timber industry and local councilors altered the terms and meanings of their discourse.

We catch another glimpse of how local residents challenged the discourse of "Scientific Japan" and "East Asian Construction" in a five-part travelogue to Sup'ung Dam written in 1940 by Kang Ik-sŏn, the Lagushao branch head for the *Mansŏn ilbo* (Manchuria-Korea Daily). The series was written as authorities made final preparations to relocate the seventy thousand residents within the dam's submersion zone. As a writer, Kang was clearly restricted as he was escorted by police officers in Ch'angsong County, warned by an official to be careful in his reporting, and banned from taking photographs in the county, which had the most productive agricultural land scheduled for submersion.[92] Although he supported the project, his entire series was tinged with sadness about the area's disappearance and the loss of homes and livelihoods. For example, Kang documented the compensation levels in one district and described landowners' complaints about the compensation levels' failure to reflect market prices, include the value of housing structures and crops, and enable maintenance of their livelihoods. He then wrote, "I believe that as citizens of a nation that is building the power plant as a glorious foundation for the construction of a new Greater East Asia . . . all of them should not feel the need to further complain."[93] Yet the article also had this rather poetic headline: "Steep Mountains and Peaks, Silky Green Farmlands in Front of Farm Houses, Submerged in an Instant—All but a Momentary Dream . . . Oh, How Powerful Science Is!"[94] Thus, as the Japanese media triumphantly celebrated science and technology's power to modernize East Asia, Kang subverted their meanings to signify destruction and impermanence.

In his last article, he described the successful efforts of the Manchukuo Concordia Association, the party charged with mobilizing the different ethnicities for pan-Asian cooperation, to convince and relocate the "listless residents" through films, lecture tours, and public meetings by proclaiming hydropower's importance for "East Asian Prosperity." But in contrast to this and his earlier descriptions of residents happily accepting compensation and relocation, Kang ended his series with the following thoughts:

For those villagers, who had to desert their beloved homes and livelihoods to join relocation groups and look for new ones, it would be impossible to understand the national policy of building a prosperous Asia. To them it was only a matter of their own life's disaster. Vacant homes demolished! Whole villages completely vacant! Truly sad scenes!

Wind was blowing towards me in the freezing cold weather. My eyes were filled with images of the vagabond villagers on their way to being relocated. At this sight, my body quickly began generating a strange heat.[95]

Thus, as colonial governments were pronouncing Japanese technology as the basis for Asian development, Kang's account revealed an attempt by the colonized to appropriate that discourse. Business interests and local residents rendered the technological imaginary unstable, an instability that is often not captured in narratives of colonial development except as bumps on the road to inevitable progress.

Colonial Power in the Purchase of Land and Transfer of Residents

The formation and implementation of Japan's civil engineering expertise at dam construction sites rested on multiple forms of colonial power—legal, police, administrative, economic, and ideological—in order to relocate the residents and mobilize workers to build the enormous structures. Residents and workers hardly ever appeared in engineers' and official accounts except as patriots who willingly parted with their lands for the greater good of East Asia or as heroes building "great pyramids" on the continent.[96] But behind the expertise of "Scientific Japan," a whole array of institutions central to the colonial enterprise had to be closely coordinated to ensure the successful completion of such large-scale projects. An examination of the more extensive documentation on Sup'ung Dam reveals the numerous institutions and techniques of power involved in making the exercise of Japanese expertise in the colonies possible.

The dam submerged 106 square kilometers in Korea and 99 square kilometers in Manchukuo, displacing 32,780 and around 40,000 people, respectively.[97] Even if authorities emphasized a spirit of cooperation as well as care in conducting negotiations with these soon-to-be-displaced residents, the actual process of compensating and transferring them involved a wider range of actors. As Kobayashi Hideo and Hirose Teizō note, colonial despotism and police power were key factors in purchasing lands and relocating residents.[98] But an overemphasis on these aspects overlooks the many other forms of legal, bureaucratic, and disciplinary power used throughout the process. From their beginnings, Manchukuo and Korea established laws and procedures for approving projects, procedures defined by an obligation to negotiate with and compensate affected parties; however, these were more streamlined than Japan's laws and lacked their checks and balances. In Korea's eminent domain law, there was no true third-party mediation, no mechanisms for dispute resolution beyond appealing to government authorities, and no language on fair value and instances requiring different compensation levels.[99] Throughout the compensation and relocation process, colonial authorities invoked the law as a means to legitimize the forced purchase of lands and diffuse resistance among residents through bureaucratic procedure.

Such different actors as the residents and their representatives, the provincial government, Yalu Hydropower, and the Government-General invoked bureaucratic procedure and laws in accordance with their respective interests. Land prices spiked when residents learned about the Sup'ung Dam project. In response, a police chief set the precedent in one district by issuing construction consent forms to 335 landlords with predetermined compensation amounts. Yalu Hydropower objected to the amounts and issued its own valuations based on tax records and land registers, which valued the lands at 15 to 20 percent lower than market value.[100] Sometimes it used the poorest quality land as the basis for determining value, which angered many residents.[101] In one 1937 sale, the company paid only one half to one third of the market price, using food and drink as tools of persuasion during negotiations.[102] Because registers did not reflect market value, lands were usually priced too high or low, causing a wave of speculation and an influx of brokers, which created more anxiety and ill will toward Yalu Hydropower.[103] Residents complained to authorities about arrogant company officials who treated the area as their private enclave, conducted hostile negotiations with landowners, engaged in their own private land speculation, and showed insensitivity toward residents about to lose their homes and livelihoods.[104] The province also criticized the company's valuations and "backroom tactics," urging the company and the Government-General in November 1937 to let it form a committee to take over the negotiation process out of sensitivity toward local opinion toward the project.[105] Thus, from the very beginning, tensions broke out among residents demanding fair compensation and rights to livelihood, the province trying to win popular support for the dam and future projects, the police pressuring the residents to settle, the company trying to minimize costs, and the Government-General mediating the entire process.

The North P'yŏng'an Province governor headed the new compensation committee, which consisted primarily of provincial councilors, provincial and local bureaucrats, and police heads.[106] It met five times before announcing its conclusions in February 1939. The meetings were contentious, with Yalu Hydropower pressuring the members to assess lands according to lower Manchukuo prices, the governor pushing for higher valuations, provincial councilors demanding direct popular representation on the committee, and the Government-General caught among all of them.[107] In the meantime, Yalu Hydropower continued using such tactics as having Korean employees act on the company's behalf to negotiate and purchase land cheaply, causing further popular distress. The province tried placating residents through lecture tours and discussions, but it also strictly banned public protests and assemblies.[108] Reacting to popular sentiment, the committee's evaluations were more detailed and higher than Yalu Hydropower's earlier ones—particularly in reference to paddy, fields, forest, and housing.[109]

In the end, total compensation was lower than the announced levels and used a more streamlined classification scheme, suggesting that company pressure to lower costs as well as increased police intimidation succeeded.[110] Although the bureaucratic process was employed to legitimize acquiring land for the project, the residents, company, and province each contested the terms and nature of that process by appropriating the language of public interest and development for their own goals. At the same time, under the guise of "fairness," "sincerity," and "sympathy" toward the residents, the province, police, and company exploited their authority, existing colonial inequalities, and lack of civil protections to pressure residents into accepting their terms when the legal process failed.[111] Meanwhile, the affected residents used whatever bureaucratic recourse available to them (official meetings, petitions to authorities and the press, provincial councilors, and sympathetic bureaucrats) to win more favorable terms and contest the discourse of development. The technological imaginary and its narrative of superior Japanese expertise generally ignored and marginalized all of these issues as minor obstacles in the larger scheme of building a prosperous, heavy industrial future. Yet it was police, bureaucratic, and legal power that in fact made the exercise and further development of Japanese expertise and the promise of development possible.

Relocating residents also involved other institutions of colonial power. The joint Manchukuo-Korea Development Committee in charge of managing the dam's socioeconomic effects decided to incorporate residents into ongoing emigration programs that resettled Koreans as colonial "pioneers" and independent farmers in Manchukuo. The arrangements were first made in 1939 and 1940 to settle nearly two thousand affected households in "collective villages" (*shūdan buraku*) managed by the Korea-Manchuria Colonization Company (*Sen-Man takushoku kabushiki gaisha*) in northern Manchuria and to issue "dispersed settler" (*bunsan imin*) permits to other residents who would cultivate land near their friends and relatives already there. Residents were also resettled in "safety villages" (*anzen nōson*) in Manchuria and north China as part of the Guandong Army's campaigns. Both governments established these large, militarized settlements with attached farmlands in 1932 to resettle Korean refugees fleeing hostilities after the Manchurian Incident the previous year. For the Manchukuo government, Korean settlers were valuable tools of colonization. For example, many of the resettled participated in building dikes and reclaiming swamplands in collective villages in northern Manchuria.[112]

As Hyun Ok Park has shown, colonization companies appealed to the hopes of poor Korean peasants by promising them land ownership, organizing them into cooperatives, and providing them with credit.[113] These together served as techniques of power as they incorporated peasants into bureaucratized regimes of debt and mobilized them into mutual aid associations that promoted values of work discipline and community. Many of these settlements, particularly the safety villages, were highly militarized. Their spaces were planned to allow for their constant surveillance by police and self-defense groups and included central squares for collective activities and drills. As they purchased land, Yalu Hydropower systematically targeted poorer tenants, whom they persuaded to pressure their more resistant landlords to settle.[114] Despite their sadness over leaving their homes and ancestral gravesites as well as anxiety over relocating, many poor peasants readily left because they had less to lose than large landowners. In his travelogue, Kang described a group of villagers in Sakchu County who were happy to sell their land to Yalu Hydropower because it was of inferior quality, demonstrating that some were perhaps attracted by promises of better prospects elsewhere.[115] Many wealthier landowners who were pressured into resettling, however, fared poorly in Manchuria, as they had almost no personal agricultural experience and were not used to the land and climate.[116] In sum, whereas engineers and officials presented the project as an achievement of Japanese expertise and technology, its completion in fact also rested on the implementation of collective emigration programs that incorporated the hopes of poor peasants into capitalist regimes of dependence, agricultural development projects that increased Japanese control over the Manchurian countryside, and counterinsurgency campaigns that disciplined collective villages into structures of mobilization.

Mobilizing and Disciplining Labor

Labor appears in engineering accounts largely as an invisible force that implemented the experts' designs or in the form of "industrial warriors" who dutifully worked and even sacrificed their lives for "constructing East Asia."[117] Although both dams were billed as representing the cutting edge of dam technology and Japanese engineers put themselves and their expertise at the center of their stories, the projects could not have been carried out without procuring large numbers of unskilled Chinese and Korean labor. At Fengman Dam, around three hundred thousand workers per month were required between November 1937 and June 1938 to build the cofferdams that closed off most of the Songhua River. At the construction's peak in 1942, fifteen thousand workers per day were utilized to quickly finish most of the dam structure in preparation for electricity production the following year.[118] From the project's beginning, securing a constant and steady labor stream proved very difficult. The Manchukuo government and private businesses usually paid Chinese labor brokers, who were usually landed elites with a network of clients, to recruit the necessary labor for large-scale construction projects. They paid lump sums to these brokers, who then paid the wages, loans, transportation, and upkeep of groups of fifteen to ninety workers (known as the *batou* system).[119]

Manchukuo on the whole constantly faced labor shortages for its many ambitious construction projects because the Guandong Army and the government's Labor Control Committee at first pursued a policy of curtailing traditional migration from north China out of concern about potential subversion, a desire to curb Manchuria's Sinicization and increase Japanese emigration, and worry over currency outflows. The Guandong Army established the Datong (*Daidō*) Company in 1932 to formalize and control the process of recruiting and supplying labor hitherto conducted by the construction companies themselves through Chinese brokers.[120] On the other hand, the Manchuria Construction Industry Association (*Manshū doboku kenchikugyō kyōkai*), a private industry group of construction companies that played an influential role in shaping labor policy, continuously campaigned for an increase in the supply of its traditional sources of Chinese labor and the creation of a system to control recruiting costs and wages. The Five-Year Industrial Plan, launched the same year Fengman Dam began construction, only increased labor demand as supply became more strained and costs rapidly rose as a result of difficulties caused by the war. In response, the National Mobilization Law was passed in February 1938, which allowed for labor service conscription, and migration restrictions were quickly loosened as well. The Manchuria Construction Industry Association also took over more control of the process through the establishment of the Manchuria Labor Association (*Manshū rōkō kyōkai*) in 1938 and its counterpart in China, the North

China Labor Association (*Hokushi rōkō kyōkai*), in 1941. Yet demand for labor in China itself resulting from new construction projects and rising wages there decreased incentives for Chinese to migrate to Manchuria, which worsened already existing supply problems. This led to an increased reliance on more inexperienced and unproductive workers, such as conscripts, prison labor, youth, and the elderly. Turnover rates were high at most construction sites as companies competed with each other for scarce labor or workers fled bad working conditions, and after the Pacific War broke out in 1941, the state turned more and more toward labor conscription and mass mobilization through such organizations as the National Prosperity Labor Association in Manchukuo and the New People's Association in north China.[121]

These problems and tensions in Manchukuo's overall labor procurement system played out acutely at Fengman Dam. Engineers reported that workers fled daily and hunting groups were formed to recapture them. To relieve the chronic labor supply problem, the Datong Company resorted to conscripting war prisoners and secured around 1,400 surrendered soldiers from the Tongzhou Incident near Beijing in July 1937 for collecting gravel. They also sought workers from as far as Shanghai in China's south to Heihe on its border with Russia in the north, but most came from their traditional sources in Shandong and Hebei Provinces. In a sign of decreasing labor quality during the war, the head of Jilin's Datong Company Office described one group of workers from Shanghai as a ragtag group of "hoodlums" who would work hard when fed.[122] A 1938 newspaper report mentioned seventeen- to eighteen-year-olds working side by side with elderly men.[123] By 1939, they began using prison labor from nearby Jilin, and in 1943 they mobilized conscripts from the Manchukuo Labor Service Corps.[124] As Fengman received the highest priority as part of the state's Five-Year Industrial Plan, laborers were worked intensely to always keep the project on schedule. For example, in order to close most of the river within the short winter time frame, laborers worked day and night in temperatures reaching minus forty degrees Celsius.[125] They dug into one to two meters of frozen earth to obtain loose gravel, sand, and stone; transported it onto the frozen river; broke through one meter of ice; and gradually filled it in. Ten thousand workers a day were required for this.[126] Engineers reported that although this method was cheap and saved on materials, approximately one person died every three days.[127] In his memoir, the head of civil engineering Uchida Hiroshi emphasized that much attention was paid to creating a satisfactory labor environment in order to maintain labor efficiency because of the project's national importance and intense state pressure to finish quickly. Housing for ten thousand workers was built by 1939 across the river from the largely Japanese town, and each unit accommodated fifty workers per structure. The food in the area was plentiful and nutritious according to Uchida—the one hundred Russian employees even brought their families and settled down in their own village.[128] The Hydropower Construction Bureau established an entertainment district with restaurants and bars operated largely by Chinese and Koreans. One newspaper report mentioned that the police post had to be upgraded to a police station because of all of the "indecent feuds over prostitutes."[129]

Workers had twelve-hour shifts with a one-hour lunch break and could take off five days a year, and foremen had two days off per month (see Figure 11). Pay was around ninety sen per day for Chinese workers, more for Koreans and Japanese, less for prisoners.[130] On average, around 70 percent of hired workers showed up each day. The Datong Company was notorious for reducing the number of hours in their wage payments and subtracting costs like travel, photo, labor management, and reception fees from workers' wages. Many workers became burdened by debt and morphine addiction—night workers received one morphine tablet during winter construction. Most casualties resulted from communicable diseases like typhoid. Uchida noted that 94 percent of the estimated one thousand total casualties died from communicable diseases because of the bad sanitation, which he expressed deep remorse for. Work accidents, such as landslides or runaway cranes, accounted for the rest.[131] Wang Kwong-Leung, a Chinese graduate of Fengtian Industrial School and an electrical technician at Fengman, also noted that disease was the number one killer. Because of the lack of hygienic facilities and lax oversight, disease spread quickly and there were only one doctor and nurse for ten thousand workers. The clinic was not very well supplied, either, and the doctor usually prescribed morphine for most cases.[132] Casualties also arose from the shooting of prisoners trying to escape, putting down prisoner revolts, and violence between Korean guards and Chinese prisoners of war.[133] A "worker memorial tower" was eventually built with the following inscription by the Manchuria Labor Association's head, Iijima Mitsuharu, carved on the back: "The hardships faced by the workers of this immense construction project pushed forward in the midst of horrible conditions such

Figure 11   Chinese "coolies" being prepared for work at Fengman Dam construction site (1939). Source: Manshūkoku suiryoku denki kensetsu kyoku, *Shōkakō dai ichi hatsudenjo kōji shashinchō*, page unnumbered.

as the freezing cold weather defy the imagination. Those workers who died at this construction site should be honored as anonymous heroes."[134] Honma also referred to the "three hundred martyrs" in a 1944 interview; however, he noted that the number of dead was not very large considering the project's massive scale.[135] As the government suppressed details about the horrific working conditions, the nameless Chinese workers were ultimately incorporated into Japan's narrative of sacrifice for the noble cause of Manchukuo's construction.

Disciplinary techniques of policing, surveillance, and mobilization were also required to efficiently manage large numbers of dam workers and increase their productivity so as to meet construction deadlines. At Sup'ung Dam, the "poor, deserted village" of Sup'ungdong in Korea grew from 90 households (480 people) to 950 households (8,782 people) in 1938—thirteen thousand people if the entire construction area was included.[136] By June 1940, the area grew to around twenty-eight thousand residents.[137] At the construction's beginning, makeshift worker huts increased exponentially and a flood of Chinese, Korean, and Japanese "fortune-seekers" rushed in to start restaurants, bars, and hostels. Out of concern over rising "indolence and sin" among workers at the national project, the district police initiated a strict permit system that limited the number of businesses based on hygiene, capital, and business durability requirements.[138] Prostitution, gambling, and theft were a great concern, and the police asked the construction companies to ban outsiders from living in their areas and conducted identity sweeps two to three times a week.[139] "Good morals and manners" were taught at neighborhood association and worker group meetings in order to prevent ethnic conflict and promote a spirit of "Japan, Manchukuo, China Coexistence and Co-prosperity" among the otherwise "nomadic" workers.[140]

Dam construction sites were heavily militarized because of the ongoing war. At Sup'ung Dam, nearby residents forced into "collective hamlets" as part of the Guandong Army's counterinsurgency campaigns to separate the general population from anti-Japanese guerrillas were employed, many of whom had no other choice, as they were deprived of their livelihoods. Learning from the Government-General's earlier tactics in combating Korean communist insurgents in the Kando border region, the Guandong Army vigorously constructed these villages throughout Manchuria after 1934—by 1939, there were 13,451 of them. Existing villages were often burned down and residents ordered to move into the hastily constructed, militarized settlements in the plains that were frequently unhygienic and did not contain sufficient means of agricultural sustenance.[141] Sup'ung Dam was also heavily guarded because Yang Jingyu's Northeast Anti-Japanese United Army and Kim Il-sŏng's division operated nearby, prompting a three-year campaign by the Guandong Army. A 1937 military study estimated that twenty thousand "bandits" operated in Manchukuo, the numbers likely being higher given their ability to blend into the sympathetic peasant population.[142] As the number of laborers from Manchukuo and China increased, Korean authorities became worried about "impure elements" posing as workers or refugees who organized, recruited, or spied for the communists.[143] Nakamura Shunjirō, the Hazama-gumi Construction Company office head, noted that police rounded up workers daily in 1938 and company officials went to work armed.[144] That year, thirty-one policemen were stationed on the dam's Korean side, and thirty-four on the Manchukuo side.[145] Two to three plainclothes or uniformed policemen patrolled the Korea construction site, bunkhouses, and nearby villages, and two officers went to the Man-chukuo side daily to exchange information about possible infiltrators.[146] Officers compiled a worker register and conducted identification checks three times a month, making more detailed inquiries into the backgrounds of suspicious persons. Bunkhouses had signs with occupants' names, and foremen were required to report all moves and changes.[147]

The government and construction subcontractors attempted to transform the largely rural workers into a disciplined industrial work-force. For example, they initiated campaigns to create "harmony between capitalist, administrator, and worker" and encourage good manners, hard work, and thrift.[148] At the beginning of concrete pouring and setting in 1939 at Sup'ung Dam, construction company employees ran around "like Nanjing rats," trying to teach "coolies" how to operate machinery, and they struggled to coordinate their workforce with the new mechanized system of conveyor belts, iron buckets, cranes, rail cars, and mixing and crushing facilities.[149] To motivate workers at the construction's height in 1941, the two subcontracted construction companies began a contest in which workers competed against each other to see who could set more concrete each day.[150]

Similar to the land compensation and resettlement process, the dam's construction required the mobilization of multiple forms of colo-nial power: the transfer and organization of colonial surplus labor through the *batou* system and Japanese labor management organizations; military campaigns that rounded up conscripts, prisoners, and peasants displaced into collective hamlets for hard labor; multiple layers of supervision rang-ing from the labor broker to the foreman to the police officer; and govern-ment and company campaigns to promote discipline, productivity, and ethnic harmony. "Scientific Japan" and its technocratic expertise thus re-lied upon multiple techniques of violence, surveillance, management, and mobilization by many different colonial organizations.

Conclusion: Naturalizing Technology's Power

Many of the same engineers, bureaucrats, and businessmen involved in colonial dam construction later utilized their expertise during Japan's high-growth era in the form of "comprehensive national land plans" and extensive multipurpose dam projects abroad as part of its influential over-seas development assistance programs. In many ways, Japanese leaders in business and government continued to invoke technology's power in the name of development and prosperity using some of the same techniques of naturalizing and legitimating technical expertise developed during the colonial era. Although several scholars have noted the continuities between wartime and postwar technocratic discourse, few have analyzed how it formed and operated as a system of power in the colonial and postcolonial context. Technology's power was produced not solely through ideological appeals to rationality, efficiency, and systematicity or promises to transcend sociopolitical divisions through development but also through attempts to naturalize these meanings of technology on the ground. This process of technology and expertise establishing itself in the colonies was clearly haphazard, messy, and contentious, thereby revealing the rather weak basis of Japan's ideological claims that technology brought about stability, progress, and order. Japan's introduction of technology and planning to the colonies required radically transforming unfamiliar and uncooperative landscapes, developing new forms of knowledge on the ground, harmonizing conflicting visions and interests, negotiating with resistant residents, mobilizing and managing labor, and creating laws and institutions to ensure cooperation and timely completion. Technology and expertise formed a system of colonial power through these specific practices and institutions rather than primarily operating at the symbolic or discursive level.

Chapter 5

Designing the Social Mechanism

The Technological Visions of the Reform Bureaucrats

Another powerful group during the war who played a major role in shaping technology's meaning were the reform bureaucrats (*kakushin kanryō*, literally "renovationist bureaucrats"). These were a close group of officials who were at the heart of Japan's economic policy making and planning administration both domestically and in the empire (particularly in Manchukuo) between 1931 and 1945.[1] At the center of their ideological program was the formation of a managed economy (*tōsei keizai*), which aimed to eliminate the inequalities of laissez-faire capitalism and establish a self-sufficient total war system. But their agenda went beyond technocratic planning from above to incorporate a fascist vision of integrating politics, culture, and society through the formation of mass corporatist parties, associations, and unions. Most of them graduated from Tokyo Imperial University during the 1920s, where they were heavily influenced by Marxist social science and radicalized by the social movements of Taishō democracy. The frequent economic crises of the 1920s and the Great Depression after 1929 in particular made people from across the political and intellectual spectrum question free-market capitalism and therefore turn to Marxism as a powerful diagnostic framework. Unlike Marxists at the time, however, these right-wing bureaucrats ultimately viewed the nation as the primary unit of political loyalty and economic analysis rather than class or the individual.[2] After graduating from university, they gained hands-on experience in economic planning in Manchukuo and China in the mid-1930s and went on to become department chiefs (*kachō*) and bureau directors (*kyokuchō*) in the Finance, Commerce and Industry, Railways, Agriculture and Forestry, Home, and Communications Ministries. Such influential bureaucrats as Kishi Nobusuke, Shiina Etsusaburō, Minobe Yōji, and Okumura Kiwao counted as among their members.

The reform bureaucrats' ideology was rooted more in notions of economic planning circulating at the time than in conceptions of technology per se, as was the case for the technology bureaucrats and engineers. In fact, the mobilization of science and technology for total war formed only one part of their larger program of establishing a managed economy. At the same time, however, reform bureaucrats shared a certain framework with engineers thoroughly rooted in emerging notions of technology who viewed society as an integrated social mechanism. During the 1920s, engineers developed a view of technology as a creative, productive force that permeated all realms of life along with an idea of themselves as "social engineers" active in planning and rationalizing society (see Chapter 2). This was part of a larger movement to elevate their status, particularly in government where engineers and technical specialists were subordinated to law bureaucrats regardless of ability or experience.

The engineers' notions of a society organized by technology and expertise converged with two influential groups who grew in power during the 1930s: the "control officers" and the reform bureaucrats. Learning from German theories of total war developed during World War I, the control officers viewed war as requiring society's comprehensive mobilization. These army officers worked throughout the 1920s to establish the institutions and plans necessary for the creation of a self-sufficient Japan capable of fighting what they saw as an inevitable war between competing imperialist powers. Because many in the army came from poor rural backgrounds and were deeply affected by the early Shōwa Era's frequent economic crises, they were also strongly anticapitalist. They firmly believed in the idea of reforming capitalism through state intervention to ensure a more equitable wealth distribution. As they grew in power after the Guandong Army's invasion of Manchuria in 1931, they developed the conception of the "Advanced National Defense State" (*kōdō kokubō kokka*)—a fully mobilized, total war economy managed by the state, which ensured that politics, economy, and culture were optimally organized not only for war but for the public's interest over those of profit-driven capitalists.[3]

With the army's support, the reform bureaucrats became the experts who designed the institutions and policies for establishing what became known as the managed economy during the war. A key component of their ideology was the notion that economic planning involved an understanding of the economy as an integrated system and of policy making as comprehensive state intervention in all areas of the economy, such as industrial organization, competition, finance, distribution, profit making, trade, labor relations, consumption, working and living conditions, and social welfare. Although they were strongly anticapitalist, their policies did not entirely eliminate capitalist institutions but relied on "voluntary" cartels and associations among businesses within each industrial area to ensure maximum productivity and efficient organization as well as to manage excess competition and profiteering. The managed economy depended more on developing a strong nationalist "ethic of responsibility" among capitalists, managers, and workers rather than on direct state control to rein in capitalism's excesses.[4] Reform bureaucrats sought to transform society into an optimal and integrated socioeconomic system geared for total war and the reorganization of capitalism; however, in the end it was the people engaged in their specific socioeconomic functions who would spontaneously and responsibly operate that system.

Over the course of the 1930s, the conceptions of state engineers and reform bureaucrats became closely intertwined. Technology associations strongly supported Japanese expansion after 1931 and campaigned vigorously for the sending of engineers to Manchukuo to help transform it into a heavy industrial national defense state (see Chapter 2). In fact, it was the Manchukuo reform bureaucrats who established such institutions as the State Council's General Affairs Agency in 1935 and drafted such laws as the Manchukuo Five-Year Industrial Plan in 1937 that enabled engineers to develop and realize their notions of "comprehensive technology" in the form of flood control, multipurpose dams, urban planning, and transportation projects. The engineers' vision of economic development through mutually reinforcing public works projects constituted an essential component of the reform bureaucrats' plans to rapidly industrialize Manchukuo and exponentially increase wartime production.

Technology bureaucrats and engineers, in turn, incorporated notions of the managed economy into their own technological imaginary, arguing that colonial management and wartime mobilization required skilled technical experts who were capable of integrating their particular projects into Japan's larger political economic goals. With the outbreak of full-scale war with China in 1937, they joined reform bureaucrats in establishing and staffing "comprehensive national policy organs" (*sōgō kokusaku kikan*), such as the Cabinet Planning Board, the Asia Development Board, and the Technology Board.[5] Modeled after Manchukuo's powerful General Affairs Agency, these transministerial institutions attempted to take charge of wartime mobilization and colonial management by consolidating research, planning, policy making, and budgetary powers under one roof. Reform bureaucrats wanted to overcome the factionalism and specialization among the established ministries by creating institutions that could formulate policy on politics, the economy, and culture in a systematic, comprehensive manner. Technology bureaucrats also firmly supported Prime Minister Konoe Fumimarō's New Order Movement in 1940, a radical effort by reform bureaucrats to strengthen the wartime mobilization system by reorganizing and increasing state control over industry, politics, labor relations, finance, and science and technology. In fact, they took the lead in designing the New Order for Science and Technology, an effort to realize their longtime objective of having technical experts plan and manage the development of a self-sufficient Japanese "science-technology" (*kagaku gijutsu*) based on the natural resources of Japan's Asian colonies.[6]

As reform bureaucrats began to implement their program of a managed economy rooted in corporatist organizations among the people, they also incorporated some of the ideological framework and language of technology being articulated by technology bureaucrats, engineers, and intellectuals. During the 1930s, technology was increasingly viewed as representing the productive mechanisms of all aspects of political, economic, and cultural life. This image of society as a dynamic, integrated system also proved attractive to reform bureaucrats as they rapidly devised policies for total war mobilization. How this new discourse on technology operated within the policies of reform bureaucrats is demonstrated by the thought and career of one of their chief ideologues, Mōri Hideoto.[7] Unlike most reform bureaucrats, Mōri published and spoke widely between 1938 and 1945, clearly articulating their ideology and program in the process. In outlining a fascist vision of a planned economy grounded in the people's productive energies, Mōri explicitly employed the discourse of technology prevalent at the time. The notion of technology not simply as the means of production but as the productive mechanisms of life easily lent itself to an articulation of a fascist program to reorganize capitalist society.

In her book on Japan's wartime reform bureaucrats, *Planning for Empire: Reform Bureaucrats and the Japanese Wartime State*, Janis Mimura aptly describes their ideology as "techno-fascism"—a fusion of technical rationality, comprehensive planning, and modern values of productivity and efficiency with ethnic nationalism and right-wing ideologies of organicism. In various ways, she shows how techno-fascism aimed to transcend traditional Japanese political divisions and incorporate them within a larger politics of technocratic planning and management, which she briefly describes as a new "mode of power."[8] The reform bureaucrats' discourse actually represented something more than a politics of technocratic planning and managerialism. In their techno-fascist ideology, we do indeed see the contours of another mode of power, but one that was based more on creation and production rather than solely on top-down management and coercion. Within the reform bureaucrats' ideas and policies, power was not simply something that organized society from above but dynamically shaped it from within through the productive practices of a whole array of institutions and people. The reform bureaucrats' ultimate goal was the construction of a fully mobilized and productive system run by creative, responsible citizens who ultimately would not require any kind of coercion from above. "Technology" therefore not only justified a technocratic ethos but also provided powerful ideological metaphors for reform bureaucrats to represent society as a dynamic, creative, concrete, and even spiritual mechanism in opposition to liberal capitalist society, which they viewed as abstract, stagnant, individualistic, and dehumanizing. Mōri had different labels for such a systematized society to describe its various aspects—the "production economy," the "national people's economy" (*kokumin keizai*), the "national life organization" (*kokumin seikatsu soshiki*), and even the "symbiotic body of East Asia" (*dai tōa kyōseitai*). Examining Mōri's conceptions on technology as well as some key reform bureaucrat policies he was involved with shows how their technological imaginary constituted a new mode of techno-fascist power.

Creative Engineers and Economic Technologies

In his writings, Mōri discussed two kinds of technology—what he described as economic and production technologies. Both were fundamentally linked to the people's creative energies and imagination and did not simply signify physical machinery or production techniques. First, he always referred to himself and other bureaucrats as "economic technicians" or "creative engineers."[9] In a 1941 roundtable with his colleagues Minobe Yōji, Sakomizu Hisatsune, and Kashiwabara Hyōtaro titled "Reform Bureaucrats Discuss the New Order," Mōri said, "We must transform ourselves from legislative bureaucrats into creative bureaucrats. Although this is a strange term, the same applies to the technical aspect [of our work]. Up to now, we have only been conservative engineers who drafted, managed, and interpreted legislation. From now on, however, we have to be 'creative engineers.' "[10] For Mōri, economic technologies were the specific policies, institutions, laws, and campaigns needed to construct and organize the "East Asian Community" (*tōa kyōdōtai*) in Japan, China, and Manchukuo. Grounded in "synthesis, planning, and science," economic technologies sought to transform Japan and its empire from a capitalist order of liberalism and free trade into a rationally planned, self-sufficient "national economy" (*kokumin keizai*).[11] However, as Mōri stated above, "economic technologies" were not merely instrumental means and techniques but required a certain type of bold, "creative bureaucrat" different from the typical drafter, administrator, and interpreter of arcane, minute legal codes. The ideal "leader" must "both grasp the deductive goals required by the nation (*minzoku*, ethnos) as a whole and the dynamic facts at work in their lives or their future potential" and then decide how to mediate and synthesize the two without doing violence to either, he added.[12] Bureaucrats must think a little less "deductively" from overall state goals and take into consideration the people's particular circumstances and interests, and those who criticized the bureaucrats, particularly the business people, should think a little less "inductively" from out of their own private interests and take into consideration the interests of the whole. Thus, the role of the economic technician was to be a mediator between the overall state goals of building the new nonliberal economic order and private economic or political gain.[13]

The target of the creative bureaucrat's economic technologies, according to Mōri, was the "liberal economic point of view" that permeated the economic lives of Japanese and Asians. Liberalism was based on the following five principles:

1. The foundation of all economic phenomena is economic man, *Homo economicus*. His material desire for fame and fortune is the motivating force of all economic phenomena.

2. Capital and profit are the pillars of economic life.

3. The essence of economic life is exchange. The market is formed through supply and demand; price is thereby constituted and dominates [society].

4. The state and taxes or any form of taxation policy arising from the state is a disturbance of the economy's natural progression.

5. The regulator of all economic phenomena is each human being's interest. This interest immediately brings about the harmony of all interests in nature.[14]

According to Mōri, Japanese social relations were thoroughly infused with these five principles of capitalism. Thus, private interest and the pursuit of profit overtook the public good, and incessant conflict rather than mutual cooperation and national development became society's essence.[15]

Until World War I, Japan held a subordinate position in the free-market world economy, which was dominated by Great Britain's abundant capital and resources. Japan's national economy was subject to this nineteenth-century system, which declared itself "natural law" worldwide.[16] Mōri wrote:

This system itself was England's national economy. The Japanese economy, which formed a link in that system, in no way promised the Japanese nation's eternal development. In sum, under this free trade system, the Japanese economy purchased raw materials from the world's cheapest places, processed them with cheap labor, and exported the finished products. By re-processing imported raw materials, Japan purchased its life necessities with the profit.[17]

Under liberalism, Japan therefore merely sought to increase its participation within Great Britain's global hegemony rather than develop a self-sufficient, independent "national economy." It only became part of the existing "international economic division of labor" whereby Japan continued to be placed among the "have-nots."[18]

In order to overcome the capitalist liberal order, new economic technologies were needed to make "the nation acquire the status of the subject rather than being an object of the economy" and make "concrete" national citizens (kokumin) the economy's directors and creators rather than "abstract universal economic man" or the "self-correcting economy."[19] Instead of market forces, the nation's "political power" guided by the new "creative engineers" would direct the economy for everyone's benefit through the "management of the total relations between natural resources, industry, finance and money, performed in accordance with the planned nature of the entire economy."[20]

As an example of "economic technologies," Mōri referred to Japanese bureaucrats who dealt with the problem of the market's erratically fluctuating prices by setting a commodity's "proper price." Ideally, this higher price would stimulate production to the point where prices would again lower as a result of overproduction. He criticized this kind of economic technology, however, as one that was still grounded in the profit principle of liberal capitalist economics—the assumption that the pursuit of profit would naturally correct the problem of high prices. Chronic price fluctuation could not be fixed by such short-term, Band Aid–type measures that were in the spirit of a "night watchman" capitalist state that largely just oversees all economic activity. Only by transforming the liberal economy into the managed production economy could this problem be fixed. Economic technologies should instead focus on expanding industrial "fixed capital" to create a firmer foundation for "expansive reproduction" rather than expanding "variable capital," which merely sought to increase demand within the market economy. Long-term "expansive reproduction" rather than the maintenance of some sort of temporary market equilibrium should be the goal of national economic technologies, Mōri argued.[21] National economic technologies should be concerned directly with people's lives rather than with corporate profit. For example, a compulsory national health insurance scheme should be set up and health clinics, occupational assistance centers, and housing facilities constructed through the imposition of a specific tax for that purpose. This would enormously improve the productivity of the people and the economy.[22] In short, life itself was supposed to be the focus of economic technologies, not capitalism's abstract market principles. "Economic technologies" were more than the technocratic restructuring of the economy by expert bureaucrats; they also suggested a mode of power aimed at stimulating creation and production in all areas of life through techniques of management and regulation.

New economic technologies were essential to encourage not only the Japanese nation's active cooperation but also the participation of colonial peoples in the project of constructing a self-sufficient economy throughout East Asia; otherwise, Japan's empire would become like a "many-storied building built on sand," according to Mōri.[23] Admitting in a speech that he was an economic technician who "was defeated by the

Manchurian and Chinese peasant," he criticized other bureaucrats who crudely divided the average Chinese person's life into abstract categories of industry, agriculture, and economics rather than taking an integrated approach more in tune with local socioeconomic relations and conditions.[24] The economies of Manchukuo and China contained a variety of productive conditions and relations ranging from feudal to capitalist to a mixture of both. Elsewhere, he stated his overall philosophy of economic technology toward China and Manchuria:

The forms of economic activity are given greater variety by the conditions of production, which are determined by particular social conditions, and probably have a variety ranging from A to Z, for instance. In such a case, when the economic technology A', which originated in Japan's commercial capitalist economic activities, is applied to the life spaces of Manchukuo and China, the economic activities of B' to X'Y'Z' are then occluded. Therefore, a total, cooperative relationship among these spaces' economic activities cannot be formed; a partial relationship of domination towards Manchukuo and China is established; and in the end, a large portion of economic life becomes isolated from a cooperative relationship.[25]

Instead of arrogantly applying Japanese policies ("capitalist economic technology") across the board in a uniform manner, bureaucrats should strive to integrate them with China and Manchukuo's complex, particular conditions. Only then could Japan truly build the "East Asian Community." Mōri urged Japanese economic technicians to understand and synthesize the various ethnicities' "life technologies" into their own economic technologies for designing the East Asian managed economy. They must be able to think of not only the "primary equations" but the "tertiary" and "quaternary" economic equations as well; otherwise, the people's "life consciousness" and the creative bureaucrats' "synthesizing consciousness" would never unite.[26]

Mōri provided several anecdotes of what he meant by a "broader economic technology" toward the colonies.[27] For example, he mentioned how Japanese "agricultural technicians" who first arrived in the colonies simply set up the same type of research stations they did in Japan and forced nearby peasants to plant their experimental seeds. But after encountering resistance, the technicians then decided to employ other techniques. Mōri recounted an instance in north China where they gathered elementary school teachers from 2,300 villages and lectured for several days on purchasing, cultivating, harvesting, and selling cotton in that region and terrain. After the technicians provided the teachers with free cottonseed, the teachers returned to their villages and asked students to plant them. A good crop would guarantee the propagation of cotton planting throughout the village and set the stage for the establishment of a local cottonseed factory, thereby further increasing cotton production.[28] In a 1941 speech to the Social Policy Institute titled "On Constructing the East Asian Economy," he recounted another instance where Japanese engineers helped build a main road to the Manchukuo capital of Xinjing using Japanese technology, only to see their roads crumble from the intensely cold temperatures. However, one provincial official was later able to make cheaper, more durable roads by working with local Chinese peasants.[29]

In the same lecture, he talked about how Home Ministry engineers used Japanese techniques of strengthening riverbeds and banks to control Manchuria's heavy flooding. This did not work because unlike in Japan, where smaller floods occurred more frequently, in Manchuria enormous floods happened once every ten to forty years. Strengthening the riverbanks was not enough to handle such floods. Thus, after three years, the Japanese engineers began to think more in terms of local conditions and built a system of artificial reservoirs and dams to siphon off floodwater in line with the emerging notions of "comprehensive technology" discussed earlier. He also talked about a Japanese coal mining company that repeatedly fired skilled Japanese technicians for beating Chinese miners because this treatment of the miners would inevitably dampen productivity.[30] Thus, in various ways, economic technicians introduced more flexible technologies that tried to employ local knowledge and were more in tune with local conditions in order to ensure cooperation and productivity rather than a "uniform technology," which Mōri argued would inevitably restrict Japan's economy.[31] In a similar way that domestic economic technologies signified a form of power designed to mobilize and reorganize all aspects of life for the nation, colonial economic technologies were designed to co-opt dissent and redirect local knowledge and skill toward constructing the New Order in East Asia.

Frustrated by the continuing inability of Japanese bureaucrats and engineers to adapt to the colonies' varied conditions and their inflexible, specialized nature, Mōri proposed at the war's height to instead train them in the particular "life cultures" of the areas in which they would serve. In a speech titled "Through the Great East Asian War," Mōri suggested that Hokkaidō University in northern Japan become a center for training leaders in the specific "life cultures" of Siberia and Manchuria and that Taipei Imperial University (Taiwan) in the south become a center for research and training on the "tropical life cultures" of the South Pacific and Southeast Asia. Only leaders with a total, integrative grasp of East Asia's particular cultures could incorporate those cultures into a prosperous, heterogeneous "Greater East Asia Co-Prosperity Sphere," not bureaucrats who specialized in minutia or who artificially divided culture or economy into specialized areas.[32]

In a 1939 discussion among colonial bureaucrats at the National Policy Research Association (*kokusaku kenkyūkai*) titled "Roundtable on Re-thinking the Managed Economy throughout Japan, Manchukuo, and China," Mōri distinguished between the "requisition economy" view, whereby Japan exploited the natural resources of China and Manchukuo for their war economy, and the reform bureaucrats' "managed economy" approach, whereby liberal capitalism was restricted and the economies of China, Manchukuo, and Japan were integrated for the people's benefit. "There is a big difference between using the Chinese economy and the well-being of China or the Chinese peoples," he said.[33] The question of "requisition economy vs. managed economy" was especially important when "China's political character today is in conflict with Japan," Mōri added. He lamented the fact that many bureaucrats were not wholly embracing their new roles as "economic technicians" of an East Asian managed economy and were instead pushing for a temporary "requisition economy" to meet Japan's immediate war needs, which would then revert back to the old liberal capitalist order after the China crisis had passed. They lacked a firm "political direction" or ethos to build an "independent" and "automatic" economy wherein bureaucrats would work with local populations to design and implement policy on the ground in accordance with the managed national economy's guiding ideals.[34]

Ultimately, the managed economy should operate like an automatic, well-oiled machine, Mōri argued, in which the individual parts worked independently with little management from above. Yet in the urgent context of war in China, administrators were still falling back on their old habits as specialized "legislative bureaucrats" rather than as "creative engineers." According to Mōri, bureaucrats from the colonies arrived at the Asia Development Board's offices daily to ask for advice on minute legislative details or on unimportant matters. In order to avoid the creation of a requisition war economy, bureaucrats had to change their behavior of monitoring and supervising the colonial economy's every detail and instead formulate bold new economic technologies that took into account China's diverse range of economic conditions. The "narrow economic technology of making one hundred things conform to one" would only spell disaster for Japan by exacerbating the conflict with the Chinese people and preventing the formation of an East Asian Community. If Japan continued to develop repressive, homogenizing technologies of control, the empire would inevitably crumble, Mōri concluded.[35]

Engineering the New Order in East Asia

Mōri's ideas on reform bureaucrats as "creative engineers" who designed East Asia's managed economy emerged from his hands-on planning experience in some of the most powerful economic institutions in Japan and its colonies. As mentioned earlier, many reform bureaucrats attended Tokyo Imperial University, where they were exposed to Marxism as a social science that addressed the gaping inequality and chronic poverty of Japanese capitalist development. Some were even caught up in the progressive atmosphere of Taishō democracy. For example, Mōri tutored workers at the Yanagashima settlement, a social welfare project to combat poverty in Tokyo's poorer neighborhoods. Students lived in settlement houses and constructed houses, schools, and day care centers.[36] Continuing this activist tendency by dedicating himself to national service, he joined the Finance Ministry's Monopoly Bureau upon graduation in 1925 through the recommendation of his distant relative Kamei Kanichirō, Foreign Ministry diplomat and later parliamentarian from the Socialist People's Party (*shakai taishūtō*). After several years of heading regional tax offices and the Finance Ministry's Tax Department, he secured a position among the second group of bureaucrats sent to establish Manchukuo's managed economy in 1933, again through Kamei's intervention.[37] He was the only one in the group who volunteered for service in Manchukuo, many of whom were fearful of the violence, the harsh conditions and the Guandong Army's tyrannical reputation.[38]

In the State Council's General Affairs Agency, the institution that effectively wielded power in Manchukuo along with the Guandong Army's Third Division, he was head of the Special Accounts Division and later the National Tax Section, which was charged with establishing Manchukuo's financial and taxation systems.[39] According to Furumi

Tadayuki, Mōri helped establish the powerful Cabinet Planning Section, which eventually became Manchukuo's central planning bureau for economic policy, and Kamei even unsuccessfully asked Hoshino Naoki, head of the General Affairs Agency, to appoint Mōri as its head.[40] Judging from the numerous reports, top-secret policy drafts, letters, and directives that Mōri had kept in his papers after the war, he had his hands in many matters in Manchukuo and China: the unification of the various Chinese currencies into a yen bloc; the establishment of banks, such as the Federal Reserve Bank of China; the planning of roads, ports, communications networks, and railroads; and the promotion of heavy industry, labor productivity, and natural resource production. As Kamei noted, Mōri was someone who "dedicated his life to planning."[41]

From its very inception in February 1932, radical military officers and reform bureaucrats dominated Manchukuo's puppet regime. In March 1933, the Guandong Army's Special Affairs Division and its brain, Mantetsu's Economic Research Association, drafted the "Outline for the Construction of the Manchukuo Economy." According to this document, Manchukuo would not conduct industrial development on the basis of the free-market capitalist system, which they viewed as not conducive to achieving Japan's national defense needs and incapable of rescuing the people from economic depression, but through the creation of a managed economy and a "Japan-Manchukuo Economic Bloc."[42] As reform bureaucrats like Mōri took over more control of Manchukuo's administrative machinery away from the more radical Guandong Army officers, they centralized policy drafting, decision making, and execution in the State Council's General Affairs Agency, particularly within its Industrial Department. Such reform bureaucrats as Kishi Nobusuke and Shiina Etsusaburō proceeded to lead the drafting and implementation of the Five-Year Industrial Plan to transform Manchukuo into a heavy industrial and chemicals base in East Asia in 1937. Rolling back the more anticapitalist Guandong Army, the plan pragmatically combined state planning with the private zaibatsu capital and expertise of Nissan's Ayukawa Yoshisuke, and it became the prototype for the total war mobilization plans they subsequently drafted for Japan's Cabinet Planning Board.[43] Concurrently, the Manchukuo Concordia Association served as the "national party" to mobilize the different peoples into the managed economy system, again paralleling similar efforts in Japan. Upon returning to Japan, such reform bureaucrats as Kishi and Mōri confidently announced that a managed national economy was being realized in Manchukuo as they pushed for a similar program to establish comprehensive national policy organs, coordinate state planning with private initiative, and form a mass national party.[44]

Mōri also acquired experience in China as an economic advisor to the China Garrison Army's headquarters in Tianjin and later to the North China Expeditionary Army's Special Affairs Division in Beijing. He developed close ties to such top military officers as Akinaga Tsukizō and Suzuki Teiichi, and he openly boasted about his good relations with Manchukuo's Guandong Army.[45] In 1935, he joined one of the first comprehensive economic study missions to north China commissioned by the Guandong and China Garrison armies to plan for its industrial development in light of their strategy of establishing cooperative regimes there and integrating them into Manchukuo's Five-Year Industrial Plan. The two-month study conducted research on existing conditions in finance, trade, politics, industry, and transportation to provide the basic data for future policy.[46] This preliminary study together with one conducted by Mantetsu set the stage for a larger-scale study on north China's economic, industrial, natural resource, transportation, and port conditions.[47] In this manner, Mōri participated in extending Manchukuo's managed economy framework to China, one that was based on the industrialization of China's natural resources for Japan's East Asian wartime system.

Much of Mōri's work as a financial official in Manchukuo and China was related to devising policies to establish and unify their banking systems. When tensions worsened between Chiang Kai-shek and the North China Expeditionary Army in 1937, Mōri played a major role in drafting the "Outline for Financial Countermeasures in North China" and later establishing the Federal Reserve Bank of China.[48] During the 1930s, there were several regional currencies in circulation throughout north China, and the North China Expeditionary Army used currency issued by Korea's Chōsen Bank to procure military supplies and natural resources. However, Chiang's Nationalist government sought to solidify control over north China's warlords, and in 1935 it issued its own yuan currency. By 1937, the Nationalist yuan became the predominant means of commercial transaction, constituting 78 percent of all money in circulation.[49] As hostilities broke out in July 1937, Chiang issued a moratorium on withdrawals from Chinese banks, which restricted the Nationalist yuan's circulation, causing inflation in north China and making goods expensive for Japanese forces. In the name of ensuring the "stability of people's lives," enabling north China's "self-sufficiency," and creating "closer relations between the economies of Japan, Manchukuo, and China through the unification of our [Japan's] economic power and that of North China," Mōri and others decided to prop up the Hebei regional currency as the predominant means of exchange and link it directly to the yen.[50] As the North China Expeditionary Army was establishing the Provisional Government of the Republic of China in Beijing in December 1937, it also drew up plans for the Federal Reserve Bank of China, which opened soon after in March 1938. Eight regional banks exchanged their silver for a new Federal Reserve yuan, and the bank was partially capitalized by Japanese bank loans and guarantees. Establishing the new yuan in such a vast area was not easy because of resistance by regional banks, the Nationalist currency's continued strength, and lack of popular recognition in interior areas.[51] However, Mōri's experience in forming a central bank from existing regional institutions, setting exchange and interest rates, regulating customs duties, and issuing and purchasing currencies in the midst of unstable wartime conditions and a complicated financial landscape may be seen as a form of what he later described in Japan as "economic technologies" designed by such "creative" reform bureaucrats as himself.

Upon returning to Japan in 1938, after a brief stint in the Finance Ministry's Deposits Bureau, he became the Economics Department chief of the newly formed Asia Development Board. The board was established to unify and centralize policy toward China under civilian control (see Chapter 2). It was headed by the prime minister and divided into four sections: political, economic, cultural, and technological affairs.[52] Its formation was the result not only of the army's need for more stable research and planning for the China front and the reform bureaucrats' desire for a comprehensive national policy organ for China but also of fierce campaigning on the part of engineering associations, such as the Japan Technology Association. The association's predecessor, the Kōjin Club, had always pushed for the incorporation of a "technical standpoint" into national policy and vigorously supported Japan's expansion into Manchuria in 1931. For them, Manchukuo represented a promised land where Japanese engineers could develop and implement their visions of comprehensive development through technology projects in alliance with reform bureaucrats who were planning a managed economy. From 1938, when the war with China was well underway, the Japan Technology Association threw its support behind the establishment of a "specialized central institution for industry and drafting of a comprehensive industrial plan rooted in the standpoint of Japan, Manchukuo, and China unity."[53] The Asia Development Board's inclusion of the Technology Department, which was in charge of conducting comprehensive research studies and laying the technical foundation for China's industrial development, represented a major victory for the engineers' movement in their quest for political power.

The Japan Technology Association's ideological leader, Miyamoto Takenosuke, became the Technology Department's head and quickly established the Asia Development Technology Committee, which he saw as a kind of "technology general staff" (see Chapter 2). Mōri served as the committee's coordinator.[54] Its mission was the "total mobilization of the essence of [Japan's] modern technology" for China's development, and it reported directly to the board's director. As Kubo Tōru notes, the board became a haven for engineers and technical experts who conducted studies on all aspects of the Chinese economy.[55] Thus, the board united the reform bureaucrats' visions of a managed East Asian economy with the technology bureaucrats' objectives of incorporating technical expertise into all areas of national policy and planning. As a departmental head, Mōri became close to Miyamoto and cooperated with him in drafting the "Outline for the Establishment of a New Order for Science and Technology." Mōri's writings on the importance of advanced technology in the wartime managed economy and the Japanese nation as well as his increasing use of technological metaphors and imagery to describe Japanese society and the empire can partially be attributed to his interaction with such technology bureaucrats as Miyamoto at the Asia Development Board.

The Economics Department clearly reflected the objectives of its reform bureaucrat designers in the Cabinet Planning Board. It supervised planning for natural resource development, economic aid to cooperative regimes in China, the management of the North China Development and Central China Promotion Corporations, regulation of private companies in China, settlement promotion, establishment of transportation and communications links between Japan and China, finance and taxation issues, and international trade.[56] As the main section's head, Mōri supervised economic planning, the implementation of development projects with the new regimes, and coordination of the other three departmental sections. Suzuki Teiichi, head of the political affairs department and later director of the Cabinet Planning Board, said Mōri was the board's rising star and his most trusted adviser.[57]

Furukawa Takahisa notes that think tanks and private gatherings were also important arenas for the development of reform bureaucrat ideology and policy, as they brought together academics, military officers, businessmen, politicians, and bureaucrats from other ministries.[58] From the time he joined the Asia Development Board, Mōri actively participated in Yatsugi Kazuo's National Policy Research Association, a think tank of influential scholars, bureaucrats, and military officers which rivaled the Shōwa Research Association for influence over national policy. Mōri delivered several talks there, including one titled "Development Toward an East Asian Cooperative Body," and participated on its National Defense Economy Committee, which compiled a "Proposal to Reorganize Economic Structures" in 1940 as part of the process of designing laws for Prime Minister Konoe's New Order movement.[59] During this time, Mōri was also a member of the "Monday Group" (getsuyōkai), a private circle of high-level reform bureaucrats and officers who met every Monday in preparation for the weekly Tuesday cabinet meeting to formulate and coordinate policy.[60]

The Cabinet Planning Board, established by Prime Minister Konoe's first cabinet in October 1937, was at the center of the reform bureaucrats' efforts to design a managed economy rooted in corporatist organizations among the people. Its purpose was to be the "economic general staff" in charge of setting up Japan's total war mobilization system. It immediately set about drafting such basic wartime laws as the Materials Mobilization Plan in October 1937, which planned and allocated resources to the military, ministries, and private sector, and the National Total War Mobilization Law of April 1938, which gave the board broad powers to impose price controls on wartime materials and consumer items, form cartels and associations in wartime industries, draft labor and monitor employment and labor conditions, and control financing and profits.[61] This paralleled the National Spiritual Mobilization Movement launched in August 1938, which placed most political and civilian organizations under centralized control and created organizations at the local level to mobilize people to cooperate with government campaigns to increase production and save materials.

Konoe's second cabinet from July 1940 to October 1941 became more aggressive in promoting the reform bureaucrats' vision of a managed economy that restricted capitalism in the name of building a "national defense economy," replacing the multiparty system with a corporatist national party, constructing the "New Order in East Asia," and promoting heavy industry and science and technology. During this period, Mōri and reform bureaucrats from other ministries received joint appointments to the Cabinet Planning Board, where they established the Deliberation Room to draft and coordinate policy for Konoe's New Order movement. Mōri transferred to the board in May 1941, where he became head of the General Affairs Division's first section. Mōri, Sakomizu Hisatsune, and Minobe Yōji were referred to in the press as the board's "three ravens." According to Sakomizu, "Mōri was the one that thought up the themes, I was the one who somehow neatly composed them, and Minobe was the one who properly arranged them into a jazz and played them for the people."[62] A list of the Deliberation Room's policy drafts reveals the range of their reformist vision: establishing "national morals" for raising wartime productivity, institutions for the managed economy, policies for small and medium businesses, systems for labor relations and "national life," laws and organizations to stabilize agriculture and rural life, and plans to promote heavy industry and chemicals. Mōri supervised the drafting of plans to establish "national morals" and a "new order for science," which referred to efforts at creating a national party and passing the technology bureaucrats' cherished New Order for Science and Technology.[63] Miyamoto, his counterpart in the Asia Development Board's Technology Department, joined the board as vice director under Suzuki Teiichi in 1941, where he finalized the New Order legislation to centralize science and technology policy, elevate Japan's technological standards, and promote science and technology throughout all areas of life.

Under Konoe, reform bureaucrats sought to greatly expand the board's powers and wartime mobilization laws, which generated strong resistance from the business community, individual ministries, and the Diet. For example, the 1940 General Plan for the Establishment of a New Economic Order originally provided for restrictions on dividends and profits, the separation of management and ownership, and the granting of official status to managers in order to increase production at the micro level of the firm rather than simply at the macro level of state regulation. However, charges of socialism from the business community resulted in a watered-down version that replaced explicit references to limiting profits and separating management from capital with the language of "organic" cooperation between capital, management, and labor and the state responsibility of managers.[64] Despite these setbacks, the reform bureaucrats achieved their basic goals of establishing industrial associations, expanding controls over production, distribution, and consumption, and encouraging a degree of cooperation between labor and management. In 1943, when defeat in the Pacific War was beginning to seem inevitable, Mōri and several other reform bureaucrats and officers drafted plans to merge the Cabinet Planning Board and most of the Ministry of Commerce and Industry into the Ministry of Munitions. This finally achieved their goal of centralizing the functions of policy planning and implementation into one gigantic institution, thereby significantly overcoming ministerial barriers to their managed economy agenda.[65] Mōri's conception of the reform bureaucrats as a committed core of "creative engineers" who actively designed and administered the managed economy in Japan and East Asia seemed to have borne fruit in the form of powerful, comprehensive national policy organs and legislation that shifted Japan and its empire away from liberal capitalism toward a productivist "national economy." Somewhat similar to the engineers and technology bureaucrats, reform bureaucrats like Mōri viewed society increasingly in technical terms, as an object for active social engineering and as a dynamic mechanism to be efficiently guided by visionary experts.

The National Life Organization and the Japanese Nation's Creative Energies

For Mōri, economic technology was not just something bureaucrats carried out from above; it needed to be grounded in a mass politics of collective mobilization both in the colonies and in Japan. In short, economic technologies had to be united with life or the "national spirit." Criticizing what he described as an "administrative control economy" in which bureaucrats merely monitored and oversaw economic phenomena as they arose, he called for an active economic technology that was united with the "multiple life functions of national life."[66] He was a vocal proponent for the establishment of a "national life organization" (*kokumin seikatsu soshiki*), which he envisioned as an "organic life system" of numerous vocations (i.e., "life functions") working for national goals. "Due to the complicated pluralization of life functions [in modern society], those who hold power should not violently squeeze the functions pursued by the people into something uniform, nor weaken the will and creativity of the people's life activities," he wrote.[67] In pursuing his or her everyday economic activity, each individual would simultaneously realize such national economic goals as strengthening wartime factories and industries, developing a productive economy based on advanced technology, and increasing agricultural output.[68]

Unlike radical military officers who emphasized complete sacrifice to the emperor and nation, reform bureaucrats continued to see an important role for the individual within the national party. Recognizing the importance of private economic initiative in promoting innovative technology and creating new management techniques and products, the state's national life organization would not completely reject the profit motive but rather curb excessive speculation by taxing dividends and encouraging reinvestment of profits into production. Capitalists would take their place within the vocationalized, national life organization as "industrial technicians" alongside workers, engineers, peasants, and bureaucrats.[69] Thus, each person would not only acquire a "concreteness" they did not have as abstract "economic individuals" in the liberal economic order but also actively participate in the national economy instead of becoming mere objects of social policy or "administrative technology."[70] In this way, the economic technologies employed by "creative bureaucrats" would be grounded in the national life organization's "political power."[71] Power would no longer be top-down and autocratic but rather expressed productively and heterogeneously in the people's multiple life functions. Mōri described this as the expressive power of the Japanese *ethnos* or nation (*minzoku*).

What distinguished the reform bureaucrats from other groups who shaped the public discourse on technology was their emphasis on Japan's national spirit as the creative fountain of technology. The national life organization's essence was the "life power of the Japanese nation."[72] Society's organization by vocation was also a concrete articulation of national life power. Mōri wrote:

The Japanese nation's development is the development of action that continues to live in eternally new life while containing something absolutely intrinsic, something immemorial of the nation that is always alive within its essence. This makes one think of the life power of the two-faced Janus who had two different life powers. The Japanese nation's life power, however, is not like Janus who had two separate life powers but rather, a unitary life power that eternally activates the intrinsic or essential into young, new life.[73]

Mōri constantly referred to this national "life power" in his essays and speeches. Japanese history from ancient times through the Meiji Restoration to the recent "Manchuria and China Incidents" was an expression of this primordial life force that also manifested itself as a fundamental adaptability, plurality, and vigor to create anew. The 1937 China Incident, according to Mōri, signified the overcoming of the liberal capitalist world order and the formation of the mobilized, self-sufficient production economy through the national life organization. It therefore represented the latest stage in the expression of the Japanese nation's creative life power. Although reform bureaucrats were modernists committed to transforming Japan into an efficient and productive social mechanism based on vocation and advanced industry, at the same time they viewed this transformation as one more expression of an eternal Japanese spirit. In this way, they reconciled their technological imaginary with the more spiritualist formulations of Japanist right-wing military officers.

The Japanese nation should thus form an organic "total life system" of dynamic "life functions" organized and expressed by the national life organization. Mōri wrote:

Although national culture becomes functionally more complex and multiple as it becomes more advanced, under the completed Japanese national order, we must be confident that no matter how complicated and multiple national culture becomes, every single life function can be vocationalized within the organic life system.[74]

In fact, vocational complexity and multiplicity within the national life organization was only a further sign of national strength and should be "expanded and developed," he argued. Vocations in advanced technological industries, such as heavy chemicals and heavy industry, would have a primary place in the national life organization, and their increasing proliferation throughout society only further demonstrated national power. However, the bureaucrats who currently ran the managed economy were instead "taming the people's will towards action in life" by "rejecting the multiplicity of the people's life functions and bringing about their uniformity." By undermining popular energies, bureaucrats were preventing the people from ultimately gaining a "firm conviction" in their own vocations, and developing and expanding newer, more advanced ones.[75] These bureaucrats' "reactionary" politics were obstructing the birth of the organic planned economy consisting of bureaucrat "economic technicians," capitalist "industrial technicians," and workers who were increasingly organized by advanced "life functions" and the organic national life organization.[76] They were stuck in an old mindset of autocratically exercising power from above rather than empowering the people in the form of an organic life system.

In a 1941 speech to engineers at the Japan Engineers Center titled "Lecture on the New Economic Order," Mōri said that through the China conflict, the "Japanese nation" was transforming itself from "citizens" (*shimin*) into "national subjects" (*kokumin*); however, they still required a further revolution in their "life consciousness" (*seikatsu ishiki*) for this to be fully realized. An essential aspect of the "New Economic Order" was the formation of a "national defense consciousness" throughout all aspects of life. This consciousness was not just a firm belief in winning the war but a determination to rein in liberalism and create a modern, independent, and self-sufficient economy in East Asia. The people's adoption of a "national defense consciousness" would be an abandonment of the selfish individualism of liberalism and the attainment of a higher individual freedom within the nation's destiny. As an individualistic "citizen," every limit on his or her freedom was felt as a restriction; however, as "national subjects" who felt the nation's destiny in every aspect of their lives, individuals would freely undergo difficulties for the higher cause. True freedom was discovered within "tremors, vibrations, and strife," Mōri noted. He urged his engineer audience to abandon the atomistic, mechanistic worldview of classical science in favor of the more total, integrated worldview of quantum mechanics at the foundation of recent technological advancements. They must also not just understand "national destiny" intellectually but actively feel it in their very being and become "builders within the storm." For example, an architect would not just plan and build worker dormitories but make them convertible into integrated military barracks as well. As national subjects, engineers must fulfill both the particular workers' needs *and* the nation's universal ones.[77]

Mōri thus not only viewed society increasingly in technological terms as a productive social mechanism to be designed and administered from above by expert technocrats. The social mechanism itself also entailed a total transformation in popular consciousness in accordance with the wartime system's goals of productivity, efficiency, and innovation.

Bureaucrats had to abandon their tendency to micromanage and instead boldly create "economic technologies" that took into account both national objectives and the people's particular conditions. Engineers must overcome their narrow, specialized worldview (symbolized by "atomism") and adopt an integrated one (symbolized by quantum mechanics) whereby their technical work simultaneously fulfilled both particular and universal goals. Finally, capitalists and workers had to correct their individualistic ways and fully take up their "life functions" or vocations within the national economy. The "national life organization" would be the collective expression of these "life functions" and operate as an organic technological system that produced the New Order in East Asia. In the new order, technology was firmly associated with the nation's multiple life energies or more specifically the national qualities of abundant creativity, efficient organization, and comprehensive management. In this way, technology was seen not only as the advanced machinery and infrastructure necessary for a total war economy but also as inextricably linked to social organization and life itself.[78]

"Humanizing" the Social Mechanism

Mōri's notions of "creative bureaucrats" who managed the economy and were firmly linked to a "national life party" had multiple origins but can be partially traced to the thought of the German Historical School economists Werner Sombart, Friedrich Göttl-Ottlilienfeld, and Othmar Spann, whom he cited periodically.[79] All three were translated prolifically into Japanese throughout the 1930s and 1940s, and their anti-Marxist, anticapitalist ideas on building a national productivist economy significantly shaped the reform bureaucrats' efforts to create a managed economy during the war.[80] Furukawa notes the strong influence of Othmar Spann, the Austrian economic philosopher and leading proponent of corporatism immediately after World War I, on Mōri's thought. Spann's most famous work, *The True State* (1921), proposed an organic society of hierarchically organized estates based on vocation. The state would be grounded in a "total spirituality" characteristic of the German nation and replace the individualistic, materialist ethics of liberal democracy and capitalism. Spiritually aware, creative leaders would guide society and each estate, which performed its specific function within the community (such as manual and skilled labor, management, or political leadership) in the interests of a higher morality.[81] Furukawa rightly notes the similarity to Mōri's notion of a society organized by function and led by expert bureaucrats; however, Spann's corporatism represented a desire to return to a perceived corporatist past in the "pure" Germanic state and a preindustrial economy of creative artisans.[82] Instead, Mōri pushed for a functionally organized society based on the development of technologically advanced heavy industry and chemicals.

Sombart's *Future of Capitalism* (1932) was also highly regarded by non-Marxist Japanese economists and bureaucrats in its diagnosis of "late capitalism," and he greatly influenced their wartime conceptions of the managed economy.[83] Sombart viewed capitalism as having moved away from a pure market system of unbridled competition and entered into a stage of increasing rationalization into cartels and trusts to regulate the market. Internationally, free-trade mechanisms were collapsing as nations focused more on building self-sufficient economies through protectionism, industrial policy, and the formation of economic blocs. For this late capitalist period, Sombart described what he called a "reformative-revolutionary" response whereby states would actively plan and systematize their economies. Such states would not impose a uniform plan but would base policy on the complexity and diversity of economic life, which differed according to region, population, culture, social structure, and so on. As such, they would mix policies that encouraged private, individualistic economic activity together with those that regulated competition and prioritized national concerns. Sombart's conception was similar to Mōri's notion of "economic technologies" outlined earlier in this chapter in "Creative Engineers and Economic Technologies," and clearly inspired efforts by reform bureaucrats to find a pragmatic third path between communism and capitalism through economic planning.[84]

Sombart and Göttl were not only influential economists but prominent theorists of technology as well. Sombart argued that technology should be viewed as systems, complexes, or totalities of means that extended throughout the economy, politics, and culture. He also recognized that advanced capitalism fostered a type of "rational technology" that was not limited by nature but rather incorporated natural science in the process of incessantly developing such newer products as synthetic natural resources, energy sources, and means of production.[85] Mōri clearly shared

Sombart's rational view of technology; however, unlike Sombart, who sought to "tame" technology and what he perceived to be the instrumentalization of cultural life, Mōri viewed advanced technology's proliferation in the form of heavy industrial and chemical industries and the concomitant social complexity as positive and tied to a primordial Japanese creative energy.[86]

Göttl was a prominent Austrian economist who later became a recognized and vocal supporter of the Nazi Party. Reform bureaucrats and intellectuals were attracted by his attempt to root economics in national life and concrete experience rather than in abstract rules and cold statistics. His 1914 work *Economy and Technology*, translated into Japanese in 1931, defined technology broadly not only as the means of production but also including the skills and knowledge associated with them, which was closely tied to the people's needs and lives.[87] In the 1920s, he introduced the term *Fordismus* into the German language and uncritically celebrated Fordism as the key to resolving capitalism's contradictions. For him, Ford's use of technology for mass production represented a spirit of "service" to the common good because the public was provided with cheap goods, workers wages were maintained at higher levels, working hours were shortened, and equal opportunities for advancement were provided. Whereas Taylorism represented the capitalist spirit of maximizing profits and worker dehumanization, the high-tech Fordist factory represented a creative social force infused with the spirit of close cooperation and social service by providing people with cheap necessities and allowing workers to therefore live more fulfilling lives.[88] Göttl's technological enthusiasm continued into the Nazi era, although it became more tempered in favor of works that grounded economics in national life and spirit, which he viewed as the most concrete form of existence, as opposed to capitalist economics and its focus on the abstract individual.[89] Similar to Göttl, Mōri also combined a firm belief in technological progress with a conviction that the resulting social complexity and rationalization arose immediately out of everyday Japanese life and the national spirit rather than being some external, abstract force.

Mōri's elaborate thought on a dynamic "national life party" based on vocation and a technologically sophisticated economy rooted in the nation's "creative energies" had domestic origins as well and manifested itself in Konoe's New Order policies during his three wartime cabinets. One important inspiration for his plan was the Foreign Ministry diplomat and later parliamentarian from the Socialist People's Party Kamei Kanichirō, with whom he had a very close relationship.[90] The Socialist People's Party was founded by conservative members of the Labor-Farmer Party in 1928 and pushed for a program to reform capitalism through such social policies as the nationalization of basic industries, land reform, and welfare legislation. Kamei was a vocal member of the party's national socialist faction, which gained more influence with the deepening economic crises of the interwar period and the rising wave of nationalism after the Manchurian Incident in 1931. He strongly supported Japanese expansion into Manchuria, and in Japan he argued for the party's transformation into a mass "consumer" party rather than one devoted solely to the proletariat or "productive" classes.[91] He was also involved in discussions with right-wing military officers in the Cherry Blossom Society (*sakurakai*) who plotted to overthrow the cabinet and "restore" the Shōwa emperor in 1930.[92] In the 1930s, he continued to push for unity between the political parties, who were increasingly seen by the public and military as corrupt and self-serving. For example, Kamei forged close relations with the army officer Nagata Tetsuzan, who led the state's efforts to establish a total war mobilization system. Upon Nagata's request for a popular tract on national defense in the "broadest sense," Kamei authored the Army Ministry's 1934 pamphlet "The Essentials of National Defense and Proposals for Its Strengthening," which fiercely indicted capitalism for creating wider wealth disparities, mass unemployment, and instability among small businesses and farmers.[93] Similar to the reform bureaucrats' program, it also called for curbing large concentrations of wealth, alleviating rural poverty, increasing state controls over the economy, developing natural resources and labor, and increasing spiritual mobilization for the establishment of a "national defense state." He continued to work with leaders of the main political parties, Diet members, the military, and labor unions to create a wartime system of national unity between the state and civil society.

After a study-tour of Germany in 1937 and 1938, Kamei became a strong admirer of the Nazi Party and its labor mobilization system and upon his return to Japan joined with the parliamentarians Asō Hisashi, Akiyama Teisuke, and Akita Kiyoshi in plans to establish a single party with Prime Minister Konoe as its head.[94] Konoe had expressed interest in establishing such a party as hostilities with China became increasingly drawn out and domestic economic problems, such as chronic unemployment, remained unresolved. Platforms and guidelines were drawn up for a centralized "Greater Japan Party" with a strong youth wing and links to labor unions, patriotic associations, consumer groups, and agricultural cooperatives as well as each level of local government. Right-wing groups were even mobilized at one point to pressure Diet members into renouncing their political affiliations and joining the national party. Despite a general agreement among the highest levels of the military and government, Konoe abandoned the plan in late 1938 because of disputes among various groups jockeying for control over the party. The idea of a national party, however, was resurrected in 1940 with Konoe's New Order movement and the formation of the Imperial Rule Assistance Association.[95] The association was established in October 1940 to replace political parties; however, as a result of resistance by the parties and the Home Ministry, reformist ideas to radically organize society along syndicalist lines were eliminated and the association primarily became an organization that managed spiritual mobilization campaigns. Kamei was instrumental in persuading political parties to dissolve themselves into the new association, and he became the head of its East Asia Division.

Kamei regarded Mōri as his most trusted personal secretary and "brain" after 1938 and involved Mōri deeply in his project to establish a national mass party. Mōri was a key member of Kamei's personal research group of intellectuals, bureaucrats, and officers who dedicated themselves to "resolving" the deepening conflict between China and Japan through the formation of an "East Asian Cooperative Body," which would be founded upon corporatist national parties similar to Japan. According to Kamei, their group was in direct communication with the military and contacts in China in planning the formation of the All Peoples Party (*Zenmintō*) to mobilize and represent the Chinese people within an East Asian community also consisting of Manchukuo's political party, the Concordia Association, and a unified national party in Japan. Mōri even drafted the party's platform and submitted it to the Army Ministry. Thus, encouraged by Kamei, Mōri continued to theorize and propagate a fascist vision of pan-Asianism during the war as part of reform bureaucrat ideology.[96]

The reform bureaucrats did not only rely on official channels to realize their vision of a managed economy based on corporatist national parties. In June 1940, leading section heads in various ministries including Mōri, Sakomizu, Okumura, Minobe, and other Monday Group members established the Wartime Daily Life Consultation Center (*senji seikatsu sōdanjo*) as a direct conduit between bureaucrats and the people. The center's avowed purpose was to listen to questions and opinions from the people about life under wartime restrictions and to direct them to the proper authorities, learn about the effects of national policies and institutions on their lives, conduct research into existing conditions of production, and elevate the quality of technology in economic life.[97] Furukawa notes how the center directly reflected Mōri's ideas on transforming bureaucrats from "administrative supervisors" into active "managers" who precisely grasped people's living conditions and shaped policy accordingly.[98] In fact, two evenings a week, Mōri and other ministry department heads would hear questions and complaints or attend meetings on specific issues. For them, the center represented a first step toward creating a "new politics" that eliminated divisive, corrupt parliamentary politics and united people with the goals of building the national defense state and developing East Asia.[99]

The government's introduction to the center in an official magazine emphasized "vocationalization" as the main direction of wartime life and the necessity to "self-consciously" adjust one's work life to meet state goals rather than pursue self-gain and profit. This did not mean negating the self for the state but "amply meeting the people's natural needs, and building institutions that freely exhibit the Japanese nation's intelligence while sufficiently expanding individuality."[100] The center was but one small step toward creating such "spontaneous" institutions to improve life for the immediate war effort and the long term. Increasing the use of advanced science and technology among the people to overcome Japan's technological backwardness was another key part of elevating wartime life. As Mōri wrote, science and technology were an essential component to increasing the nation's productivity rooted in the "energy of the people," and it was the job of the bureaucrat or "creative engineer" to "synthesize" the people's actual living and working conditions with the requirements of advanced science and technology.[101] The center's actual work consisted primarily of advising small to medium businesses that were being abolished or shifted over to wartime production as a result of state directives. Its reports also covered such issues as food shortages, rising cost of living, rationing of surplus materials, and the organization of industrial cartels.[102] Although it is unclear how much if any of these reports and meetings affected actual policy, for Mōri and the reform bureaucrats, the center represented an important step toward radically renovating all areas of Japanese life into a fascist social mechanism.

In October 1944, Mōri joined Suzuki Teiichi and several others from the recently dissolved Cabinet Planning Board to lead the Greater Japan Industrial Patriotic Association (*dai Nihon sangyō hōkokukai*; hereafter, Sanpō). Established in November 1940 from out of the semiprivate Industrial Patriotic League (*sangyō hōkoku renmei*), it centralized the management of industrial relations under the ideology of "dedicated labor" (*kinrō*). First developed in the second Konoe cabinet's Outline for the Establishment of a New Order for Dedicated Labor in November 1940, the ideology institutionalized the reform bureaucrats' vision of work and vocation as the nation's central organizing principle. Blue-collar and white-collar workers would overcome their class identities to gain true membership in the nation as citizens actively and purposefully engaged in their specific occupations, which now had public significance. According to *kinrō* ideology, work "expressed one's full personality" and was understood as "creative" and "spontaneous" rather than as "drudgery." As Saguchi Kazurō notes, five-member Sanpō units were formed in factories to encourage cooperation, innovation, and productivity, and the principle of a minimum living wage was institutionalized during the 1940s through worker appeals to this idea of dedicated labor.[103]

Suzuki, Mōri, and other ex–Cabinet Planning Board bureaucrats were brought in to revitalize Sanpō, which was criticized as bureaucratic and inefficient and increasingly seen as nothing more than a "consciousness-raising movement" for workers. Mōri was appointed Sanpō's managing director and later headed its planning room from January 1945. On the eve of Japan's defeat, Sanpō was largely an ineffective organ torn by divisions between regional offices, industrial control associations, and the Home, Welfare, and Munitions Ministries, all of which had competing jurisdictions.[104] Nevertheless, in the context of the worsening war situation, it was felt that the reform bureaucrats could rationalize labor-management relations with their extensive planning experience, and in October 1944 Mōri drafted the "New Direction for the Industrial Patriotic Movement" for Sanpō workers, which was published widely in newspapers. It began by affirming the idea that those who were engaged in production were not divided into "managers, engineers, and workers" but became "imperial dedicated laborers" (*kōkoku kinrōsha*) who did not merely engage in physical labor but used all of their abilities to employ productive facilities, materials, transport, and so on for the purpose of exhibiting "total productive power." The document then pushed for an economic thought centered entirely on production and the elimination of capitalist profit-centered production. Advanced science and technology should form the nucleus of this "national defense production system" with a strong focus particularly on shipping and aircraft production and the wide array of industries associated with them. Avenues for vertical and horizontal communication between workers, managers, and bureaucrats needed to be improved in order to improve national productivity, innovation, and morale, and production leaders had to incorporate the rapid rise of Korean, female, and student labor into the industrial patriotic movement.[105] Despite the increasingly hopeless war situation, the reform bureaucrats continued to insist on their fascist vision of a society organized as an advanced productive mechanism driven by creative, responsible workers dedicated to the nation as the only path to victory.

Pan-Asian Nationalism

For reform bureaucrats, wartime mobilization was more than about achieving victory on the battlefield. It represented the total renovation of Japanese politics, economics, and culture in accordance with their fascist visions. Reorganizing the economy and mobilizing the people at home were closely tied to similar efforts in Japan's expanding East Asian empire, such as the establishment of the "Japan-Manchukuo-China Economic Bloc," and promoting such populist parties as the New People's Association in north China to win the cooperation of various peoples for Japan's goal of creating a self-sufficient, prosperous New Order in East Asia. For reform bureaucrats, pan-Asianism was more than empty sloganeering or a cynical justification of Japanese imperial rule; it entailed a radical program of change to unite Japan and East Asia.

For Mōri and the reform bureaucrats, the national life organization that would build the national economy was inseparable from the construction of a similar political order in East Asia. Along with the encouragement of more pragmatic economic technologies to co-opt Chinese resistance, Mōri argued for the grounding of these in a new political order as well that would unify various peoples into an "East Asian nation" (tōa minzoku). He criticized those who saw Chinese nationalism as a recent response to chronic warlordism and did not recognize its long history dating back to the overthrow of the Qing Dynasty in 1911 and popular revolts against the European powers as well as those who thought that Chinese peasants were apolitical and had no national or ethnic consciousness.[106] If the Japanese continued to narrowly reject Chinese nationalism, "history would teach us a lesson," Mōri warned.[107] Instead, the Japanese people should "affirm the fact that the ethnic pursuit of the Chinese nation to realize its unity is the motivating force of their political power."[108]

Japanese and Chinese nationalism should not remain at the level of "simple ethnic emotion" but rather mutually develop under the rubric of a higher multiethnic East Asian nationalism that would cooperatively build the noncapitalist, self-sufficient order in East Asia.[109] In fact, Mōri argued that such East Asian nationalism was an essential part of the Japanese nation, which he described as a "plural nation."[110] Japan's multiethnic nationalism would "fertilize" the Chinese national instinct and "make possible the East Asian nation's greater and faster construction" and thereby the "Symbiotic Body of East Asia." East Asian nationalism thus formed the political ideology or "will" for the "liberation from China's international capitalist and communist orders" and the establishment of a "total life order" in Japan, Manchukuo, and China along the lines of the national life organization.[111]

Mōri saw world history as moving away from the particular nationalisms institutionalized by the Versailles treaty system and toward the formation of large, multiethnic nations that incorporated "smaller and weaker" ones. The five great plural nations—the Japanese, Slavs, Germans, Chinese, and Indians—would shape the course of world history.[112] Plurality in terms of ethnicity, culture, lifestyle, and economic activity was not a sign of weakness but rather of strength and superiority, he said in a 1939 speech to the Social Policy Institute.[113] This affirmation of multiplicity and the need for bureaucrats to understand and synthesize such difference instead of homogenizing it was a recurring theme in Mōri's totalitarian philosophy. As the world's leading "plural nation," Japan would affirm, synthesize, and revitalize the weaker China, and in the process, each nation would transform the other into a higher East Asian nation. The Chinese and Japanese peoples would also transform themselves from "private citizens" of a liberal capitalist order into "national subjects" within a corporatist "total life order." This new East Asian nation would resolve persistent contemporary questions of ethnic minorities and class conflict, which the nineteenth century world gave birth to and the twentieth century failed to resolve—this was Japan's "world-historical task."[114] For Mōri, the East Asian nation's "political power" was to be expressed not as the homogenization of Asia's cultures and economies but rather as a productive synthesis and expression of their multiple energies and skills.

Toward the "Economic Construction of East Asia"

Mōri's pan-Asianist ideology developed hand in hand with the reform bureaucrats' efforts to create a managed economy in East Asia and beyond. In November 1938, as hostilities expanded across China, Prime Minister Konoe announced Japan's mission of forming a "New Order in East Asia" in Japan, Manchukuo, and China based on a platform of anticommunism and "mutual cooperation" in politics, economics, and culture. Konoe's second cabinet passed the more detailed "Outline for Japan-Manchukuo-China Economic Construction" in October 1940, and reform bureaucrats in the Cabinet Planning Board immediately began drafting policy for its realization. The outline called for the creation of an "organic self-sufficient sphere" that "comprehensively united" the politics, culture, and economics of all three nations. Policies would be rooted in the particular "life stages" of each region and ethnic group. Japan was to become the New Order in East Asia's center and focus on "expanding national power by mobilizing the people and reorganizing the economy to guide and manage the construction of Manchukuo and China" and promoting science and technology as well as advanced industry at home. Manchukuo was envisioned as a base for mining and electricity production and some heavy and chemical industries. China would provide natural resources for heavy industry in the other two nations and develop light industry and agriculture.[115]

Because the goal was to collectively transform Japan, Manchukuo, and China into Asia's industrial center, technology played a significant role in the outline. Japan would train more engineers to help develop industries, educate skilled workers, and introduce "labor technologies" to mobilize workers in accordance with each nation's assigned function within the new order. Investment into improving transportation infrastructure and the communications network between the three nations was also prioritized. The reform bureaucrats firmly tied their renovationist conceptions to transform Japan into an advanced, managed economy rooted in the people with concurrent efforts to incorporate Manchukuo and China into that same order. This bore fruit in the large-scale projects examined in Chapters 3 and 4, where the reform bureaucrats were also heavily involved with planning and implementation.

The Cabinet Planning Board's "three ravens"—Akinaga, Minobe, and Mōri—were involved in setting up the main planning organ to realize the outline's conception, the Japan, Manchukuo, and China Economic Conference (*Nichi-Man-Shi keizai kyōgikai*). The conference began its activities in 1941 under close management by board bureaucrats. It consisted of midlevel officials from such relevant ministries as the Cabinet Planning Board, Asia Development Board, and Manchukuo Office, as well as leading representatives from business, academia, and the media. Divisions were formed to draw up five-year and ten-year production targets in industry, mining, agriculture, livestock, and forestry and to make plans to establish necessary institutions and infrastructure to expand production, procure materials, ensure sufficient capital, facilitate trade, mobilize labor, and build the necessary transportation and electricity networks. Of particular note were the conference's aggressive and optimistic plans to expand heavy industry in Manchukuo, especially mining and electricity production. This controversial reform bureaucrat plan to shift some of Japan's heavy industrial production to Manchukuo near abundant natural resources in China was carried over into the conference's next incarnation, the Greater East Asia Construction Council (*dai tōa kensetsu shingikai*).[116]

After war was declared with the United States in December 1941 and Japan expanded farther into Southeast Asia, Prime Minister Tōjō Hideki announced the formation of the "Greater East Asia Co-Prosperity Sphere," thereby formalizing earlier military plans to "advance south" and secure wartime natural resources that were then being subject to Western blockades. As the main organ for planning resource allocation and materials mobilization, the Cabinet Planning Board again took over the leadership in determining policy for its realization. The board established and ran the Greater East Asia Construction Council in February 1942, an expanded version of the earlier conference with added divisions on economic policy, agriculture and fisheries, education, and ethnic policy. Mōri and three others from the board were appointed to run the council, and he was heavily involved in drafting its initial guiding document, the "Basic Policy for Greater East Asia Economic Construction."[117] With the army's backing, the reform bureaucrats pushed to make Japan, Manchukuo, and north China into the heavy industrial nucleus for the development of a larger economic sphere. Setting the tone for the rest of the "Co-Prosperity Sphere," this region would be firmly based on the principles of the colonial managed economy—rational distribution of industries based on local conditions, comprehensive planning of natural resources, electricity, and labor and the reorganization of firms on the basis of productivity rather than profit. This conception was later watered down, however, as a result of competing visions from the Ministry of Commerce and Industry, which wanted to centralize economic control in Japan and prioritize Japanese industries, and from the navy, which wanted to invest more in Southeast Asian development.[118]

Furukawa notes the influence of Mōri and board reform bureaucrats in shaping the language of Tōjō's famous January 1942 Greater East Asia Co-Prosperity Sphere policy speech to the Diet. His statement that Japan would become the core of "an order of co-existence and co-prosperity . . . whereby all nations and peoples of Greater East Asia will be enabled to find their true place" came directly from the board's basic policy draft on "constructing the Greater East Asia Co-prosperity Sphere" that Mōri largely wrote. From 1938, as a vocal proponent of establishing a "symbiotic sphere" in East Asia, Mōri envisioned an extended version of the corporatist national party whereby each nation and people were united under a higher "East Asian" nationalism and had a specific "place" or role in building a strong, prosperous community in accordance with their specific economic, social, and cultural conditions. The basic policy contained several of Mōri's other pan-Asianist visions, such as mobilizing the Japanese people to eliminate the Western "self-centered" and "materialistic" worldview and fulfilling Japan's "destiny" to lead in the development of a managed and closely coordinated national defense economy in East Asia rooted in "geo-political" considerations. Despite resistance from other ministries, the navy, and business, much of the reform bureaucrats' anticapitalist and techno-fascist language made it into the board's final policy document. Although much of this planning was never realized because of the war, the reform bureaucrats' conceptions of a "symbiotic" and highly integrated East Asian community entered into the very highest levels of cabinet decision making, demonstrating that the pan-Asianist technological imaginary was more than empty rhetoric.[119]

Conclusion: Technology, Fascism, and Power

Technology, imperialism, and fascism merged in Mōri's thought and the reform bureaucrats' ideology and policies. The expansion of technology's meaning in Japan's 1930s intellectual discourse toward representing a society's entire productive mechanisms as well as more subjective values of creativity, efficiency, and systematicity became incorporated into their visions of a managed economy tied to the nation's popular energies. Mōri's thoughts on technology and society closely resemble what Jeffrey Herf refers to as "reactionary modernism"—the reconciliation of "antimodernist, romantic, and irrationalist ideas present in German nationalism and the most obvious manifestation of means-ends rationality, that is, modern technology."[120] In a similar fashion to German thinkers, Mōri rooted modern technology in the primordial energies of the Japanese nation, who were destined to overcome liberal capitalism and lead the development of a "Greater East Asia Co-Prosperity Sphere." Herf, however, employs his own preestablished definition of technology as "means-ends rationality" based on a liberal Enlightenment tradition, which leads him to conclude that the synthesis of "technics and unreason" in Nazi thought was "paradoxical."[121] By predefining technology in this manner, Herf not only ignores the term's historicity and how different groups articulated it but also does not question how technology and rationality were able to operate as a system of power and mobilization.

The conceptions and policies of Mōri and the reform bureaucrats signified the emergence of a new mode of power through the articulation of fascism with technology. They appropriated emerging notions of technology for their project of building a managed economy where every aspect of life was rationally planned and mobilized to exhibit its maximum potential and productivity. Even their "irrational" notions of the Japanese nation and spirit were subsumed within their vision of a rationally organized corporatist party composed of "national citizens" who acquired an ethic of responsibility, hard work, and creativity for the nation. In their conceptions and policies, power did not merely structure social life from above through state institutions but infused consciousness and the totality of social relations in all of their multiplicity. Power was fundamentally restrictive *and* productive.

For example, economic technologies sought to employ local knowledge and socioeconomic relations in order to fully mobilize the "energies" of the colonial peoples. This notion of bureaucrats as "creative engineers" employing economic technologies emerged out of their experiences of establishing a managed economy in Manchukuo and China through such policies as the Manchukuo Five-Year Industrial Plan and efforts to create the necessary political, financial, economic, and technical infrastructures. The formation of such comprehensive national policy organs as the Cabinet Planning Board further institutionalized their conception of an "economic general staff" managing an interrelated and complex social mechanism through a wide array of policies ranging from industrial organization to colonial management, social welfare, science and technology, and "national morals." At a pan-Asian level, this manifested itself in plans to transform Japan, Manchukuo, and north China into a technically advanced, integrated industrial zone at the center of a Greater East Asia Co-Prosperity Sphere. Such comprehensive laws as the 1938 State Total Mobilization Law, the 1940 New Order outlines on the economy, labor, and science and technology, and the establishment of the Japan, Manchukuo, and China Economic Conference in 1941 and the Greater East Asia Construction Council in 1942 realized many of their techno-fascist conceptions at the state's highest levels despite resistance from other ministries, business, sections of the military, and the Diet.

Mōri and the reform bureaucrats tied their technological imaginary to the level of daily life by proposing the formation of a plural "national life organization" that was efficiently organized based on technologically advanced vocations. In the colonies, these efforts were mirrored in attempts to incorporate Chinese nationalism into pan-Asianism through mass parties. Despite Konoe's failed campaign to establish a syndicalist national party, reform bureaucrats succeeded in forming the Industrial Patriotic Federation (later, Greater Japan Industrial Patriotic Association) for labor in 1938, the Imperial Rule Assistance Association for political parties in 1940, and the Wartime Daily Life Consultation Center in 1940. All of these organizations utilized techniques to cultivate an ethics of responsibility or spirit of creativity in order to mobilize people's active participation in the managed economy. In China and Manchukuo, the New People's Association (1938) and Concordia Association (1932) functioned similarly with an organizational emphasis on promoting social welfare and "enlightenment."

Technology also played a key role in their ideological agenda of transforming liberal capitalism into the managed national economy. By promoting technology's creative essence, the managed economy would liberate technology from its materialist meanings under capitalism as the instrumental means for extracting profit. With the second industrial revolution, the reform bureaucrats argued, technology had escaped nature's

limits to the extent that it could now reproduce natural products and resources by utilizing innovations in energy production, physics, and chemistry, for example.[122] In the process of reorganizing capitalism, technology's creative essence would be harnessed to further the development of
vital wartime industries and an advanced economy based on heavy industry, electricity, and chemicals. Such laws as the 1941 Outline for the Establishment of a New Order for Science and Technology were passed to promote scientific and technical research, rapidly industrialize its results, and
nurture a "scientific spirit" among the populace. The Cabinet Planning
Board and the National Policy Research Association drew up blueprints
and studies on developing colonial resources and improving industrial
and technical infrastructure. For example, the National Policy Research
Association issued a report, *Theory of the Order of Technology for the
Greater East Asia Co-Prosperity Sphere*, which suggested policies to advance technical training for non-Japanese workers, increase investment in
technical schools, promote children's science education in the colonies,
and expand public hygiene programs to eliminate epidemics.[123] Such blueprints were grounded in the reform bureaucrats' technical view of space
whereby the managed economy constituted a concrete "space for living"
(*Lebensraum*). The abstract view of space behind the capitalist exploitation
of natural resources would be replaced by a "technological conscience"
that developed new resources and productive industries for the Greater
East Asia Co-Prosperity Sphere.[124] In sum, "technology" was utilized by
reform bureaucrats not only as a way to immediately mobilize Japan for war
and increase the exploitation of colonial natural resources but also as an ideological trope to reenvision and reshape society into a managed social system
of mobilized subjects who actively participated in capitalism's renovation.

In the end, rather than ensuring individuality, autonomy, and creativity, the techno-fascist utopia envisioned by Mōri and the reform bureaucrats greatly contributed to establishing a dystopian social mechanism
that harnessed the lives and labor of millions of Asians during the war while
entrenching Japan's wartime state and business interests. Their technological imaginary greatly rationalized Japanese war crimes (often blamed on
irrationalist right-wing military officers and ideologies) by seeking to establish a new mode of power that operated at the level of subjectivity and
life. Rather than therefore dismissing their formulations merely as empty
rhetoric or delusional, it is important to take them seriously, as they produced a range of institutions, policies, and projects that continuously

invoked powerful discourses of technology and modernization. These initiatives continued into the postwar era as the Japanese and other Asian regimes mobilized their peoples to exert all of their energies for high-speed growth through appeals to technology's possibilities.[125] In ways continuous with the wartime and colonial eras, creativity, responsibility, and independence were once again encouraged under the rubric of technology in connection to postwar goals of mobilizing citizens to build strong, prosperous nation-states. In wartime Japan, the articulation of technology with fascism (i.e., techno-fascism) signified a new mode of power designed to organize the multiplicity of subjects in Japan and East Asia into an organic "national life system" that would completely eliminate political antagonism and therefore any chance of democratically transforming social relations of subordination. This new mode of power survived relatively intact after Japan's surrender—hidden beneath the popular postwar myth of the wartime era as the dark valley of unquestioning spirituality and mindless irrationality, which Japan had supposedly departed from after 1945.

Epilogue
Legacies of Techno-Fascism and Techno-Imperialism in Postwar Japan

Several hours after Emperor Hirohito's broadcast on August 15, 1945, announcing Japan's surrender to the Allies, Prime Minister Suzuki Kantarō addressed the nation for the first time. He noted how the enemy had utilized science and technology with devastating effect in the form of the atomic bombs. He also lamented the weakness of Japan's wartime science and technology and urged all Japanese to make extra efforts to develop them for the nation's postwar reconstruction:

It is essential . . . that the people should cultivate a new life spirit of self-reliance, creativity and diligence in order to begin the building of a new Japan, and in particular should strive for the progress of science and technology, which were our greatest deficiency during the war.[1]

Technology during the wartime era constituted a new system of power designed to mobilize the same "self-reliance, creativity, and diligence" that Suzuki implored the Japanese people to once again cultivate with science and technology after Japan's defeat. As Tessa Morris-Suzuki argues, "the vision of technology as the basis of the Greater East Asia Co-Prosperity Sphere was transformed into a vision of technology as the basis of the new Japan."[2] Janis Mimura notes how "wartime techno-fascism" quickly transformed into "postwar managerialism" as many of the same figures, such as Kishi Nobusuke and Shiina Etsusaburō, adapted their technocratic "managed economy" approach to Japan's postwar goals of building a middle-class consumer society.[3] This "techno-fascist" system was closely connected to "techno-imperialism," specifically in the form of comprehensive technical projects designed to integrate the empire into Japan's wartime economic system and mobilize colonial peoples. These two components of the technological imaginary—techno-fascism and techno-imperialism—adapted to Japan's postwar context of building a prosperous nation at home and exercising soft power abroad through overseas development assistance. Technology continued to serve as a system of power and mobilization as Japan shifted to a peacetime economy and a nonaggressive foreign policy after 1945.

Japan's new leaders adapted the earlier wartime technological imaginary to the context of postwar reconstruction and high-speed economic growth. In the conclusion to his book *Miyamoto Takenosuke*, Ōyodo Shōichi discusses the career of the former wartime engineer and bureaucrat Ōkita Saburō, who took on upper-level positions in the Economic Planning Agency in the 1950s. This agency drafted some key laws for Japan's high-speed economic development, such as the 1960 National Income Doubling Plan and the 1963 National Land Comprehensive Development Plan. Ōkita was part of an influential group of economists led by Arisawa Hiromi who wrote the 1946 document "Basic Questions for Reconstructing Japan's Economy," which foreshadowed some of the directions that Japanese economic planning would take. The document gave a prominent position to science and technology as factors that were essential to improving Japan's living standards and economic conditions. Similar to wartime efforts to spread technical values of rationality and productivity throughout everyday life, the plan called for more science and technical education as part of developing the new "economic man" [*sic*] who was hard working and dedicated to his vocation—not the capitalistic, individualistic economic man who was concerned only with money and status.[4] This image of the ideal highly skilled worker dedicated to the company and nation would form a pillar of Japan's rapid development into a global economic power.

Former bureaucrats from the Cabinet Planning Board, Technology Board, and other wartime organizations immediately called for the reestablishment of the same types of institutions that they had designed to encourage and manage technological development during the war. They argued that the "power of science had to be effectively utilized for Japan to rebuild itself as a healthy cultural nation-state."[5] It was therefore necessary to organize scientists and engineers under firm national goals so that scientific values would eventually be reflected throughout administration, industry, and national life. They viewed the spread of such values and the promotion of science and technology as absolutely essential to Japan's reconstruction. Former wartime bureaucrats, intellectuals, and engineers, such as Ōkita, Aki Kōichi (Home Ministry engineer), Tsuru Shigeto (economist and advisor), and Taira Teizō (Mantetsu researcher), established the Resources Investigation Association (*shigen chōsakai*) in 1948 to formulate a policy of "comprehensive resource use grounded in science-technology" in light of Japan's dire postwar economic situation. This transministerial, interdisciplinary organization formulated new notions of comprehensive resource planning and regional development, which were inspired by the Tennessee Valley Authority (TVA) chairman David Lilienthal's philosophy and what his translator, the engineer Wada Koroku, described as Japan's failure to comprehensively plan and integrate technology during the war.[6] In the 1950s, former wartime bureaucrats joined together with business leaders and politicians to successfully lobby the government to establish the Science and Technology Agency in 1956, which promoted research for national development and coordinated science and technology policy among the different government ministries. This movement went hand-in-hand with the establishment of the Agency of Industrial Science and Technology under the Ministry of International Trade and Industry (MITI), which was to play a central technocratic role in promoting Japan's accelerated economic growth.[7] Whereas proposals to systematize technology put forth by such intellectuals as Aikawa and technology bureaucrats as Miyamoto largely failed during the wartime era, these proposals succeeded immensely during the postwar era.

The technological imaginary also extended itself once more to the level of workers' lives on the factory floor. William Tsutsui examines the continuity of wartime industrial management techniques after 1945. For example, with the aid of the United States and Japanese governments, midlevel business leaders and management experts established the Japan Productivity Center (*Nihon seisansei honbu*). The center popularized the latest management techniques as part of "rationalization" campaigns and promoted "the ideology of productivity" based on "harmonious labor relations, a compliant working class, and a social consensus that legitimized a managerial, technocratic order."[8] They promoted human relations management and quality control, both of which had their roots in the wartime period. The human relations approach emphasized "humanizing the workplace" or meeting the social and psychological needs of workers rather than using more heavy-handed techniques of increasing productivity.[9] Quality control circles gave small groups of workers responsibility over improving quality, efficiency, and productivity as a way to allow for more worker self-expression and creativity. These efforts to align the interests of workers with management resonated with the proposals of bureaucrats and intellectuals on encouraging worker innovation to increase wartime production, for example, and clearly manifested themselves in earlier state policy. For example, the New Order for Dedicated Labor (1940) emphasized the ideology of *kinrō* (dedicated labor), whereby attention was paid to the workers' "full personality," which was considered to be "creative and autonomous," so that workers would be totally involved in their work.[10] National goals instead of wages would motivate workers. In these and other ways, postwar management techniques often continued some of the methods used for wartime mobilization.

Japan's transformation into a mass consumer economy in the postwar era is often cited as a major distinction from the prewar and wartime periods. Simon Partner has examined this transformation through a study of Japan's rise to become a consumer electronics giant in the postwar era. Confronting the absence of consumer demand in the immediate postwar period when most Japanese barely earned enough to meet their living expenses, electronics makers invoked the discourse of scientific and technical rationality in their attempts to persuade Japanese to purchase expensive television sets, washing machines, and refrigerators. Partner goes beyond the conventional narrative of how Japanese "electronics samurai" engineers imported and adapted actual American technologies to examine the broader mass production and marketing technologies that engineers and managers had learned and employed to establish a large domestic consumer market for their expensive products. For example, electronics firms sold the image of the "Bright Life" to Japanese consumers, which presented a nuclear household of refrigerators and washing machines that rationalized the housewife's work, thereby giving her more leisure time to spend with her family, and televisions that brought more culture into the Japanese home. Thus, as Partner shows, the story of Japan's postwar transformation into a mass consumer, high-tech society involved the invocation of a broader definition of technology as the rationalization of the workplace, household, and consumer behavior through the employment of management and marketing technologies.[11] This vision of society organized according to technological principles was not simply imported from the United States in the postwar era but, as shown throughout this book, had wartime precedents. The technological imaginary constituted a powerful ideology that sought to shape people's everyday values and behavior for the cause of war and empire. As Sheldon Garon has shown, broad technologies of "moral suasion" designed to make people internalize a range of values (such as thrift and diligence) were strengthened during the wartime era, and various social organizations continued to propagate these values well beyond the postwar era.[12] Even the distinction between the "consumerist" postwar era and the "productivist" wartime era has been somewhat exaggerated, as state intellectuals, such as Ōkochi Kazuo, also argued about the importance of encouraging consumption and bureaucrats concerned themselves with utilizing private initiative for wartime mobilization.[13] In a similar manner, many of the basic policies designed to mobilize Japanese in the postwar era were also rooted in the emergence of the technological imaginary and a new mode of power during the war.

The ideological concept of "comprehensive technology" also continued into the postwar era in the form of "comprehensive national development plans" for Japan's reconstruction and high-speed economic growth. With the enactment of the 1950 Comprehensive National Development Law (*zenkoku sōgō kaihatsu keikaku hō*) drawn up by such technocrats as Ōkita on the Economic Stabilization Board, Japan began to construct multipurpose dams in specially designated regions with U.S. assistance. The purpose of this law, which remains in effect today, was as follows:

By considering the natural conditions of national land, to integrally utilize, develop and conserve national land from the general viewpoint of policies concerning economy, society or culture, to attempt the appropriate selection of industrial sites, and at the same time to improve social welfare.[14]

Similar to the colonial and wartime eras, a wide range of such matters as the "utilization of natural resources, disaster prevention, river management, urban and rural relationships, the location of industrial zones, and even matters of culture" were placed under the rubric of comprehensive national land planning (*sōgō kokudo keikaku*).[15] However, different from the wartime era, projects during the 1950s were more inspired by the TVA's grassroots development philosophy, which was enthusiastically received by Japanese bureaucrats and intellectuals at the time.[16] In this context, many former colonial dam engineers joined the Electric Power Development Corporation (Denpatsu), Japan's public hydropower development company, or remained with the construction giant Hazama-gumi, where they used their expertise to complete Sakuma Dam (1956) on the Tenryū River in only three years. Sakuma was billed as the "largest in the Orient" and was Japan's first large-scale development project after the war. Similar to the cases of Sup'ung and Fengman Dams, Sakuma was presented by state and company officials as an example of Japanese science and technology's power to improve life in the region through flood control, irrigation, and power production. Despite the fact that this infrastructure development was rebranded under the banner of postwar democracy and peacetime economic growth, the dam's social effects nevertheless bore strong resemblance to the effects associated with other dams constructed during the wartime era. The sacrifices made by residents affected by these dams' construction were deemed necessary, insofar as Japan had a pressing need for electricity to rebuild its shattered industrial infrastructure. Although Sakuma district boomed during the dam's construction, in the years afterward, its lumber and fishing industries declined, residents increasingly moved away to the cities, and tax revenues dried up. In the end, Sakuma Dam's electricity would benefit industrial growth in the booming industrial metropolises of Tokyo and Nagoya more than in the Sakuma region itself.[17]

When the First Comprehensive National Development Plan was passed in 1963, the Japanese government shifted away from a policy of developing specific regions and more toward the "balanced development of national land."[18] The plan's objective was to coordinate and develop adequate infrastructure to match the high-growth-rate targets of Prime Minister Ikeda's 1960 Income Doubling Plan and check trends toward industrial overconcentration on Japan's Pacific belt. Although the plan succeeded at building "new industrial cities" in the regions, it still focused primarily on projects geared toward urban industrial development, especially low-margin industrial materials factories that shipped goods to the coastal metropolitan cities, rather than on infrastructure construction for rural residents and the relocation of more high-margin assembly and processing factories that would contribute more to rural development. Air and water pollution became major issues during this period and prompted the rise of citizens' movements that raised questions about the overall direction of Japanese high-speed economic growth and its effects on the quality of life.[19] Thus, similar to the wartime era, comprehensive development projects were far from balanced and were skewed more toward the interests of big business and the government's economic growth targets than toward the people. As Gotō Kunio notes, "The root of the problem was more in the excessive centralization that occurred during the war, and which established a pattern that was maintained throughout the postwar redevelopment of Japan."[20] Despite these problems, the Japanese government continued to invoke the ideology of comprehensive development as a way to mobilize an increasingly critical public. For example, Prime Minister Tanaka Kakuei's 1973 grand plan to "Remodel the Japanese Archipelago" entailed building large-scale industrial complexes, extending the highway and high-speed railway network, and constructing integrated industrial, cultural, and administrative cities with sophisticated communications networks in the regions.[21] In the end, as Gavan McCormack notes, these projects that had once again invoked technology's power to resolve socioeconomic problems simply contributed to the rise of the "construction state" (*doken kokka*)—the Japanese government's ongoing massive investment in large-scale engineering and public works projects that prioritize the interests of big business, contribute to ecological devastation, worsen the urban-rural imbalance, create an unsustainable public debt, centralize power at the national level, and strengthen the corrupt nexus between bureaucrats, politicians, and businessmen.[22]

As citizens' movements focused attention on the social and environmental costs of Japan's high-speed heavy industrial growth in the late 1960s and early 1970s, such Japanese theorists as Umesao Tadao, Koyama Kenichi, Hayashi Yūjirō, and Masuda Yoneji began to popularize the concept of the "information society" (*jōhōka shakai*) as a vision for "postindustrial" Japan. Similar to the theories put forth by American futurists like Daniel Bell, information society theorists saw the computer as revolutionizing industrial production through the automation and integration of the office, factory, and consumer (e.g., "just-in-time production").[23] Production would shift away from the manufacture of material goods and become information-intensive, and "innovation, planning, design, and marketing would represent an integral and increasing share in the value of goods and services." Masuda went so far as to envision a "computopia," where the increasing availability of information and leisure would result in a kind of spiritual renaissance, the formation of egalitarian civic virtues, and the elimination of class conflict.[24] In the aftermath of the 1973 Oil Shocks, the Japanese government mobilized the nation by reviving the prewar and wartime slogan of "technological nation-building" (*gijutsu rikkoku*) in their campaign to shift Japan to a more high-tech information society that relied less on heavy industry.[25] Even during the post-bubble era, expensive urban planning projects that combined rapid transportation links, intelligent buildings and control centers, fiber-optic communication links, and grand leisure and entertainment areas were implemented throughout Japan, regardless of whether they provided any benefits for its citizens.[26] Similar to the wartime era, technology continually served as a powerful horizon to mobilize people's hopes and dreams and diffuse socioeconomic discontent.

In the realm of technology theory, Marxist standpoints reemerged and became dominant after 1945. Theories, such as Aikawa's, that emphasized technology as a comprehensive social system were dismissed as fascist in the new democratic environment of the American occupation and immediate postwar period. The *Kōza-ha* Marxist belief that Japan's semifeudal economy and social structure held back scientific and technological progress continued to hold sway after 1945 and shaped postwar theories of technology.[27] As mentioned in Chapter 1, in the late 1940s, the physicist Taketani Mitsuo put forth a new theory of technology that resonated with contemporary movements among scientists and engineers to democratize their workplaces—"the conscious application of rule-governedness in human (productive) praxis." In this way, Taketani emphasized technology's subjective aspects, in particular the essential role of scientific rationality in improving human welfare, which he believed the wartime period had failed to develop. According to him, the spirit of free scientific inquiry would continuously elevate Japan's productive forces, which in turn would enter into conflict with outmoded capitalist relations of production and ultimately advance Japan toward a more rational socialism where science and technology united with the lives of the proletariat.[28] Taketani was so confident in technology's progressive nature that he became a leading proponent of the peaceful use of nuclear power. In a 1952 article for a prominent women's magazine, he argued that nuclear energy would not only help resolve Japan's electricity shortage but also urbanize and industrialize natural wastelands, end starvation through agricultural expansion, and lead to major medical advancements, among other things.[29] His strong faith in technology's "rational" employment for the people's benefit represented a strong postwar ideological current among not only such Marxists as himself but also the government, which went on to promote the institutions of high-speed industrial growth.

As Gotō notes, postwar Marxists, such as Taketani, possessed a "naive confidence in science and rationalism."[30] His influential follower, Hoshino Yoshirō, extended his arguments by arguing that democratically minded scientists would lead technological innovation in areas that were key to Japanese economic growth, such as automation, synthetic chemicals, and nuclear power. Some Marxists, such as Nakamura Seiji and Oka Kunio, continued to argue for a "systems theory of technology" or the importance of understanding how capitalism structurally organized technology rather than focusing solely on its subjective aspects.[31] The intellectual debate remained at a rather simplistic level of emphasizing either technology's subjective or objective elements and became increasingly irrelevant to Japanese society's rapid technological transformation during the 1960s and 1970s. Although Marxists produced sophisticated scholarship on Japan's new structure of "state monopoly capitalism" that formed the backdrop to its technological revolution, Marxist orthodoxy largely prevented an analysis of technology's social and political aspects. The Marxists viewed technology as an ultimately neutral means that could be used instrumentally by the working class to achieve socialism, and they therefore failed to explore how technology operated as a particular system of power. Ironically, many wartime figures, such as Aikawa, Mōri, and Miyamoto, in their own ways possessed a greater understanding than postwar Marxists of the complicated dynamics between technology, subjectivity, and power. But Japan's dominant postwar narrative of technical rationality and progress, which many Marxists shared albeit for different political reasons, foreclosed an analysis of these dynamics.[32] In fact, their failure to vigorously critique the overall framework of technical progress contributed to the strength of the predominant "wartime as a dark valley of irrationality and spiritualism" narrative, a narrative that hid the wartime and colonial origins of Japan's postwar technological imaginary.

The techno-scientific regimes of expertise that legitimized wartime ideologies of technology also adapted and evolved in the postwar context. The loss of Japan's colonial empire by no means ended the involvement of Japanese engineers in building comprehensive infrastructure development projects overseas. For example, Kubota Yutaka, Yalu Hydropower's president from 1940 until the war's end, returned from Korea after the war to start the development consultancy, Nippon Kōei. From the 1950s, its core business centered on the application of Japanese foreign aid assistance to large-scale overseas development projects. Missing the excitement of developing "virgin frontiers," Kubota was instrumental in persuading the Japanese government to provide wartime reparations payments in the form of massive infrastructure projects that he argued would not only help rapidly industrialize newly independent Asian nations but also provide lucrative contracts and export opportunities to Japanese firms.[33] In a striking reinvocation of the earlier discourse of "constructing Asia," Kubota argued to Japanese government officials that Japan's leadership in undertaking development projects in developing nations would help create Asian "co-prosperity and co-existence."[34] As Japan became an overseas development power, Kubota's company became known domestically as the "vanguard of Japan's industrial advance overseas," and Kubota himself was later called the "Shōgun of the Mekong" in international development circles for his vital role in Southeast Asia's Mekong River Development Project.[35] In attempting to win overseas contracts, Kubota often boasted to Southeast Asian leaders of his vast experience in building dams in coordination with heavy chemical industrial development in colonial Korea. He even noted that although most Koreans hated the Japanese because of the colonial past, Koreans welcomed him because of his "great work" during that era. The Korea Power Company therefore asked him to bring the same engineers to aid in the completion of unfinished colonial hydropower projects as well as the planning of new projects to promote South Korea's economic development during the 1960s.[36] In the late 1960s, Kubota convinced Indonesia's president Suharto and the Japanese government to build the Asahan River Dam, a Dutch-conceived project in northern Sumatra that Kubota had surveyed during the war for the Japanese military. Japanese equipment suppliers and construction firms benefited from the $2 billion project, in which local smelters provided cheap aluminum to manufacturers in Japan. The 2,600 smelting jobs created by the project went to migrant Javanese laborers, and the project failed to promote the additional development of industrial infrastructure in that region. Thus, as in the colonial period, such development projects never seriously incorporated the interests of local residents but subsumed them under grand visions of comprehensive national development through technology.[37]

There is a great need to more closely analyze how ideologies of technology backed up by specific techno-scientific regimes of expertise with strong roots in Japanese imperialism continued to operate because they have so significantly shaped postwar Japan's history. Former colonial networks of businessmen, bureaucrats, and engineers revived and reinvented wartime techno-scientific regimes of expertise under Japan's new banner of developmentalism in order to legitimate their projects and activities.[38] For example, utilizing expertise acquired in colonial Korea at its enormous chemical complex at Hŭngnam, Nitchitsu (now Shin-Nitchitsu) reemerged as a giant in Japan's heavy chemical industry after the Japanese government began focusing its energies on creating a mass consumer society (which required such industrial materials as plastics) and later petrochemicals as part of its national resource policy. Executives and engineers from colonial Korea returned to take up top managerial positions in Shin-Nitchitsu, which quickly rose once again to become the "head temple" in Japan's heavy chemical industry by 1950. In a renewed affirmation of its status as one of Japan's technological leaders, Emperor Hirohito visited its Minamata factory for the second time in 1949, the first time having been in 1931 when Nitchitsu was establishing itself in Korea as an essential industry to imperial Japan.[39] Yet as a stark reminder of the technological imaginary's brutal underside, Shin-Nitchitsu was also responsible for one of Japan's worst industrial pollution cases, the widespread mercury poisoning around its Minamata plant and the outbreak of the severe and often fatal Minamata disease among thousands of residents. Hashimoto Hikoshichi, Nitchitsu engineer and head of the Minamata factory during the war, invented a mercury-based procedure to increase acetaldehyde production that led to the dumping of deadlier toxins from the 1930s, which manifested itself in the form of Minamata disease after 1951.[40] With the cooperation of local government (including Hashimoto, who was Minamata's mayor in the 1960s), Shin-Nitchitsu's executives covered up the pollution, actively obstructed investigations, evaded responsibility, and prevented a comprehensive settlement with the victims for decades. Because the chemical industry was an essential component of policies to stimulate high-speed economic growth, national ministries, such as MITI—the institution that resurrected and adapted wartime economic planning techniques in the postwar era—also participated in the cover-up and obstruction. As the economist Fukai Junichi notes, the corporate and governmental culture that facilitated the Minamata pollution disaster was nothing new and had some of its origins in efforts to put down local pollution and compensation struggles during the prewar, wartime, and colonial eras as Nitchitsu began constructing hydropower dams and heavy chemical factories in Japan and Korea with the state's full support.[41] Minamata was only the first of several prominent industrial pollution cases during the 1960s and 1970s that galvanized civil society to question the technocratic system they had helped create to achieve Japan's "economic miracle." But even as it became clear that postwar economic growth was being achieved at the expense of people's health, lives, and communities, former colonial bureaucrats and engineers went on to take up top-level positions in government and business, where they continued to justify Japan's path of technology-led development. For example, Honma Norio, who headed Fengman Dam's construction and established a development consultancy after the war, once said, "Recently, everyone is saying pollution, pollution but in reality there has been even more public good. As evidence, haven't people's average life spans greatly increased?"[42]

In addition to its role in shaping Japan's overseas development policy, Nippon Kōei became a key member of the "construction state" (*doken kokka*) and was involved in the implementation of the Japanese government's comprehensive national development projects.[43] Kubota's former construction industry partners in colonial Korea, Nishimatsu-gumi and Hazama-gumi, were founding investors in his company, and his wartime bureaucracy and business contacts enabled him to gain financing and win precious reconstruction projects in the immediate aftermath of Japan's surrender.[44] As Kawamura Masami shows, the same network of top bureaucrats, businessmen, and engineers involved in Sup'ung Dam's construction—such as Kishi Nobusuke (former head of Manchukuo's Industrial Department and prime minister from 1957 to 1960), Kobayashi Ataru (wartime financier and first head of Japan Development Bank), and Takahashi Tatsunosuke (head of Manchurian Heavy Industries and first president of Denpatsu)—were instrumental in pushing for big dam development overseas using Japanese foreign aid money. Domestically, river basin management and hydropower projects constituted much of Nippon Kōei's business.[45] As Daniel Aldrich notes, Japan's 2,734 existing dams and 373 dams under construction make Japan one of the most heavily dammed nations in the world.[46] Under the banner of Japan's comprehensive national development plans, the government and construction industry vigorously promoted multipurpose dams as the key to promoting regional prosperity because they promised to control floods, irrigate new lands, and provide cheap water and electricity for industry. Filmmakers also enthusiastically responded with Iwanami releasing the 1954 award-winning documentary *Sakuma Dam*, for example, and Mifune Productions even producing a dramatized feature film, *The Sun in Kurobe*, in

1968, about Japan's largest dam project at the time. These films portray dam construction as a heroic battle between humans and nature and represent such large-scale projects as symbolic of Japan's struggle to rapidly rebuild and become prosperous after the war.[47]

Yet in a demonstration of similar power dynamics involving Nitchitsu and the promotion of the heavy chemical industry, dam construction has also led to the marginalization of localities and ecological devastation in the name of national economic development. The Ministry of Construction, the postwar reincarnation of the wartime Home Ministry's engineering sections, intensified the prewar and wartime agenda of colonizing nature under the banner of environmental "improvement" (*seibi*)—the straightening and reinforcement of rivers through concrete dam and waterway structures—rather than planning projects more in tune with the natural capacities of Japan's environments. As McCormack notes, dams have largely failed to fulfill their functions of flood control because planners have not addressed root causes of deforestation and environmental neglect. They have also quickly filled with silt and muck, required expensive repair or demolition, and created a needless oversupply of water at enormous public expense. This continued colonization of nature has been accompanied by "entrenched, centralized, bureaucratic systems of water supply, sewerage, and flood control, with their complex of huge dams, aqueducts, and tunnels" as well as the monopolization of Japan's electricity market by ten regional power companies.[48] Techniques of appropriating land and co-opting local resistance against dam construction also continued well into the postwar era. Aldrich describes the Construction Ministry's approach to dam projects as "Decide, Announce, Defend" and notes that they increasingly turned to such social control tactics as invoking eminent domain laws, exploiting legal ambiguities, slashing local subsidies, and using police force. In the event of fierce resistance, they created local deliberation committees filled with dam supporters, increased and aggressively pushed incentive schemes, held public meetings, and used the media and local leaders to advertise dam benefits.[49] Once again, the failure to critically evaluate the wartime technological imaginary resulted in unchecked ecological devastation and the continued mobilization and subjugation of local communities in the name of national progress.

These events and trends highlight the continuing need to denaturalize and critique ideologies of technology and regimes of techno-scientific expertise that have survived the colonial era in reinvigorated forms. Although Japan's wartime ideologies on technology (and the legacies of different projects and policies) should be examined more closely in terms of how they changed in relation to specific historical contexts, different confluence of actors, and changing political and economic circumstances, the continuities in the deployment of technology as a system of power after 1945 require further study because socioeconomic conflict in Japan continues to be subsumed under an ideology of science and technology as the foundation for national progress.[50] Understanding and directly confronting the undemocratic legacies associated with technology that originated in the wartime era may open up more possibilities for Japan to manage such twenty-first-century challenges as chronic economic recession, an energy policy dangerously overcommitted to nuclear power, the continued predominance of unaccountable bureaucrats, and overseas development assistance programs that fail to truly promote local empowerment—issues that the state and its various allies have continually addressed by uncritically appealing to technology as a long-term solution.

Reference Matter

Notes

Introduction

1. Commission on the History of Science and Technology Policy, *Historical Review of Japanese Science and Technology Policy*, i, iii.

2. Tatsuno, *Technopolis Strategy*; Bloom, *Japan as a Scientific and Technological Superpower.*

3. Kodama, *Analyzing Japanese Advanced High Technologies*, 173–74.

4. With the burst of the Japanese economic bubble in the early 1990s, the discourse around Japanese technology gradually shifted to a U.S. neoliberal paradigm aggressively pursued by Prime Minister Koizumi Junichirō, which viewed the free market as the best determiner of technological innovation and development. Although the Japanese government has pursued some degree of market liberalization, privatization, and spending cuts, the tradition of state-sponsored technology development still remains strong. For a recent example on Prime Minister Abe Shinzō's Innovation 25 program to remake Japanese society using the latest technologies (particularly household robots), see Robertson, "Robo Sapiens Japanicus."

5. Katada, "Why Did Japan Suspend Foreign Aid to China?" 47–48.

6. On the "flying geese model," see Hatch, *Asia's Flying Geese.*

7. Samuels, *"Rich Nation, Strong Army."*

8. On the issue of omitting empire in writing the history of modern Japan, see Schmid, "Colonialism and the 'Korea Problem,'" 951–76.

9. "Imagineering" is a term originally popularized by Disney referring to the combination of engineering and imagination in the theming of goods, services, and places. Archer, "Limits to the Imagineered City," 322. I do not use this term in the same sense of techniques of commodification but more as a way to capture the powerful wartime ideology of technological utopianism among Japan's elites that bore fruit in specific policies and projects.

10. Morley, "Introduction: Choice and Consequence," 9. See the other essays in Morley, *Dilemmas of Growth in Prewar Japan*, for examples of the modernization theory approach.

11. Johnson, *MITI and the Japanese Miracle*; Dower, "Useful War," 49–70; Garon, "Rethinking Modernization and Modernity in Japanese History," 350; Garon, *Molding Japanese Minds*. Works that analyze how modernization contributed to wartime mobilization include Garon, *State and Labor in Modern Japan*; Fletcher, *Search for a New Order*; Kasza, *State and the Mass Media*.

12. Weber, *Protestant Ethic and the Spirit of Capitalism*, 181.

13. Ibid., 181, 182; Feenberg, *Critical Theory of Technology*, 68.

14. Marcuse, *One Dimensional Man*, 158.

15. Habermas, "Science and Technology as 'Ideology,'" 107, 109.

16. Ibid., 103. Emphasis is in original. Brackets are mine.

17. Ibid., 103, 104.

18. Herf, *Reactionary Modernism*, 16.

19. Maier, "Society as Factory," 26–29, 30, 32, 35, 40.

20. Roger Griffin's definition of fascism as a "palingenetic form of populist ultranationalism" interprets fascism's antimodern, tradition-affirming, and spiritualist elements as part of a forward-looking, revolutionary transformation of society rather than the affirmation of some idyllic past or preservation of the status quo. Griffin, "Palingenetic Core of Generic Fascist Ideology," 2–3.

21. Maruyama, "Ideology and Dynamics of Japanese Fascism."

22. Peter Duus and Daniel Okimoto propose "corporatism" as a more appropriate framework than "fascism," arguing that fascists were only a "minor side current" in wartime policy making. Duus and Okimoto, "Fascism and the History of Prewar Japan," 65–76. After comparing 1930s Japan with Europe, Kasza uses the term "renovationist authoritarian right" to describe the wartime political system instead of "fascism." The "renovationist authoritarian right" lay between the more status quo–oriented "conservative right" and a fascism demanding sweeping sociopolitical revolution. Kasza, "Fascism from Below?" 625. Herbert Bix employs the term "emperor-system fascism" to describe the wartime political system. Bix, "Rethinking Emperor-System Fascism," 20–32. Walter Skya avoids the term "fascism" in favor of "radical Shintō ultranationalism" as wartime Japan's hegemonic ideology. Skya, *Japan's Holy War*. There has also been some work analyzing Japan's "cultural fascism," which had strong interactions with concurrent trends in Europe. See Harootunian, *Overcome by Modernity*, and the essays in Tansman, *Culture of Japanese Fascism*. Finally, although the essays in E. Bruce Reynolds's edited volume do much to position Japan within global currents of fascism, they do not devote much attention to the modernizing, rational, and technocratic aspects of Japanese fascism. Reynolds, *Japan in the Fascist Era*.

23. McCormack, "Nineteen-Thirties Japan," 29.

24. Or as Andrew Gordon notes, fascism was "a set of common ideas that justified the new regimes, and the common programs they adopted." Gordon, *Labor and Imperial Democracy in Prewar Japan*, 334.

25. Mizuno, *Science for the Empire*, 13.

26. Ibid., 3.

27. Harootunian, *Overcome by Modernity*, xxviii–xxix. He borrows "molecular power" from Gilles Deleuze and Félix Guattari.

28. Tsutsui, *Manufacturing Ideology*, chapters 1–3.

29. Kawahara, *Shōwa seiji shisō kenkyū*, 58, 64–67. For an overview of Ōkochi's philosophy, see Cusumano, "'Scientific Industry.'" For an analysis of different articulations of technocracy by reform bureaucrats and their allies, see Mimura, *Planning for Empire*. On Miyamoto's life and thought, see Ōyodo, *Miyamoto*. Chapter 2 of *Constructing East Asia* focuses on Miyamoto and the engineers' movement. Chapter 5 analyzes the reform bureaucrats.

30. Barnhart, *Japan Plans for Total War*; Mimura, *Planning for Empire*, 15–21.

31. Robertson, "Mobilizing for War, Engineering the Peace," 62–84. As head of the army's research apparatus, Tada supervised the development of several prototypes in radar and remote-control technology during the war. The most comprehensive study in English on how Japan mobilized science and technology to build advanced weapons during World War II is Grunden, *Secret Weapons and World War II*.

32. Kawahara, *Shōwa seiji shisō kenkyū*, 68–70.

33. Ibid., 203–34; Yamanouchi, "Total War and System Integration," 23–26.

34. Kawahara, *Shōwa seiji shisō kenkyū*, 74–79; Koschmann, "Rule by Technology/Technologies of Rule," 5–10.

35. Naoki Sakai notes how Nishida perceived of technology as "subjective technology" (*shutaiteki gijutsu*), by which "the subject manufactures itself through praxis" and not just as the lifeless instruments used by an epistemological human subject (*shukan*) to achieve its goals. Sakai, *Translation and Subjectivity*, 24–25, 198–99.

36. Miki, "Gijutsu tetsugaku," 220.

37. Quoted in Calichman, *Overcoming Modernity*, 113.

38. Kimoto, "Tosaka Jun and the Question of Technology."

39. Nakai, "Bigaku nyūmon," 111; Moore, "Para-Existential Forces of Invention," 127–57.

40. Harootunian, *Overcome by Modernity*, 95, 106–18.

41. Quoted in Nornes, *Japanese Documentary Film*, 91. For a comprehensive account of how film technology was discursively employed to create hegemonic notions of spectatorship and subjectivity, see Gerow, *Visions of Japanese Modernity*.

42. Harootunian, *Overcome by Modernity*, 104; Weisenfeld, *MAVO*, 125–38.

43. Silverberg, *Erotic Grotesque Nonsense*.

44. Kawahara, *Shōwa seiji shisō kenkyū*; Mimura, *Planning for Empire*; Mizuno, *Science for the Empire*; Ōyodo, *Miyamoto*.

45. Silverberg, *Erotic Grotesque Nonsense*, 8.

46. Yamanouchi, "Total War and System Integration," 22.

47. Lo, *Doctors Within Borders*, 35–37.

48. Zaiki and Tsukahara, "Meteorology on the Southern Frontier of Japan's Empire," 187–88.

49. Other works that examine Japan's colonial infrastructure include Yang, "Japanese Colonial Infrastructure in Northeast Asia," 90–107; Tucker, "Building 'Our' Manchukuo"; Koshizawa, "Sangyō kiban no kōchiku to toshi keikaku," 183–242; Molony, *Technology and Investment*; Matsusaka, *Making of Japanese Manchuria*. Yang's book on telecommunications in the Japanese empire also illustrates a more explicit and systematic discourse on technology and empire among communications engineers emerging after the formation of Manchukuo in 1932, which culminated in the planning and construction of a "comprehensive East Asian telecommunications network" to integrate the "New East Asian Order" after the war's outbreak in 1937. Yang, *Technology of Empire*, part 2.

50. Ibid., 5.

51. Scott, *Seeing Like a State*, 4. On the modernism behind Japan's infrastructure projects in Manchuria, see Sewell, "Rethinking the Modern in Japanese History," 313–58.

52. Scott, *Seeing Like a State*, 348–49. Yang's *Technology of Empire* also puts conflict and negotiation at the heart of the engineering process with regard to telecommunications.

53. Reuss, "Seeing Like an Engineer," 545–46.

54. The technological imaginary also developed in relation to the rise of "comprehensive" domestic infrastructure projects in the 1930s. However, these never reached the scale and vision of some colonial projects, and they were plagued by more bureaucratic conflict, political maneuvering, local resistance, and legal restraints than was the case in the empire. Thus, the most ambitious and idealistic engineers often worked in the colonies rather than in Japan. On domestic social infrastructure projects, see Matsuura, *Senzen no kokudo seibi seisaku*.

55. Skya, *Japan's Holy War*, 25.

56. Many groups, such as fiction writers, lower-level engineers, the military, farmers, workers, and consumers, and how they thought about technology are unexplored in this book because of space restrictions and a desire to more fully articulate the thought of some of the primary actors who have not received very much attention in English. On literature and science fiction, see Mizuno, *Science for the Empire*, 143–72, and Nakamura, "Making Bodily Differences," 169–90. On managers and workers, see Tsutsui, *Manufacturing Ideology*. On farmers in relation to the state's rural revitalization programs and public works projects, see Smith, *A Time of Crisis*. On emerging middle-class consumers in relation to new ideas of the modern urban house and home, see Sand, *House and Home in Modern Japan*.

57. Williams and Wallace, *Unit 731*, 37–38. Also see Grunden, *Secret Weapons and World War II*, 165–96.

58. Driscoll, *Absolute Erotic, Absolute Grotesque*, 266, 272–73, 276–77, 314. Also see Kratoska, *Asian Labor in the Wartime Japanese Empire*.

59. Driscoll, *Absolute Erotic, Absolute Grotesque*, 247.

60. Watanabe, "Militarism, Colonialism, and the Trafficking of Women," 3–17.

61. Dower, *War Without Mercy*; Tanaka, *Hidden Horrors*.

62. For an analysis of Japanese imperialism in terms of how capitalism incessantly operated at the level of extracting surplus value from life in three distinct forms—the

biopolitical, neuropolitical, and necropolitical—see Driscoll, *Absolute Erotic, Absolute Grotesque.*

63. For example, in his analysis of "techno-nationalism," or Japan's ideology of linking technological development to national security, which originated in the prewar and wartime eras, Richard Samuels concludes that it was "admirable, rational, flexible, and ought to be embraced—mutatis mutandis—in the United States as well." Samuels, *"Rich Nation, Strong Army,"* x. Nakamura Takafusa pioneered the scholarship into the "useful" wartime origins of the postwar Japanese "economic miracle." Nakamura, *Postwar Japanese Economy.*

Chapter 1

1. Saigusa, "Gijutsu no shisō," 147.

2. Ibid., 127.

3. The comprehensive bibliography and commentary in Aikawa's "Modern Theory of Technology" provide an idea of the wide range of works on technology in each discipline available to Japanese. See Aikawa, *Gendai gijutsuron,* 287–326.

4. Kawahara, *Shōwa seiji shisō kenkyū,* 256–57; Mizuno, *Science for the Empire,* 98–101, 109–11.

5. An example of a strong supporter of the more utopian view of technology was Miyamoto Takenosuke (see Chapter 2).

6. Koschmann, "Rule by Technology/Technologies of Rule," 1.

7. Barshay, *Social Sciences in Modern Japan*; Fletcher, *Search for a New Order.*

8. Koschmann, "Rule by Technology/Technologies of Rule," 5.

9. Quoted in Koschmann, *Revolution and Subjectivity,* 144.

10. Ibid., 146. Koschmann borrows from Herbert Marcuse's view of the industrial world as a vast "technical apparatus" that systematizes political struggle and rationalizes submission. See Marcuse, *One-Dimensional Man,* 158.

11. Yamanouchi, "Total War and System Integration," 1–42. On techniques of winning popular support for nuclear power, see Aldrich and Dusinberre, "Hatoko Comes Home," 683–705.

12. For more on these postwar debates on the development of a "free and democratic subjectivity" and their continuities with the wartime period, see Koschmann, *Revolution and Subjectivity.*

13. Härd and Jamison, "Conceptual Framework," 2–3, 7.

14. Härd, "German Regulation," 33–67.

15. Maier, "Society as Factory."

16. Härd and Jamison, "Conceptual Framework," 4.

17. Marx, "Idea of 'Technology,'" 248; Schatzberg, "*Technik* Comes to America," 486.

18. Ibid., 486, 494–96.

19. Marx, "Idea of 'Technology,'" 249.

20. Iida, *Gijutsu,* 372–73; Mizuno, *Science for the Empire,* 26.

21. Ōyodo, *Miyamoto,* 208.

22. On the development of the social institutions of science and technology in Japan after World War I, see Hiroshige, *Kagaku no shakaishi*. For more on the early engineer movement, see Mizuno, *Science for the Empire*, 19–42.

23. For a detailed history of the debate, see Nakamura, *Gijutsuron ronsōshi*.

24. Mizuno, *Science for the Empire*, 80. For more on Marxism's introduction to Japan, see Hoston, *Marxism and the Crisis of Development* and Barshay, *Social Sciences in Modern Japan*, 36–71.

25. Sugiyama, "World Conception of Japanese Social Science," 209.

26. Nakamura, *Gijutsuron ronsōshi*, vol. 1, 6–10.

27. Rubinstein, "Relations of Science, Technology, and Economics under Capitalism, and in the Soviet Union," 20–21.

28. Aikawa, *Gijutsuron*, 8.

29. Sugiyama, "World Conception of Japanese Social Science," 208–9, 212–13.

30. Barshay, *Social Sciences in Modern Japan*, 81; Aikawa, *Gijutsuron*, 221–22.

31. On Tosaka's theory of technology, see Kimoto, "Tosaka Jun and the Question of Technology." On Oka, see Kimoto, "Tosaka Jun and the Question of Technology."

32. Aikawa, *Gendai gijutsuron*, 3.

33. Aikawa, *Gijutsuron nyūmon*, 8, 14–15.

34. Marx, "Theses on Feuerbach," 400. Aikawa openly quoted from the "Theses" in his earlier work on technology in discussing the inseparability of truth or thought from praxis. See Aikawa, *Gijutsuron*, 122. Although all references to Marx drop out of his wartime works, he still insisted that "praxis, actual praxis, and the results of praxis" constituted the "primary meaning of the human spirit." Aikawa, *Gijutsuron nyūmon*, 190.

35. Although Aikawa emphasized praxis as technology's essence, he was careful to note that technologies always took on specific historical form based on the historical mode of production (e.g., tools, instruments, machines, machine systems). Much of his work in fact was on the historical development of technology. Aikawa, *Gijutsuron nyūmon*, 26–50.

36. Aikawa, *Gendai gijutsuron*, 47–48.

37. Aikawa, *Gijutsuron nyūmon*, 59–60.

38. Nakamura, *Gijutsuron ronsōshi*, vol. 1, 60, 65–66.

39. Aikawa, *Gijutsuron nyūmon*, 85. Figure 1 is a more elaborate version of an earlier diagram in Aikawa, *Gendai gijutsuron*, 95.

40. Aikawa, *Gendai gijutsuron* , 85.

41. Similar to Aikawa, many Marxists committed themselves to Japan's wartime effort by conducting official studies or writing reports and articles for state institutions. For example, Mizuno discusses how Marxists were some of the most enthusiastic proponents of science during the war, as they hoped that the spread of a rational scientific worldview would lead to a critical and independent spirit among the population. Mizuno, *Science for the Empire*, 71–142. Louise Young analyzes leftists who conducted agrarian studies in Manchukuo in the hopes that these would lead to radical agrarian reform of

semifeudal conditions. Young, *Japan's Total Empire*, 291–302. Sugiyama Mitsunobu examines Marxists who conducted economic studies of Japan's "Greater East Asia Co-Prosperity Sphere" with the goal of "liberating" or "modernizing Asia from imperialism. Sugiyama, "World Conception of Japanese Social Science."

42. On "overcoming modernity" and capitalism's "spiritual crisis," see "Naze gijutsu o kangaeru ka" in Aikawa, *Gijutsuron nyūmon*, 5–16. Also see "Gendai gijutsu kara gijutsuron he" in Aikawa, *Gendai gijutsuron*, 5–11.

43. Aikawa, *Sangyō gijutsu*, 211; Ōyodo, *Miyamoto*, 406–11. Engineers and technology bureaucrats are discussed more in Chapter 2.

44. Aikawa, *Sangyō gijutsu*, 79–80.

45. Quoted in Morris-Suzuki, *Technological Transformation of Japan*, 148.

46. For more on the New Order of Science and Technology, see Mizuno, *Science for the Empire*, 63–68 and Pauer, "Japan's Technical Mobilization."

47. Aikawa, *Sangyō gijutsu*, 39–43. Aikawa often criticized the term "science-technology" (*kagaku gijutsu*) for privileging science over technology and therefore being abstract. For example, Aikawa, ibid., 53. For him, science was only "potential technology," not actual technology. Aikawa, *Gijutsuron nyūmon*, 104.

48. Aikawa, *Sangyō gijiutsuron*, 210, 236.

49. Aikawa, *Gijutsu no seisaku to riron*, 135, 142–43, 223–24.

50. In 1941, Aikawa and other technology bureaucrats in the Japan Technology Association drafted a proposal for a corporatist organization of scientists and engineers that would not only work to develop science and technology for wartime production but also thoroughly spread technical values of efficiency, optimality, and rationality outside the factory throughout all areas of life. The proposal was submitted to the Imperial Rule Assistance Association, the Cabinet Planning Board, and the All-Japan Federation of Science and Technology Organizations, but it was rejected as too "idealist." Ōyodo, *Miyamoto*, 411. On the whole, Aikawa's proposals strongly resembled Stakhanovite visions of technical mobilization and productionism. In his only postwar theory of technology essay, Aikawa criticized another intellectual's lack of understanding of Stakhanovism's significance, which he described as a grassroots cultural movement that was giving birth to a technological intelligentsia responsible to the proletariat. Stakhanovism represented a type of technological humanism whereby technically trained workers freely employed their creativity, originality, and research to help create new technologies that would enable the Soviet Union to overtake productivity in capitalist countries and eventually form the foundation for communism. See Aikawa, "'Dai niji sangyō kakumei' setsu no hihan," 16. Thus, one can view Aikawa's wartime push for mass technological mobilization rooted in vocational organizations and community institutions as evidence of his belief that rapid modernization through the strengthening of a total war system could lay the foundations for socialism. On Stakhanovism, see Siegelbaum, *Stakhanovism*.

51. Aikawa, *Gijutsu no seisaku to riron*, 234–35.

52. Aikawa, "Hitachi seisakujo," 43–45.

53. Ibid., 44; Aikawa, *Gijutsu no seisaku to riron*, 235–38.

54. Aikawa, "Hitachi seisakujo Hitachi kōjō," 40–43.

55. Aikawa, *Sangyō gijutsu*, 115–32.

56. Aikawa, *Gijutsu oyobi ginō kanri*, 60–77.

57. Aikawa, *Sangyō gijutsu*, 112–14.

58. Mizuno, *Science for the Empire*, 51.

59. Aikawa, *Sangyō gijutsu*, 152.

60. Ibid., 154–55, 175–76, 189–91.

61. For more on establishing a "Japanese agricultural technology," see ibid., 253–85.

62. Aikawa's Southeast Asia study was published as Aikawa, *Tōna Ajia no shigen to gijutsu*. It originally appeared serialized in the institute's journal, *Shin Ajia* (New Asia), which focused on Southeast and West Asia. The institute drafted numerous research reports on natural resource, industrial, and agricultural development in Manchuria, China, and other parts of Japan's empire. They also published translations and conducted anthropological, historical, and sociological studies. On Mantetsu's research institutes, see Kobayashi, *Mantetsu—"Chi no shūdan."*

63. An example of such a military study is Tada, *Nanpō kagaku kikō*.

64. Aikawa, *Tōna Ajia no shigen to gijutsu*, 5, 8–9, 46.

65. Ibid., 57, 76. On Japanese policy in wartime Southeast Asia, see Part 3 of Mendl, *Japan and Southeast Asia*.

66. Aikawa's "rubber pyramid" chart is in Aikawa, *Tōna Ajia no shigen*, 168. Also see ibid., 157–202 for his full analysis of the pyramid, which included land ownership patterns, types of technologies used, wage systems, and processing facilities for each area.

67. Ibid., 264, 340, 387.

68. Ibid., 253.

69. Ibid., 412.

70. Ibid., 408–9.

71. Ibid., 445.

72. Ibid., 448.

73. Ibid., 451.

74. Aikawa, *Gendai gijutsuron*, 255.

75. Ibid., 250–57.

76. Ibid., 259–60.

77. Ibid., 272.

78. Ibid., 264.

79. Ibid., 277.

80. Ibid., 276.

81. Ibid.

82. Ibid., 276–77.

83. Ibid., 277.

84. Ibid., 276–79.

85. Nornes, *Japanese Documentary Film*, 63, 95.

86. Aikawa, *Bunka eigaron*, 16.

87. Nornes, *Japanese Documentary Film*, 6.

88. All of these are located at the Yamagata International Documentary Film Festival office in Tokyo.

89. Imamura, *Kiroku eigaron*.

90. Aikawa, *Bunka eigaron*, 15.

91. Ibid., 11.

92. Ibid., 15–16.

93. Ibid., 21–22.

94. Ibid., 31, 38, 46, 66–67. Aikawa's "neorealism" should not be mistaken with the formal neorealism movement in cinema.

95. The series consisted of Mizuki Sōya's *Production for Victory* (*Shōri he no seisan*, 1942), Imaizumi Zenju's *The Shipbuilding Corps* (*Zōsen teishintai*, 1942), and the film analyzed here, Nakayama Yoshio's *The Present Battle*. While making the film, Aikawa frequently talked about the difficulty of "turning thought into image" (*shisō no eigaka*), particularly the difficulty of linking workplace solidarity to national solidarity and the greater mission of "cultural construction." See, for example, Aikawa, *Bunka eigaron*, 30–32. See also Kubo, "Kōjō to bunka eiga," 36 and Asano, "Konnichi no tatakai: Seisaku hōkoku," 30–31. Aside from depicting national solidarity, the filmmakers were also interested in trying to capture the worker's point of view within mechanized factory work and shooting in an enormous, dark, and confusing factory.

96. Unfortunately, the film no longer exists. The description of the film's content comes from Asano, "Konnichi no tatakai," 66–71.

97. Aikawa, *Bunka eigaron*, 30–31.

98. Ibid., 32.

99. Ibid., 33, 57.

100. Aikawa borrowed the phrase "engineers of the human soul" from Maxim Gorky and used it in his early works on theory of technology. Aikawa, *Gijutsuron*, 179; Gorky, "Literature and the Soviet Idea," 53–54, 64–67. Stalin had famously used this term in a speech, too, at Maxim Gorky's house in October 1932 in preparation for the first Congress of Soviet Writers, where he described the "production of souls" as more important than the "production of tanks." It became one of the central tenets of "socialist realism," whose mission was to accurately portray the proletarian revolution and the proletariat's construction of a socialist culture.

101. See Aikawa, "Zaiso minshū undō no ichi kessan," 271–97.

102. See Chūō kōron jigyō shuppan, *Aikawa Haruki shōden*, 217–33 for more details.

103. Aikawa, "Sobieto-teki ningen no keisei," 8–16.

104. *Aikawa Haruki shōden*, 19, 78.

105. Steinhoff, *Tenkō*; Tsurumi, *Intellectual History of Wartime Japan*.

106. The economist Sakairi Chōtarō noted that Aikawa helped establish and lead the Center for Technology Culture (*gijutsu bunka kenkyūjo*) as a haven for leftist and Marxist intellectuals who were increasingly being rounded up or sent to war between 1942 and 1945. Prominent leftist intellectuals such as Saigusa Hiroto, Oka Kunio, Taketani Mitsuo,

and Yamazaki Toshio participated in the center's activities, which were partially funded by Kamei Kanichirō, one of the theoretical forefathers of the reform bureaucrats and Prime Minister Konoe Fumimaro's New Order policies. A few months after its establishment, Aikawa was drafted, and the center was bombed toward the war's end, destroying all of its records. See Chūō kōron jigyō shuppan, *Aikawa Haruki shōden*, 84–85. Thus, he never abandoned his former leftist colleagues while conducting his statist intellectual work during the war, which suggests a continuing commitment to his earlier, more explicit objective of proletarian revolution.

107. Samuels, *"Rich Nation, Strong Army"*; Dower, "Useful War," 49–70; Nakamura, *Postwar Japanese Economy*. Mimura's *Planning for Empire* challenges this earlier scholarship through her characterization of wartime reform bureaucrats as "techno-fascists." She mainly focuses on the managerial and technocratic tendencies within fascism rather than on how techno-fascism constituted a new mode of power that sought to comprehensively mobilize subjectivity and life.

Chapter 2

1. "Technology bureaucrats" (*gijutsu kanryō*) constituted a new type of bureaucrat "with both specialized technical knowledge and broad administrative experience." Mimura, "Technology Bureaucrats and the 'New Order for Science-Technology,'" 98. For another study of technology bureaucrats (defined more narrowly as bureaucrats with an engineering background), see Ōyodo, *Gijutsu kanryō no seiji sankaku*.

2. Hiroshige, *Kagaku no shakaishi*, vol. 1, 34.

3. Yoneda, "Ozawa Kyūtarō naimu gishi to naru ki," 107. According to Kubota Yutaka, a government engineer who later quit to build dams and hydroelectric plants in Korea, they were placed under the category of "construction costs," thereby confirming the popular view of engineers as objects or "means." Nagatsuka, *Kubota*, 59.

4. Ōyodo, *Miyamoto*, 46–52. On the U.S. engineers' movement, see Layton, *Revolt of the Engineers*. For more on the rise of science-based industry, see Noble, *America by Design*.

5. On Taylorism in Japan, see Tsutsui, *Manufacturing Ideology*.

6. Naoki, *Gijutsu seikatsu yori*; Ōyodo, *Miyamoto*, 51–52, 54.

7. Ōyodo, *Miyamoto*, 110.

8. Schatzberg, *"Technik* Comes to America," 498, 503.

9. Veblen, *Engineers and the Price System*, 43–44.

10. Ibid., 47.

11. Ibid., 86–104.

12. Mizuno, *Science for the Empire*, 19–42.

13. During his 1924–25 study tour to Europe and the United States, Miyamoto met a Labor Party leader in England named Burns, who told him to read Veblen's *Engineers and the Price System* and Frank Taussig's *Inventors and Money Makers* after hearing Miyamoto describe the Kōjin Club's program of improving the social position of engineers and increasing their role as social managers. Ōyodo, *Miyamoto*, 119. When an engineer-

ing professor in New York told him the same thing, he immediately purchased a copy and read it on the boat back to Japan. Miyamoto, "Tekunokurashī no kenkyū," 25–26. Miyamoto had previously heard of Veblen's ideas while establishing the Kōjin Club. Yamaguchi Noboru, another founder and a professor of civil engineering at Tokyo Imperial University, had heard discussions of Veblen's ideas among socialists and intellectuals who periodically visited his home. Ōyodo, *Miyamoto*, 110.

14. Quoted in Ōyodo, *Miyamoto*, 112. Also in Mizuno, *Science for the Empire*, 24.

15. Quoted in Ōyodo, *Miyamoto*, 112.

16. Mizuno, *Science for the Empire*, 19–42.

17. Ōyodo, *Miyamoto*, 117–20, 148, 152.

18. Mizuno calls the ideology of mobilizing people through science and technology "scientific nationalism," which she traces among such different social actors as engineers, intellectuals, bureaucrats, and fiction writers. For an introductory discussion of this ideology, see Mizuno, *Science for the Empire*, 8–11. Samuels also employs the term "techno-nationalism" to describe the Japanese state's emphasis on defense technology and national security from Meiji into the postwar era. Samuels, *"Rich Nation, Strong Army."*

19. For more on the "rationalization" movement as well as its links to earlier scientific management ideas, see Tsutsui, *Manufacturing Ideology*, 58–89.

20. Miyamoto, "Manmō mondai to gijutsuka," 33.

21. Ōyodo, *Miyamoto*, 56.

22. Ibid., 69–70. For a historical account of the various pan-Asianist discourses, see Hotta, *Pan-Asianism and Japan's War.*

23. Miyamoto, "Manmō mondai to gijutsuka," 32.

24. Ibid., 31. On the widespread discourse of the "Manchurian lifeline," see Young, *Japan's Total Empire*, 88–95.

25. The Guandong Army held supreme power in Manchukuo and determined most of the regime's basic policy directions in addition to controlling its military affairs. The General Affairs Agency in the State Council (largely staffed by Japanese bureaucrats) was a comprehensive institution that planned, drafted, and implemented most of the Manchukuo government's policies and laws in consultation with the Guandong Army. Mantetsu was often consulted, too, not only because of its significant business holdings but also for its extensive research expertise as represented by its Economic Research Section.

26. Ōyodo, *Miyamoto*, 190.

27. Naoki, "Sōsa ni Manshū he," 3.

28. In 1922, the Manchuria Engineers Association, an organization formed by Kōjin Club and Kōseikai engineers in Manchuria, asserted a notion of "comprehensive technology" in their charter. Similar to the case in Japan, in Manchuria they were pushing for including more engineers in administrative positions. But "comprehensive technology" back then simply meant closer coordination between engineers, bureaucrats, and managers in planning and constructing development projects rather than any notion of comprehensive development and mobilization. Ōyodo, *Miyamoto*, 162.

29. McCormack, "Modernity, Water, and the Environment in Japan," 445–47.

30. Ōyodo, *Miyamoto*, 182–86. Mikuriya Takashi gives a detailed history of attempts at formulating "comprehensive" policies among Japan's conflicting ministries. Mikuriya, *Seisaku no sōgō to kenryoku*.

31. Ōyodo, *Miyamoto*, 186; Matsuura, *Senzen no kokudo seibi seisaku*, 5–9. Japan's first multipurpose dam (Ikari Dam) was planned from 1926, and construction began in 1931; however, it was not completed until after the war because of construction difficulties and wartime materials shortages. Kuwabara, "Kawa kaihatsu no shisō," 38.

32. Billington and Jackson, *Big Dams of the New Deal Era*, 71–102.

33. Rassweiler, *Generation of Power*.

34. Pietz, "Controlling the Waters in Twentieth Century China."

35. Ōyodo, *Miyamoto*, 186. The emerging idea of "comprehensiveness" in the 1930s was rather different from the Tokugawa conception of achieving an organic balance between accommodating and using water. Comprehensiveness was still largely based on total control over nature and actively shaping it to achieve optimal results for the different "functions" of flood prevention, irrigation, and electricity production (rather than adjusting to a local ecosystem's rhythms).

36. Scott et al., *Introduction to Technocracy*. For a contemporary bibliography of work on Technocracy in Japan, see "Tekunokurashī ni kansuru bunken," 47.

37. "Tekunokurashī zadankai," 23–45.

38. Ōyodo, *Miyamoto*, 208, 210.

39. Miyamoto, "Tairiku hatten to gijutsu," 115. "Scientific industry" was the philosophy of Ōkochi's Riken, an industrial combine dedicated to patenting and industrializing inventions from its own research laboratory. Scientific industry, as opposed to "capitalist industry" dedicated to short-term profit and capital preservation, "meant the continuous development of new products and production methods or equipment, and the raising of wages while lowering commodity prices to stimulate demand." High wages were necessary to increase consumer purchasing power and to motivate workers. Higher wages, Ōkochi maintained, would pay off in the long term through greater productivity and quality, and "improvements in production technology would allow firms to cut prices to meet the ultimate objective of scientific industry . . . 'low-price, high quality goods.'" Thus, scientific industry would encourage skilled, technologically advanced, productive workers who would produce not for profit but for the good of the firm *and* society as a whole (by producing cheap, high-quality goods). Cusumano, "'Scientific Industry,'" 284. On "scientific industry," see Ōkochi, *Shihonshugi kōgyō to kagakushugi kōgyō*. Ōkochi even suggested that China required "scientific industry" because Chinese workers had not sufficiently developed an "industrial spirit" after years of being paid low, exploitative wages by Western and Chinese capitalists. Ōkochi, "Hokushi no kōgyō," 68–76.

40. Miyamoto was criticizing the "rural revitalization movement," the government's employment and public works programs in response to the Great Depression during the early 1930s. On this, see Smith, *Time of Crisis*.

41. Miyamoto, "Tōhoku shinkō keikaku hihan," 65–66, 68, 70–71.

42. Ōyodo, *Miyamoto*, 120–28, 130–33. For a comprehensive study on industrial education in Japan, see Ōyodo, *Kindai Nihon no kōgyō rikkokuka*. On Japan's attempts to acquire "technological independence," see Morris-Suzuki, *Technological Transformation of Japan*, 105–42.

43. Matsuura notes that "comprehensive" public works projects officially became institutionalized in the Home Ministry around this time as well (1936–37). Matsuura, *Senzen no kokudo seibi seisaku*, 5–9.

44. Nakamura, *Senji Nihon no Kahoku keizai shihai*, 9–87.

45. Ōyodo, *Miyamoto*, 239.

46. Chapter 3 deals with some of these projects in detail.

47. Miyamoto, "Shina kaihatsu to gijutsu," 3; Ōyodo, *Miyamoto*, 239–43.

48. Miyamoto, "Shina kaihatsu to gijutsu," 1–3.

49. See the November 1937 statement of the Engineers Assembly for Establishing a Nation of Technology in Ōyodo, *Miyamoto*, 218–20.

50. Shinohara, "Kagaku no sōgōka to gijutsu no sōgōka," 21–29.

51. Shinohara, "Chōki kensetsu kokumin sai soshiki to 'gijutsu' no igi no kakujū," 31.

52. Mizuno, *Science for the Empire*, 58–59.

53. Sakai, "Subject and Substratum," 462–530. Japanese Orientalism has a longer history dating from the early Meiji era. For an analysis of the emergence of the "Orient" (*tōyō*) as a discursive object that positioned Japan as Asia's leader, see Tanaka, *Japan's Orient*.

54. Miyamoto, "Shina Manshū bekken ki," 226–27, 230. Also see Miyamoto, "Nikka shinzen to gijutsu-teki teikei," 307–16. At the time of the Oriental Engineering Congress, the Association for Civil Engineering established an "East Asia Coordination Committee" within its East Asia section to attract, guide, and keep track of exchange students from China and elsewhere as well as an "East Asian Investigation Committee" to examine further possibilities for academic exchange. Their activities ended when Chiang Kai-shek took over from the more accommodating Wang Jingwei as China's premier at the end of 1935. Ōyodo, *Miyamoto*, 200–201.

55. Miyamoto, "Nikka shinzen to gijutsu-teki teikei," 312.

56. Miyamoto, "Shina Manshū bekken ki," 225, 240–41; Miyamoto, "Ō Chō Mei to Kō Yū Nin," 119–23.

57. Miyamoto, "Nanman," 286–87.

58. Buck, *Good Earth*, vii.

59. Miyamoto, "Nanman," 286–87. "Preserve the borders and pacify the people," "Kingly Way Paradise," and "Harmony of the Five Ethnicities" were Manchukuo's official slogans, which were rooted in pan-Asianism and Confucianism. On these attempts at cultural synthesis to win popular support, see chapter 5 of Komagome, *Shokuminchi teikoku Nihon no bunka tōgō*.

60. Miyamoto, "Nanman," 298–99, 316–17.

61. Miyamoto, "Shina Manshū bekken ki," 271.

62. Miyamoto, "Nanman," 303.

63. Sakai, "Subject and Substratum," 482–86.

64. On the establishment of the Asia Development Board, see Hensha, "Hajime ni," 1–20, and Shibata, "Kōain to Chūgoku senryōchi gyōsei," 21–46.

65. Hensha, "Hajime ni," 10.

66. Mizuno, *Science for the Empire*, 56–57.

67. Hōseikyoku, *Kōa gijutsu iinkai kansei o sadamu*, August 25, 1939. Japan Center for Asian Historical Records. www.jacar.go.jp. (Accessed August 14, 2009.)

68. Ōyodo, *Miyamoto*, 262–63.

69. Kubo, "Kōain," 44–64.

70. Miyamoto, "Jogen," 1.

71. Miyamoto, "Kōa gijutsu no mittsu no seikaku," 178–79.

72. Miyamoto, "Kōa gijutsu no shidō seishin," 9–14. Zischka's works on resource scarcity, international conflict, and the role of science and technology in total war and alleviating resource dependence received much attention in Japan in the late 1930s. In addition to this work, several of his books on the global struggle for oil, rubber, cotton, and foodstuffs were translated into Japanese between 1938 and 1943.

73. Miyamoto, "Kōa gijutsu no shidō seishin," 9, 11.

74. Miyamoto, "Kōa gijutsu no konpon genri," 147.

75. On the movement for a "New Order for Science and Technology," see Sawai, "Kagaku gijutsu shintaisei kōsō no tenkai to gijutsuin no tanjō," 367–95.

76. Miyamoto, "Kōa gijutsu no konpon genri," 148, 152–53.

77. On Nazi influences, see Mimura, "Technology Bureaucrats and the 'New Order for Science-Technology,'" 110–11. For an example of Nazi-inspired work by a prominent technology bureaucrat, see Morikawa, *Natchisu Doitsu no kaibō*.

78. Miyamoto, "Kōa gijutsu no shidō seishin," 15–16.

79. Miyamoto, "Tairiku hatten to gijutsu," 118.

80. Miyamoto, "Kōa gijutsu no mittsu no seikaku," 182.

81. Miyamoto, "Kōa gijutsu no kihon seikaku," 122–24.

82. Miyamoto, "Kōa gijutsu no mittsu no seikaku," 182–83.

83. Miyamoto was citing Sun Yat Sen's famous 1924 speech in Kobe on pan-Asianism, in which he challenged Japan to be the representative of the Asian tradition of the "rule of right" rather than the Western tradition of the "rule of might." Miyamoto, "Kōa gijutsu no kihon seikaku," 124.

84. Miyamoto, "Shin tōa kensetsu to gijutsu no shimei," 196–98.

85. Miyamoto, "Kōa gijutsu no kihon seikaku," 127–28.

86. Nakamura, *Senji Nihon no Kahoku keizai shihai*, 237–49. Also, Hanwell and Bloch, "Behind the Famine in North China," 63–68.

87. Ōyodo, *Miyamoto*, 283.

88. "Hokushi sangyō gokanen keikaku an tekiyō," December 1941. Japan Center for Asian Historical Records. www.jacar.go.jp. (Accessed September 3, 2009.)

89. Ibid.; "Kahoku no sangyō kaihatsu."

90. Miyamoto, "Kōa gijutsu no konpon genri," 145.

91. Miyamoto, *Miyamoto Takenosuke nikki*, April 23, 1939 entry, 45.

92. Ibid., June 16, 1940 entry, 69.

93. Ibid., September 30, 1940 entry, 118.

94. Ōyodo, *Miyamoto*, 277–78.

95. For more cultural conceptions of pan-Asianism, see Hotta, *Pan-Asianism and Japan's War*; Skya, *Japan's Holy War*. Also see the essays in Koschmann and Saaler, *Pan-Asianism in Modern Japanese History*.

Chapter 3

1. *Nessa no chikai*. The film was the second most popular film in Japan when first released, and the most popular upon second release. Akaeda, " 'Nessa no chikai' no omoide," 143. For a more detailed reading of this film and another "construction film" (*The Green Earth*) about a Japanese canal construction project in wartime China, see Baskett, *Attractive Empire*, 33–40.

2. Akaeda, " 'Nessa no chikai' no omoide," 142–43. An agency employee, Sakaba Nobuō, helped write the scenario. Kitamura, "Kodoku no tensai," 327.

3. Ueno, "Ishizu unga no kaisō." http://www.shin-nihon.net/main/forum/unga .htm. (Accessed October 27, 2008.)

4. "Kensetsu no uta," 137.

5. Kobayashi, "Shokuminchi keiei no tokushitsu," 3–26.

6. For example, see Mizuno, *Science for the Empire*; Mimura, *Planning for Empire*.

7. Young, *Japan's Total Empire*, 241–303.

8. For example, the closely coordinated planning of flood control, electricity production, agricultural development, urban planning, and transportation improvement projects as a way to most rapidly and efficiently promote balanced economic development.

9. Schmid, "Colonialism and the 'Korea Problem,' " 958.

10. Home Ministry engineers and reform bureaucrat planners in Japan also pushed for the Tōhoku Industrial Promotion and Development Program using the same language of "comprehensive development." Approved by the Diet in 1936, this program established the Tōhoku Industrial Promotion and Development Company, which oversaw the development of enterprises ranging from vegetable oil pressing to aluminum refining and oversaw the operation of eleven hydropower stations in the region. It focused largely on improving Tōhoku's industrial and transportation infrastructure. See Dinmore, "Small Island Nation Poor in Resources," 167–70. But state engineers and planners in Japan faced much more bureaucratic jurisdictional infighting and conflict with business, which thereby hindered the development of "comprehensive technology" domestically. Thus, they increasingly turned to the empire as the "new frontier." For a detailed account of the bureaucratic battles that hindered the formation of "comprehensiveness" domestically during the 1930s and 1940s, see Mikuriya, *Seisaku no sōgō to kenryoku*.

11. Kuzuoka, "Shinkyō kōgakuin," 145–46; Nan, "Gakuchō kakka," 147–49; Satō, "Shinkyō no yūjintachi," 136.

12. "Chikoku no daihon," 34.

13. A major flood in the northern Songhua River region in 1932 also made the Guandong Army quickly pay attention to flood control from Manchukuo's very beginning.

14. "Chisui o kanete hatsuden tekichi o tosa," *Manshū nippō*," February 5, 1935. The Liao River was an essential transport artery and flowed into the important port of Yingkou. Frequent floods often disrupted transportation, communications, and economic activity.

15. Haraguchi, "Ryōga no kaishū ni tsuite 1"; Haraguchi, "Ryōga no kaishū ni tsuite 2"; Haraguchi, "Ryōga no kaishū ni tsuite 3," 390–93; "Chikoku no daihon," 33–34.

16. *Ryōga chisui keikaku shingikai gijiroku*, 6.

17. "Chikoku no daihon," 33. Ōyodo Shōichi also notes that the Liao River plan was Japan's first "comprehensive" flood control project that combined flood control and water use and therefore proved to be a valuable experience for Japanese civil engineers. Ōyodo, *Miyamoto*, 194.

18. *Ryōga chisui keikaku shingikai gijiroku*, 19.

19. The first Japanese river studies of Manchuria were based on earlier Mantetsu and Chinese ones. For example, see Haraguchi, "Manshū no kasen ni tsuite (sono ichi)," 416–23; Manshūkoku shi hensan kankōkai, *Manshūkoku shi (Kakuron)*, 197, 963; Minami Manshū tetsudō keizai chōsakai, *Manshū chisui hōsaku*. For the Russian engineer's speech, see Unikowski, "Manshūkoku no kaihatsu to chisui mondai," 205–15.

20. "Chikoku no daihon," 33.

21. Akigusa, "Kasen," 69. For more on concurrent Chinese Nationalist attempts at comprehensive river control, see Pietz, *Engineering the State*.

22. On incorporation of Chinese techniques and knowledge, see Aki, "Chūgoku no kasen ni manabu," 99–102; Hirao, "Omoide," 265–67.

23. Miyamoto, "Kōa kensetsu no suiri mondai," 103–14; Igarashi, "Ryūga chisui kōji ni tsuite," 5.

24. *Ryōga chisui keikaku shingikai gijiroku*, 17.

25. Manshūkoku shi hensan kankōkai, *Manshūkoku shi (Kakuron)*, 963.

26. Minami Manshū tetsudō keizai chōsakai, *Manshū chisui hōsaku*, 40–88.

27. *Ryōga chisui keikaku shingikai gijiroku*, 25.

28. Ibid., 49, 57, 59.

29. Ibid., 28, 30, 86. There is also evidence of tension between local residents and provincial officials as villagers argued for the construction of wider dikes, whereas the province worried about rising land compensation costs. Ibid., 84.

30. Ibid., 34–36, 37–39, 77, 81–82, 85–89, 91–92.

31. Kōtsūbu ryōga chisui chōsasho ed., *Ryōga suikei shiryō*. Also see Manshūkoku shi hensan kankōkai, *Manshūkoku shi (Kakuron)*, 967–69.

32. Kōtsubu Shōbu chisui kōteisho, *Ryūga chisui kōji no taiyō*.

33. Igarashi, "Ryūga chisui kōji ni tsuite."

34. Ibid., 9; Ryōga chisui chōsasho, "Ryōga suiri shiken no kinkyō," 24–29; Manshūkoku shi hensan kankōkai, *Manshūkoku shi (Kakuron)*, 979.

35. Manshūkoku shi hensan kankōkai, *Manshūkoku shi (Kakuron)*, 979–87. This labor statistic likely reflects total manpower required rather than total number of laborers.

36. Miyamoto, "Nanman oyobi hokushi zakkan."

37. Koshizawa, "Nihon senryōka no Pekin toshi keikaku," 266.

38. "Hokushi kensetsu sōsho ni kinin shitaru Miura Shichirō hakase to sono ichigyō," 34.

39. Henshū iin Ogawa, "Soshiki no hensen to nisseki shokuin no posuto," 7–8. The agency's total number of staff in March 1938 was 528; in April 1941, it was 1,270. Ibid., 8.

40. Horii, "Shinminkai to Kahoku senryō seisaku (chū)," 3.

41. Taniguchi, *Tairiku no kyokusen*, 97; Emori, "Kensetsu sōsho rekidai gikan ni tsuite," 14. On Yin Tong, see Tōa mondai chōsakai, *Saishin Shina yōjin den*, 2–3. After returning from studying abroad in Japan, Yin served in the Nationalist Army and later joined one of Mantetsu's coal companies. He later became a key negotiator between the Nationalists and the Guandong Army after the Manchurian Incident in brokering the infamous Tanggu Truce of 1933 that formally ended the Japanese invasion of Manchuria and resolving railway disputes between China and the new state of Manchukuo. Under the North China Provisional Government, he not only served on its political affairs committee but also headed the North China Transportation Corporation in addition to taking up various other posts.

42. For memoirs of "pacification officers" who worked in rural China, see Kōshinkai zaika gyōseki kiroku henshū iinkai, *Ōdo no gunzō*. For an introduction to the ideology and activities of the New People's Association, see Iriye, "Toward a Cultural Order," 254–74.

43. Satō and Yamazaki, "Dōro," 55; Miura, "Hokushi ni okeru doboku kensetsu jigyō."

44. KA-sei, "Miura Shichirō o mukaete," 144.

45. For more on the Yellow River flooding disasters, see Lary, "Drowned Earth," 191–207. Apparently, the main breach was never entirely fixed as a result of obstruction from the North China Expeditionary Army, which gained more administrative territory as a result of the Yellow River's course shift southward at the expense of Japan's Central China Expeditionary Army. Tabuchi, *Aru doboku gishi no han jijoden*, 149.

46. Miura, "Hokushi ni okeru doboku jigyō no gaiyō," 1385; Akigusa, "Kasen," 68.

47. On specific agency projects, see Kōain, *Dokuryū nyūkai genka shori hōsaku* (September 1941); "Kōga shori yōkō (an)—kyokumitsu." For a series of anecdotes on these projects and how these ideas arose in China, see *KB*, 67–106.

48. Miura, "Hokushi ni okeru doboku jigyō," 1387.

49. Shiobara, *Toshi keikaku ten to sen*, 5; Koshizawa, "Nitchū sensō ni okeru senryōchi toshi keikaku ni tsuite," 386. On Satō's career, see Koshizawa, *Harupin no toshi keikaku*. For an overview of Japanese urban planning activities in Manchuria, see Koshizawa, *Shokuminchi Manshū no toshi keikaku*. On the design philosophies of Manchukuo's urban planners, see Tucker, "Building 'Our' Manchukuo." I discuss some of the relationships between urban planning and what would become known as "national land planning" (*kokudo keikaku*) during the war later in this chapter.

50. For an introduction to the Garden City and City Beautiful movements, see Hall, *Cities of Tomorrow*, 87–141, 188–217.

51. Urban planners in Manchuria did try to incorporate what they saw as indigenous ideals of the "Kingly Way" and "ethnic harmony." For example, the plans for Manchukuo's capital included a palace residence for its chief executive and emperor, Pu Yi. However, the Capital Construction Bureau continuously clashed with Mantetsu planners about its location, with the bureau insisting on a south-facing imperial palace along the lines of Beijing's Forbidden City and Mantetsu arguing for a palace smoothly integrated into the modern city. Ultimately, they agreed on a site for the palace in the western part of the city and decided to build a temporary palace in the center. This central site soon became the permanent one, and construction began in 1938 but was halted in 1943 because of wartime materials restrictions. All this time, Pu Yi lived in a small, cramped residence outside the urban planning area in an industrial neighborhood near the railway. Upon his coronation in 1933, the Capital Construction Bureau hastily built an outdoor Temple of Heaven modeled after the one in Beijing at the empty palace site. In the end, the imperial residence project took a back seat to the modernist aspects emphasizing "civilization" and "progress." Planners of Manchukuo's capital also implemented what Tucker calls the "ritual Japanization of space" by building monuments and shrines firmly linked to Japan's war efforts and state Shintō. Tucker, "Building 'Our' Manchukuo," 361–63, 394–409.

52. Pekin Nihon shōkō kaigisho, *Pekin Nihon shōkō meikan*, 119.

53. Tucker, "Building 'Our' Manchukuo," 211, 376.

54. Shiobara, *Ten to sen*, 5, 83.

55. Shiobara, "Toshi," 109.

56. Shiobara, *Ten to sen*, 6, 83, 86–87.

57. Ibid., 39. The population of the Beijing area in 1941 was around 1,500,000. See Kōain, *Pekin toshi keikaku gaiyō*, 7–10 for more detailed population statistics and breakdowns by ethnicity.

58. Kōain, *Pekin toshi keikaku gaiyō*, 18–19.

59. Koshizawa, "Nihon senryōka no Pekin toshi keikaku," 268. Japanese residents and companies would purchase or rent old Chinese houses and clumsily renovate them into hybrid Japanese-Chinese residences, causing problems with Chinese landlords and residents. Sosiroda, "Hokushi manmō ni okeru Nihonjin no jyūkyo kenchiku kōzō chōsa hōkoku," 32–38.

60. Shiobara, *Ten to sen*, 8.

61. Koshizawa, "Nihon senryōka no Pekin toshi keikaku," 268.

62. Kōain, *Pekin toshi keikaku gaiyō*, 2, 19, 28, 35; Shiobara, *Ten to sen*, 7, 9–10; Pekin Nihon shōkō kaigisho, *Pekin Nihon shōkō meikan*, 124.

63. Kōain, *Pekin toshi keikaku gaiyō*, 19.

64. Koshizawa, "Nihon senryōka no Pekin toshi keikaku," 270; Kensetsu sōsho toshi kyoku, *Pekin toshi keikaku yōzu*.

65. Shiobara, "Toshi," 113.

66. Kōain, *Pekin toshi keikaku gaiyō*, 16, 27–37.

67. Pekin Nihon shōkō kaigisho, *Pekin Nihon shōkō meikan*, 124–25; Shiobara, *Ten to sen*, 10–11; Kōain, *Pekin toshi keikaku gaiyō*, 37. Wartime urban planning in China and Manchukuo was also closely tied to the historical experience with urban planning in Japan, particularly after the 1923 Great Kanto Earthquake. On the history of urban planning in Japan proper, see Hanes, *City as Subject*.

68. Tucker, "Building 'Our' Manchukuo," 227.

69. Shiobara, *Ten to sen*, 42.

70. Hokushina hōmengun sanbōchō, "Pekin seikō shin shigai kensetsu keikaku ni kan suru ken." Japan Center for Asian Historical Records. www.jacar.go.jp. (Accessed July 23, 2009.)

71. Kōain, *Pekin toshi keikaku gaiyō*, 24–25; Shiobara, *Ten to sen*, 43–44.

72. Beijing was placed in a proposed Beijing-Tianjin Unified Region that included Tangshan and Cangzhou. Beijing would be the political and cultural center; Tangshan, an industrial center surrounded by abundant mineral resources; Tianjin, a main shipping and trading city; and Cangzhou, an agricultural center of the proposed region. Shiobara, *Ten to sen*, 109–15.

73. Shiobara, "Kahoku ni okeru toshi kensetsu," *Toshi kōron* 27, no. 6 (1944): 32.

74. Shiobara, *Ten to sen*, 84–85.

75. Satō, "Hokushi ni okeru hōjin jūtaku mondai no bekken," 731.

76. Ibid., 729; Shiobara, "Kahoku ni okeru toshi kensetsu," *Toshi kōron* 27, no. 8 (1944): 23. In Tianjin, having five to six Japanese in one room was not rare. The average number of Japanese per household in Beijing was 3.4. Often one "household" consisted of single men rather than families, which contributed to the social problems. See Ōmura, "Zai tairiku hōjin," 762; Satō, "Hokushi ni okeru hōjin jūtaku mondai," 732. Twenty-six percent of Japanese residents in Beijing shared houses with Chinese. Ōmura, "Zai tairiku hōjin," 762. The male-female ratio among Japanese in Beijing was 61.3 percent to 38.7 percent. Satō, "Hokushi ni okeru hōjin jūtaku mondai," 733. On insensitivity and the need to educate Japanese, see Ōmura, "Zai tairiku hōjin," 763.

77. Ōmura, "Zai tairiku hōjin," 762; Satō, "Hokushi ni okeru hōjin jūtaku mondai," 732.

78. Koshizawa, "Nihon senryōka no Pekin toshi keikaku," 268.

79. Kōain, *Pekin toshi keikaku gaiyō*, 27–28.

80. Shiobara, "Kahoku ni okeru toshi kensetsu," *Toshi kōron* 27, no. 8 (1944): 23.

81. Kōain, *Pekin toshi keikaku gaiyō*, 28.

82. Pekin Nihon shōkō kaigijo, *Pekin Nihon shōkō meikan*, 124.

83. On ethnic separation in Manchuria, see Perrins, "Doctors, Diseases, and Development," 108–9. The Russians also created separate neighborhoods in their earlier urban planning projects in Harbin and Dalian. Koshizawa, *Shokuminchi Manshū no toshi keikaku*, 61.

84. Tucker, "Building 'Our' Manchukuo," 380.

85. On the concept of "hygiene towns," see Shiobara, *Ten to sen*, 8, 11. According to a study, 65 percent of Japanese residents in Beijing's walled city had to buy their water on

the street and 72 percent used Chinese-style, nonflushable toilets. Ōmura, "Zai tairiku hōjin," 762.

86. Shiobara, "Kahoku ni okeru toshi kensetsu," *Toshi kōron* 27, no. 8 (1944): 24.

87. Those Chinese who applied to reside in the New Town had to be of some means, as housing prices were out of the range of most. The discourse of hygienic separation clearly had a class aspect as well, mainly targeting Chinese day laborers and the lower classes.

88. Inose, "Pekin Tōseikō," 116–17.

89. Shiobara, "Kahoku ni okeru toshi kensetsu," *Toshi kōron* 27, no. 6 (1944): 32–33. In China and Manchukuo, "special budgets" also included funds from a range of illegal activity, such as the opium trade, which financed a large portion of the Guandong Army's industrialization programs. See Driscoll, *Absolute Erotic, Absolute Grotesque*, 227–62.

90. Inose, "Pekin Tōseikō," 116.

91. Yamasaki, "Sen kyūhaku sanjūhachi nen no Kahoku toshi keikaku no omoide," 298.

92. For a sample ad, see Shiobara, *Ten to sen*, 103–8.

93. Inose, "Pekin Tōseikō," 117.

94. Ibid., 116.

95. In Suzhou, another city where they conducted urban planning, the agency branch worked with the army and arbitrarily assigned prices to housing classified as "lower, middle, or upper," then proceeded to compensate and transfer the residents. Umezawa, "Soshū no omoide," 167.

96. Nakamura, "Kahoku no omoide no ki," 247.

97. Minami Manshū tetsudō kabushiki kaisha chōsabu, *Shina toshi fudōsan kankō chōsa shiryō*, 1–36, 70–72. The report also noted many "uncertainties" in the whole process, such as the amounts of the various fees and how exactly to coordinate and time the transfer from original landowner to lessee. Ibid., 21. In Tanggu, officials were faced with unclear property records and land registers, and cases where Japanese used Chinese intermediaries in the past to acquire land. See Minami Manshū tetsudō kabushiki kaisha chōsabu, *Tōko ni kansuru hōkokusho*, 97–98.

98. For examples of how Chinese labor was used in agency projects, see Iwamiya, "Sainan no omoide," 155; Kohiro, "Hotei sekōjo, jimukanjo tōji no omoide," 194.

99. Orisaka, "Pekin shi kōcho kōmukyoku," 40.

100. Yin Tong, North China Construction Agency head, was the first president of the North China Labor Association; however, he died in 1942 soon after its establishment. For an overview of labor recruitment in North China, see Ju, "Northern Chinese Laborers and Manchukuo," 61–80.

101. On construction problems, see Satō, "Hokushi ni okeru hōjin jūtaku mondai," 731. Inflation rose from a price index of 261.85 in December 1939 to 354.12 in May 1940.

102. "Rikushi mitsuden gojūgo-go." Japan Center for Asian Historical Records. www.jacar.go.jp. (Accessed July 23, 2009.)

103. Orisaka, "Pekin shi kōcho kōmukyoku," 38–40.

104. Inose, "Pekin Tōseikō," 117–18; Koshizawa, "Nitchū sensō ni okeru senryōchi toshi keikaku ni tsuite," 388. North China House Building Corporation (*Hokushi bōzan kabushiki gaisha*) constructed three residential blocks for the North China Construction Agency, the Japanese Residents Association, and International Transportation Corporation employees by May 1941. Eight different types of houses between 12.2 and 45.5 tsubo (1 tsubo = 3.3 square meters) were constructed with those over 22 tsubo including Western living rooms. Tomii, "Pekin ni okeru kyū Nihonjin jūtakuchi ni kansuru kenkyū," 5–8.

105. Chapter 4 analyzes Sup'ung Dam in more detail.

106. Son, *Nihon tōchika no Chōsen toshi keikakushi kenkyū*, 175–76; Yang, "Japanese Colonial Infrastructure in Northeast Asia," 99–100.

107. Sugai, *Kokudo keikaku no keika to kadai*, 2.

108. Suzuki, "Chōsen ni okeru toshi keikaku to kokudo keikaku," 5–7.

109. Ibid., 8–9.

110. Kuroda, "Futōkō 'Daitōkō' ga mondai ni naru made," 11–12.

111. Kōtsūbu daijin kanbō shiryōka, *Kotsūbu yōran*, 222.

112. Kuroda, "Futōkō 'Daitōkō' ga mondai ni naru made," 12–13.

113. Ibid.; Manshūkoku shi hensan kankōkai, *Manshūkokushi (Kakuron)*, 995.

114. Kuroda, "Futōkō 'Daitōkō' ga mondai ni naru made," 15.

115. Kōtsūbu daijin kanbō shiryōka, *Kotsūbu yōran*, 241.

116. Naoki, "Daitōkō futō jōkyo shisatsu kōen, 2; Manshū jijō annaijo, *Daitōkō to hōko Tōhendō*, 34.

117. Kuroda, "Futōkō 'Daitōkō' ga mondai ni naru made,"14; Kōtsūbu daijin kanbō shiryōka, *Kotsūbu yōran*, 222.

118. "Ōryokkō guchi ni okeru Sen-man chikukō to toshi keikaku," 59.

119. Koshizawa, "Daitōkō no keikaku to kensetsu," 227.

120. Yamada, "Chihō keikaku ni tsuite," 3.

121. Kōtsūbu daijin kanbō shiryōka, *Kotsūbu yōran*, 224.

122. Ibid.; Yokoyama, "Daitōkō toshi kensetsu jigyō to enchi," 111–13.

123. Yoneda, "Daitōkō kensetsu jigyō ni tsuite," 49–50.

124. Manshū jijō annaijo, *Daitōkō to hōko Tōhendō*, 45. Ford's application was approved by the Dadong Port Construction Agency but rejected by the Guandong Army. Koshizawa, "Daitōkō no keikaku to kensetsu," 232.

125. Manshū jijō annaijo, *Daitōkō to hōko Tōhendō*, 45.

126. Kōtsūbu daijin kanbō shiryōka, *Kotsūbu yōran*, 225; Yokoyama, "Daitōkō toshi kensetsu jigyō to enchi," 114.

127. Yokoyama, "Daitōkō toshi kensetsu jigyō to enchi," 114–16; Kōtsūbu daijin kanbō shiryōka, *Kotsūbu yōran*, 228.

128. Kōtsūbu daijin kanbō shiryōka, *Kotsūbu yōran*, 236; Tochika keikaku mata, "Daitōkō toshi keikaku no naiyō," 80.

129. Manshū jijō annaijo, *Daitōkō to hōko Tōhendō*, 44.

130. Kōtsūbu daijin kanbō shiryōka, *Kotsūbu yōran*, 228.

131. Tochika keikaku mata, "Daitōkō toshi keikaku no naiyō," 82.

132. Yokoyama, "Daitōkō toshi kensetsu jigyō to enchi," 116.

133. Kōtsūbu daijin kanbō shiryōka, *Kotsūbu yōran*, 223.

134. "Ōryokkō guchi ni okeru Sen-man chikukō to toshi keikaku," 42.

135. Kuroda, "Futōkō 'Daitōkō' ga mondai ni naru made," 9–11.

136. Naoki, "Daitōkō futō jōkyō shisatsu kōen," 6; Manshū jijō annaijo, *Daitōko to hōko Tōhendō*, 48.

137. "Ōryokkō guchi ni okeru Sen-man chikukō to toshi keikaku," 67.

138. Naoki, "Daitōkō futō jōkyō shisatsu kōen," 4.

139. Antō shi kōsho, *Yakushin Antō*, 61.

140. Kōtsūbu daijin kanbō shiryōka, *Kotsūbu yōran*, 226, 233.

141. Tochika keikaku mata, "Daitōkō toshi keikaku no naiyō," 53; Kōtsūbu daijin kanbō shiryōka, *Kotsūbu yōran*, 230.

142. Kōtsūbu daijin kanbō shiryōka, *Kotsūbu yōran*, 230–32.

143. Koshizawa, "Daitōkō no keikaku to kensetsu," 231–35. The rental and sale revenue was to be used to pay off the bonds issued for construction. Manshū jijō annaijo, *Daitōkō to hōko Tōhendō*, 36–38.

144. Various other projects were also begun, such as flood control, basic water provisioning, and park formation. Kōtsūbu daijin kanbō shiryōka, *Kotsūbu yōran*, 230–41.

Chapter 4

1. Hoshino, "Honma Norio-kun," 55.

2. 'Manshū dengyō shi' henshū iinkai, *Manshū dengyō shi*, 578–80 (hereafter, *MDS*); Manshū jijō annaijo, *Kitsurin jijō*, 199 (hereafter, *Kitsurin jijō*); Uchida, *Hōman damu*, 3, 13 (hereafter, *Hōman damu*).

3. *Kitsurin jijō*, 145.

4. *Hōman damu*, 7, 11, 42–44; Honma, "Manshūkoku suiryoku denki jigyō ni tsuite," 35. One *koku* of rice is 278.3 liters, or what was considered to be enough to feed one person for a year.

5. *MDS*, 575; Yamamoto, "Honma moto suiden kyokucho o shinobu," 146; *Hōman damu*, 3.

6. For a record of the many different visitors to the dam site, see Kitsurin shōkō kōkai, *Kitsurin shōkō kōkai jigyō hōkokusho 1940*, 96–104. Kitsurin shōkō kōkai, *Kitsurin shōkō kōkai jigyō hōkokusho 1941*, 82–88.

7. 'Suihō damu shōjo hatsu sōden," *Keijō nippō*, August 26, 1941; Harada, *Suihō hatsudenjo kōji taikan*, 15.

8. Kodaira, *Shingishū shōkō yōran shōwa jūshichi nenkan*, 130. For more on these projects, see Koshizawa, "Daitōkō no keikaku to kensetsu," 223–34, and Son, *Nihon tōchika no Chōsen toshi keikakushi kenkyū*, 175–212.

9. Chūō Nikkan kyōkai, *Chōsen denki jigyōshi*, 415, 535; Hazama-gumi hyakunen-shi hensan iinkai, *Hazama-gumi hyakunen shi*, 636. The figure of twenty-five million is the statistic for total manpower used over the course of construction, not the sum total of laborers.

10. Mimura, *Planning for Empire*; Mizuno, *Science for the Empire*.

11. Young, *Japan's Total Empire*, 303.

12. Fukai, *Minamatabyō no seiji keizaigaku*, 79–80.

13. Harada, *Suihō hatsudenjo kōji taikan*, 22, 63–64, 84, 120, 123, 125–26; Kubota, "Watakushi no rirekisho," 241–321. Molony, *Technology and Investment*, also largely centers the story of Nitchitsu in Korea on Noguchi, Kubota, and other managers. For a similar type of account on Fengman Dam, see *Hōman damu*.

14. For an overview of Nitchitsu's development in Korea, see Kan, *Chōsen ni okeru Nitchitsu kontserun*, 81–107.

15. Ibid., 23, 96, 100, 145, 157, 162.

16. Quoted in Molony, *Technology and Investment*, 156.

17. White, *Organic Machine*, 76. For other work that focuses on how environments are "rationalized," see Mitchell, *Rule of Experts*, 19–53.

18. Korea's first hydropower study only measured minimum streamflow to ensure constant, stable electricity production for "conduit-style" dams, which diverted river flow over steep inclines through large pipes rather than storing and controlling water through large reservoirs and high dams. The second study was essential for its "paradigm shift" in reading the environment, and its conclusions of abundant and cheap electricity spurred the wave of high dam construction and surge in heavy industrialization in 1930s Korea. By the end of the 1930s, 82 percent of Korea's electricity came from hydropower, as opposed to 56 percent for Japan. Kawai, "Dai niji suiryoku chōsa," 304. The study's value also lies in its descriptions of how Japanese engineers measured and negotiated the environment instead of simply providing data. The third hydropower study (1936–41) conducting more detailed studies of the Yalu River, and therefore more directly relevant to Sup'ung Dam, has been lost.

19. Chōsen sōtokufu teishinkyoku, *Chōsen suiryoku chōsasho*, 15, 22, 171. Similar studies of the Fengman Dam site that exhibit this effect of transparency can be found in Suiryoku denki kensetsu kyoku, *Manshūkoku suiryoku dengen chōsa nenpō*.

20. Chōsen sōtokufu teishinkyoku, *Chōsen suiryoku chōsasho*, 12, 75, 161, 183, 194–96; Kawai, "Dai niji suiryoku chōsa," 314–15.

21. Dams built before Sup'ung Dam in Korea were of the "conduit type," although they were much larger than those built in Japan.

22. 'Honma-san suiden o kataru," 14. Previous Mantetsu and Communications Department studies of the Songhua River suggested that dams would only have the capacity to produce 30,000 to 50,000 kilowatts of sustained hydropower.

23. Hoshino, "Honma Norio-kun," 46–49.

24. Manshūkoku shi hensan kankōkai, *Manshūkokushi (Kakuron)*, 1062.

25. *MDS*, 151–52; Hoshino, "Honma Norio-kun," 49.

26. *MDS*, 238. The first period aimed at establishing basic political and economic institutions and creating an independent economy, and the second aimed at the formation of a national defense economy subordinated to Japan. Mimura, *Planning for Empire*, 94–95.

27. Manshūkoku shi hensan kankōkai, *Manshūkokushi (Kakuron)*, 971–74.

28. Nan, "'Manshūkoku' ni okeru Nihon no shokuminchi tōchi," 113.

29. *MDS*, 362.

30. Naoki, "Manshū no suiryoku denki jigyō," 4.

31. Hori, "'Manshūkoku' ni okeru denryokugyō to tōsei seisaku," 18. The Manchukuo Five-Year Industrial Plan focused heavily on increasing mining and metals production but also included plans for improving agricultural productivity and transportation infrastructure and increasing emigration. For more on the Five-Year Plan and other aspects of economic policy, see Nakagane, "Manchukuo and Economic Development."

32. *MDS*, 252–57.

33. Sunaga, "Manshū ni okeru denryoku jigyō," 97–98.

34. Manshūkoku shi hensan kankōkai, *Manshūkokushi (Kakuron)*, 1063–66.

35. *Kitsurin jijō*, 154–55.

36. "Honma-san suiden o kataru," 14–15.

37. Hoshino, "Honma Norio-kun," 52; Manshūkoku shi hensan kankōkai, *Manshūkokushi (Kakuron)*, 1068.

38. In 1921, American engineers who built the world-famous Arrowrock Dam advised Japanese engineers in the construction of Komaki Dam. Japanese civil engineers also participated in the World Dam Congresses of 1933 (Stockholm) and 1936 (Washington, D.C.) and were regular visitors to most of the major dams in Europe and the United States in the 1920s and 1930s. For information on the development of Japanese dam-building expertise in relation to the United States, see Hirose, "Gunju keiki to denryoku kensetsu kōji."

39. Ibid., 142–43; Matsuura, "Konkureeto damu ni miru senzen no sekō gijutsu," 570–71.

40. Kuga, *Beikoku ni okeru kōentei shisatsu hōkokusho*, 3, 11, 14–15, 23–26, 34–36; Matsuura, "Konkureeto damu ni miru senzen no sekō gijutsu," 575.

41. For more on the New Order for Science and Technology, see Mizuno, *Science for the Empire*, 60–68. For more on various Japanese efforts at technological autonomy during the war, see Morris-Suzuki, *Technological Transformation of Japan*, 143–60; Yang, *Technology of Empire*; Grunden, *Secret Weapons and World War II*.

42. *Kitsurin jijō*, 144–46; Manshū denki kyōkai, *Shōkakō suiryoku hatsuden keikaku gaiyō*, 27; Yamamoto, "Honma moto suiden kyokucho," 147.

43. Manshūkoku tsūshinsha kaisha, "Shōkakō hatsudenjo kensetsu no kushin o kiku zadankai," 37; "Honma-san suiden o kataru," 15.

44. Manshū denki kyōkai, *Shōkakō suiryoku hatsuden keikaku gaiyō*, 35–36; "Hōman damu," 58–60.

45. As a result of the war, generator and turbine deliveries from abroad were either delayed or blocked. The Swiss order made it to New York in May 1940 but was turned away as a result of the ban on Japanese ships through the Panama Canal—only one large turbine and two internal turbines were delivered by way of the Trans-Siberian Railway. Three generators were delivered from the United States, and the German order arrived in

three separate shipments. One German generator and turbine was transported through Siberia in 1939; a camouflaged destroyer secretly delivered another pair in 1941; and the last delivery of another pair by destroyer in 1942 survived an English air attack en route along the coast of Africa. Nan, "'Manshūkoku' ni okeru Nihon no shokuminchi tōchi," 130; *MDS*, 581.

46. 'Honma-san suiden o kataru," 15.

47. Nan, "'Manshūkoku' ni okeru Nihon no shokuminchi tōchi," 130.

48. Manshūkoku suiryoku denki kensetsu kyoku, *Shōkakō dai ichi hatsudenjo kōji shashinchō*; *Hōman damu*, 58–59.

49. 'Honma-san suiden o kataru," 15.

50. Sup'ung Dam also faced fierce institutional conflict as the Manchukuo government, Korea Government-General, and Nitchitsu disagreed over such issues as project feasibility, private versus public management, industrial policy, electricity rates, and electricity cycles. See Moore, "Yalu River Era of Developing Asia."

51. Hirose, "Chōsen sōtokufu no doboku kanryō," 284.

52. Honma, "Manshūkoku suiryoku denki jigyō," 31; "Honma Norio ni taisuru—tōjitsu no shitsugi ōtō sokkiroku," 42.

53. *MDS*, 577.

54. Hydrological studies around the Fengman Dam site can be found in Suiryoku denki kensetsu kyoku, *Manshūkoku suiryoku dengen chōsa nenpō*.

55. The highest recorded stream flow was ten thousand cubic meters per second. *MDS*, 578.

56. Nan, "'Manshūkoku' ni okeru Nihon no shokuminchi tōchi," 124.

57. Yamamoto, "Honma moto suiden kyokucho," 138.

58. *MDS*, 589–91.

59. 'Honma-san suiden o kataru," 15.

60. Yamamoto, "Hōman chosuichi," 16–17.

61. Nan, "'Manshūkoku' ni okeru Hōman suiryoku hatsudenjo no kensetsu," 12, 14–15.

62. 'Honma-san Suiden o kataru," 16–17.

63. For more on this point about "hybrid expertise" in a different context, see Mitchell, *Rule of Experts*, 37.

64. Nakayama, *Yakushin Kitsurin*, 8

65. Kitsurin shōkō kōkai, *Suiryoku hatsuden*, 8–18.

66. Ibid., 51, 160–61.

67. *Kitsurin jijō*, 134–35, 167–77, 179–80.

68. Ibid., 138; Nakayama, *Yakushin Kitsurin*, 29–30.

69. Nakayama, *Yakushin Kitsurin*, 34, 39–43.

70. 'Honma-san suiden o kataru," 18.

71. *Kitsurin jijō*, 186–92; Matsumoto, *Kitsurin ringyōshi*, 266. In 1940, the Jilin Chamber of Commerce, in cooperation with the provincial and city government, the county governments of Huadian, Jiaohe, and Bingji Counties, the Hydropower Construction

Bureau, Northern Manchuria Transport Company, Harbin's Shipping Bureau, Manchukuo's Forestry Department, and Manchuria Development Company launched a "comprehensive investigation" into the dam's effects on transportation, fishing, and commerce. They planned to release a limited edition pamphlet on their study, which has been lost. See Kitsurin shōko kōkai, *Kitsurin shōkō kōkai jigyō hōkoku sho 1940*, 83–84.

72. Nakayama, *Yakushin Kitsurin*, 11.

73. *Kitsurin jijō*, 192.

74. Kitsurin shōkō kōkai, *Suiryoku hatsuden*, 3.

75. Matsumoto, *Kitsurin ringyōshi*, 258; *Kitsurin jijō*, 152. Honma briefly described "naïve" and superstitious Chinese residents who agreed to sell their lands because they did not believe the Japanese could successfully complete the project, which would therefore enable them to keep their lands as well as the already paid compensation money. They also thought that the resident water gods would eventually smash the structure to pieces. "Honma-san suiden o kataru," 15; Manshūkoku tsūshinsha kaisha, "Shōkakkō hatsudenjo kensetsu no kushin," 34–35.

76. Fukai, *Minamatabyō no seiji keizaigaku*, 89–92.

77. Chūō Nikkan kyōkai, *Chōsen denki jigyōshi*, 280, 282, 305–6; Hirose, "Shokuminchiki Chōsen ni okeru Suihō hatsudenjo kensetsu," 42.

78. Nōrinkyoku iin, "Ryūbatsu taisaku" (1938?), in *Ōryokkō kaihatsu iinkai kankei*, 247–48. These papers are hereafter referred to as *OKIK*.

79. Chūō Nikkan kyōkai, *Chōsen denki jigyōshi*, 297.

80. Nōrinkyoku iin, "Ryūbatsu taisaku" (1938?), in *OKIK*, 247–48.

81. Nōrinkyoku iin, "Ryūbatsu taisaku" (1939?), in *OKIK*, 142–43.

82. Nōrinkyoku iin, "Ryūbatsu taisaku" (1941?), in *OKIK*, 25–26.

83. Hirose, "Shokuminchiki Chōsen ni okeru Suihō hatsudenjo kensetsu," 47.

84. Ibid., 48.

85. 'Suihō damu chikuzō to tsūbatsu shisetsu mondai," 11.

86. 'Suihō damu chikuzō to Ōryokkō ryūbatsu mondai," 28.

87. Hirose, "Shokuminchiki Chōsen ni okeru Suihō hatsudenjo kensetsu," 50.

88. Ibid.; "Ryūbatsu mondai masu masu jūdaika," 10–11.

89. 'Ryūbatsu mondai masu masu jūdaika," 11.

90. Hirose, "Shokuminchiki Chōsen ni okeru Suihō hatsudenjo kensetsu," 50–51.

91. Hyōya, "Ōryokkō mokuzaikai no tenbō," 17–18; Hirose, "Shokuminchiki Chōsen ni okeru Suihō hatsudenjo kensetsu," 52.

92. Kang, "Parabononi, simsan yugok—Chŏnch'ang chungsŏkkwang chidae," *Mansŏn ilbo*, January 29, 1940.

93. Kang, "Hŏmsan chullyŏng, munjŏn oktap, ilcho e sugukhwa rani hanbat'ang kkum iroda," *Mansŏn ilbo*, January 28, 1940.

94. Ibid.

95. Kang, "Sumolchi rŭl ttŏnanŭn tongp'o yŏ," *Mansŏn ilbo*, February 1, 1940.

96. Harada, *Suihō hatsudenjo kōji taikan*, 105.

97. Chūō Nikkan kyōkai, *Chōsen denki jigyōshi*, 298–99; Hirose, "Manshūkoku ni okeru Suihō damu no kensetsu," 8.

98. Hirose, "Suihō hatsudenjo kensetsu ni yoru suibotsuchi mondai," 19–20; Kobayashi, *Dai tōa kyōeiken no keisei to hōkai*, 347–52. Unlike Kobayashi, Hirose discusses the various legal and administrative maneuvers employed during the dam's construction.

99. Hirose, "Chōsen ni okeru tochi shūyō rei," 2–9.

100. Heianhokudō iin, "Tochi kaishū taisaku" (1938?), in *OKIK*, 217–18.

101. Tanaka, "Ōryokkō suiden kaisha no yōchi kaishū," 391.

102. Heianhokudō iin, "Tochi kaishū taisaku" (1938?), in *OKIK*, 219.

103. Tanaka, "Ōryokkō suiden kaisha no yōchi kaishū," 393; Heianhokudō iin, "Tochi kaishū taisaku" (1938?), in *OKIK*, 221, 223.

104. Tanaka, "Ōryokkō suiden kaisha no yōchi kaishū," 394–95.

105. Heianhokudō iin, "Tochi kaishū taisaku" (1938?), in *OKIK*, 219.

106. Ibid., 221.

107. Hirose, "Suihō hatsudenjo kensetsu ni yoru suibotsuchi mondai," 10–12.

108. Heianhokudō iin, "Tochi kaishū taisaku" (1938?), in *OKIK*, 219–20, 223.

109. Hirose, "Suihō hatsudenjo kensetsu ni yoru suibotsuchi mondai," 17; "Ōryokkō Suiden suibotsuchi happyō saru," 92–94.

110. Hirose, "Suihō hatsudenjo kensetsu ni yoru suibotsuchi mondai," 18.

111. "Ōryokkō suiden suibotsuchi happyō saru," 91.

112. Hirose, "Suihō hatsudenjo kensetsu ni yoru suibotsuchi mondai," 23, 25; Gaimubu iin, "Suibotsuchi jūmin no shori" (1938?), in *OKIK*, 228; Naimukyoku iin, "Tochi kaishū taisaku" (1941?), in *OKIK*, 31.

113. Park, *Two Dreams in One Bed*, 162–97.

114. Hirose, "Suihō hatsudenjo kensetsu ni yoru suibotsuchi mondai," 19.

115. Kang, "Paeksŏl ŭl kŏdŏch'ago paesuryŏng ŭl nŏmŏsŏni Hwangbuk' en odumach'a pangŭl sori yoranhago chŏgi chigetkun ŭi susimga hŭnggyŏpta," *Mansŏn ilbo*, January 27, 1940.

116. Naimukyoku iin, "Tochi kaishū taisaku" (1941?), in *OKIK*, 32.

117. Harada, *Suihō hatsudenjo kōji taikan*, 59.

118. Kitsurin kōteikyoku, "Shōkakkō entei hatsuden kōji gaikyō," 210; Manshūkoku shi hensan kankōkai, *Manshūkokushi (Kakuron)*, 584.

119. Eda, Kai, and Matsumura, *Mantetsu rōdōshi no kenkyū*, 219–20.

120. Tucker, "Building 'Our' Manchukuo," 306–10. Since Mantetsu began contracting labor for construction projects in 1907, they employed the flexible *batou* system, whereby Chinese foremen were paid a lump sum to transport, pay, and maintain workers for short periods of time. This system continued into the Manchukuo era but became more centralized through the Datong Company. See Tucker, "Building 'Our' Manchukuo," 295–97. For more on the Manchukuo labor procurement system, see Tucker, "Labor Policy and the Construction Industry in Manchukuo."

121. Tucker, "Building 'Our' Manchukuo," 319, 330, 334–35.

122. Manshūkoku tsūshinsha kaisha, "Shōkakō hatsudenjo kensetsu no kushin," 32–33.

123. Tagami, "Nidai suiden kōji o miru 2," *Manshū nichi nichi shinbun*, September 22, 1938.

124. Nan, "'Manshūkoku' ni okeru Hōman suiryoku hatsudenjo," 6.

125. Kitsurin kōteikyoku, "Shōkakō entei hatsuden kōji gaikyō," 210.

126. *Hōman damu*, 47–48.

127. Manshūkoku tsūshinsha kaisha, "Shōkakō hatsudenjo kensetsu no kushin," 34.

128. *Hōman damu*, 62–64.

129. Tagami, "Nidai suiden kōji o miru 2."

130. One hundred sen equals one yen. Manshūkoku tsūshinsha kaisha, "Shōkakō hatsudenjo kensetsu no kushin," 33. See *Hōman damu*, 68, for 1939 statistics on worker pay.

131. *Hōman damu*, 4, 65, 67.

132. Nan, "'Manshūkoku' ni okeru Hōman suiryoku hatsudenjo," 6.

133. Manshūkoku tsūshinsha kaisha, "Shōkakō hatsudenjo kensetsu no kushin," 34.

134. Nan, "'Manshūkoku' ni okeru Nihon no shokuminchi tōchi," 138.

135. Honma only mentioned deaths from falling and dynamite accidents and praised the fact that there were no deaths from floods or bandit attacks. "Honma-san suiden o kataru," 15–16. In 1970, a burial pit was discovered near Sup'ung Dam containing around one thousand bodies, some tied together with wire, handcuffed, or with bayonet wounds in their chest. Wen, "Genba de no shōgen," 51–52. Similar pits were discovered at Feng-man Dam. On the "holes in the ground filled with ten thousand bodies" (*manninkō*) at Fengman, see "Hōman damu manninkō." http://www.ac.auone-net.jp/~miyosi/db3.htm. (Accessed July 27, 2012.)

136. Heianhokudō iin, "Keibi oyobi hōan torishimari taisaku" (1938?), in *OKIK*, 235.

137. Chūō Nikkan kyōkai, *Chōsen denki jigyōshi*, 418.

138. Heianhokudō iin, "Keibi oyobi hōan torishimari taisaku" (1938?), in *OKIK*, 235–36.

139. Keimukyoku iin, "Keibi oyobi hōan torishimari taisaku" (1939?), in *OKIK*, 133.

140. Heianhokudō iin, "Keibi oyobi hōan torishimari taisaku" (1938?), in *OKIK*, 236; Keimukyoku iin, "Keibi oyobi hōan torishimari taisaku" (1939?), in *OKIK*, 135.

141. Oida, "Shiryō: 'Shūdan' buraku ni tsuite," 86–88.

142. Kajimura, "Sen kyūhaku sanjū nendai Manshū ni okeru kōnichi tōsō," 35.

143. Heianhokudō iin, "Keibi oyobi hōan torishimari taisaku" (1938?), in *OKIK*, 239.

144. Harada, *Suihō hatsudenjo kōji taikan*, 89.

145. Keimukyoku iin, "Keibi oyobi hōan torishimari taisaku" (1939?), in *OKIK*, 131.

146. Heianhokudō iin, "Keibi oyobi hōan torishimari taisaku" (1938?), in *OKIK*, 233.

147. Keimukyoku iin, "Keibi oyobi hōan torishimari taisaku" (1939?), in *OKIK*, 132.

148. Heianhokudō iin, "Keibi oyobi hōan torishimari taisaku" (1938?), in *OKIK*, 239–40.

149. Harada, *Suihō hatsudenjo kōji taikan*, 90.

150. Hazama-gumi hyakunen shi hensan iinkai, *Hazama-gumi hyakunen shi*, 638. Koreans and Chinese were not only workers or "coolies." According to Tamaoki Masa-haru, head of Yalu Hydropower's Electricity Division, most of the Koreans who worked in the more skilled positions at Sup'ung Dam were technicians rather than engineers and were not in any positions of higher leadership. Tamaoki, "Chōsen no suiryoku ni tsuite,"

24. A list of some of the main members of the hydropower and construction companies involved in the Sup'ung Dam project shows that Koreans (and some Chinese) had positions primarily in their supplies, construction, machinery, engineering works, repair, concrete mixing, and power station divisions. Harada, *Suihō hatsudenjo kōji taikan*, 92–99. Although the Government-General had a long-standing policy of restricting scientific and technical knowledge among the colonized, thereby channeling them into lower-paid unskilled labor, Koreans increasingly took advantage of lower-level technical schools in Korea and Manchuria, which were established to meet Japanese demand for skilled workers as Korea pursued a policy of rapid industrialization from the 1920s. Thus, despite the Government-General's policy of policing scientific and technical knowledge, Koreans were beginning to appropriate that knowledge to a certain degree for their own social advancement. For more on Korean engineers and technicians under Japanese colonial rule as well as their educational backgrounds, see Hirose, "Chōsen sōtokufu no doboku kanryō," 260–332.

Chapter 5

1. My historical profile of Mōri and the reform bureaucrats comes from the following sources: Itō, "Mōri Hideoto ron oboegaki," 235–39; Mimura, "Technocratic Visions of Empire: The Reform Bureaucrats in Wartime Japan"; Mimura, *Planning for Empire*; Hata, *Kanryō no kenkyū*; Furukawa, *Shōwa senchūki*, 121–25. For more on the reform bureaucrats, see Spaulding, "Japan's New Bureaucrats"; Spaulding, "Bureaucracy as a Political Force, 1920–1945"; Weiner, "Bureaucracy and Politics in the 1930s."

2. Gao, *Economic Ideology*, 70.

3. Barnhart, *Japan Prepares for Total War*; Furukawa, *Shōwa senchūki*, 16; Mimura, *Planning for Empire*, 15–21, 26–27. For an example of the Control Officers' thought on the "national defense state," see Rikugunshō shinbunhan, *Kokubō no hongi to sono kyōka no teishō*.

4. Koschmann, "Spirit of Capitalism as Disciplinary Regime," 97–116; Gao, *Economic Ideology*, 97.

5. Furukawa, *Shōwa senchūki*, 18–19.

6. As Mizuno demonstrates in detail, technology bureaucrats coined the term "science-technology" (*kagaku gijutsu*)—"planned and systematized science for the purposes of industrialization and practical applications"—in campaigning for and designing the New Order for Science and Technology. Mizuno, *Science for the Empire*, 60–68.

7. Furukawa, *Shōwa senchūki*, 18, 91. For a detailed sense of the reform bureaucrats' diversity, see Mimura, *Planning for Empire*.

8. Mimura, *Planning for Empire*, 3.

9. Mōri, "Shina no sangyō kaihatsu," 73; Miki and Mōri, "Ashita no kagaku Nihon no sōzō," 196.

10. Mōri, Sakomizu, et al., "Kakushin kanryō," 54.

11. Kamakura, " "Jiken dai yonki wa seiji o tenkai su,","79; Mōri, "Tai shi keizai gijutsu no sōzō," 100. Kamakura Ichirō was Mōri's penname.

12. Mōri, Sakomizu, et al., "Kakushin kanryō," 54.

13. Ibid., 55–56.

14. Kamakura, "Nihon kokumin keizai no keisei to seiji," 27.

15. Mōri cited the Austrian economist Othmar Spann, who pushed for the establishment of an authoritarian, corporate state; the Austrian economist Friedrich Göttl-Ottlilienfeld, follower of Spann and leading advocate of the Nazi totalitarian economy; and the nineteenth-century German economist Friedrich List, who argued against free-trade classical economics and for the construction of a national economy, as influential to his thought on the antiliberal "national economy." Nihon hyōron shinsha, *Yōyōko*, 176. All of these economists were translated into Japanese. Mōri specifically cites List's *The National System of Political Economy* in Mōri, "Nihon kokumin keizai no keisei to seiji,'" 31. For more on the intellectual influences and background of the reform bureaucrats, see Furukawa, "Kakushin kanryō no shisō to kōdō."

16. Kamakura, "Tōa kyōdōtai to gijutsu no kakumei," 6–7.

17. Mōri, "Shina no sangyō kaihatsu," 76.

18. Kamakura, "Tōa kyōdōtai to gijutsu no kakumei," 5–6.

19. Kamakura, "Nihon kokumin keizai no keisei to seiji," 26.

20. Kamakura, "'Tōa ittai' toshite no seijiryoku," 11.

21. Kamakura, "Chūshōteki bukka to gutaiteki bukka," 5–6, 8.

22. Mōri, "Seisan keizai no konpon rinen," 26–27.

23. Kamakura, "Tōa kyōdōtai to gijutsu no kakumei," 12. Mōri always employed the term *minzoku* (nation, ethnicity) in a plural fashion. The "Japanese nation" therefore consisted of Taiwanese, Koreans, and the peoples of the Japanese archipelago. Mōri never used a "blood nationalism" argument in his writings but rather a more plural and incorporative ethnic nationalism. For Mōri's idea of the Japanese nation as one of the world's "superior plural nations," see the speech, Mōri, "Dai tōa sensō o tsūjite." Suzuki Teiichi associated Mōri with pan-Asianism in contrast to Mutō Akira, a senior officer in Manchuria and China who was an uncritical follower of Nazism. Mutō later became an active figure in the more reactionary and spiritualist Imperial Rule Assistance Association in the early 1940s. Mimura, "Technocratic Visions of Empire: The Reform Bureaucrats in Wartime Japan," 229–30. Mōri's idea of *minzoku* will be discussed in more detail later.

24. Mōri, "Dai tōa sensō o tsūjite," page unnumbered.

25. Mōri, "Tai shi keizai gijutsu," 100, 105.

26. Mōri, "Dai tōa sensō o tsūjite," page unnumbered. Aikawa Haruki also used this term "life technologies" in his 1942 work *Introduction to a Theory of Technology*. See chapter 2, "Seikatsu no naka no gijutsu," in Aikawa, *Gijutsuron nyūmon*, 26–50.

27. Mōri, "Haihin kaishū roku," 164.

28. Mōri, "Shina no sangyō kaihatsu," 79.

29. Mōri, "Tōa keizai kensetsu ni tsuite, dai ikkō."

30. Mōri, "Shina no sangyō kaihatsu," 79.

31. Kamakura, "Tōa kyōdōtai to gijutsu no kakumei," 11.

32. Mōri, "Dai tōa sensō o tsūjite," pages unnumbered.

33. Ishimoto, Kubo, Shiina, et al., "Nichi-Man-Shi o tsūzuru keikaku keizai no sai ginmi zadankai," 2–3.

34. Ibid., 3.

35. Ibid., 3–4.

36. For more on the contemporary student movement, see Smith, *Japan's First Student Radicals*.

37. Mimura, "Technocratic Visions of Empire: The Reform Bureaucrats in Wartime Japan," 223.

38. Ibid., 136; Hata, *Kanryō no kenkyū*, 130.

39. Akinaga et al., "Manshūkoku keizai no genchi zadankai," 36.

40. Furumi, *Wasureenu Manshūkoku*, 101–2; Kamei, *Kamei Kanichirō shi danwa sokkiroku*, 196. Furumi was the right-hand man to Hoshino Naoki, the head of Manchukuo's General Affairs Agency.

41. Kamei, *Kamei Kanichirō shi danwa sokkiroku*, 35.

42. Furukawa, *Shōwa senchūki*, 40.

43. Mimura, "Technocratic Visions of Empire: The Reform Bureaucrats in Wartime Japan," 65.

44. Furukawa, *Shōwa senchūki*, 43–44.

45. Hata, *Kanryō no kenkyū*, 131.

46. Nakamura, *Senji Nihon no Kahoku keizai shihai*, 28–29.

47. Kobayashi, "Kahoku senryō seisaku no tenkai katei."

48. Mōri already had experience in unifying Manchukuo's currencies and issuing a new one based on silver in 1932 and subsequently shifting Manchukuo to the gold standard in 1935. See Kobayashi, "Shokuminchi keiei no tokushitsu," 21.

49. Nakamura, *Senji Nihon no Kahoku keizai shihai*, 147.

50. Kuwano, *Senji tsūka kōsaku shi ron*. See pp. 14–18 for the texts of both outlines.

51. The preceding information on currency stabilization is from Nakamura, *Senji Nihon no Kahoku keizai shihai*, 148–54. Apparently in the beginning, bank officials had to visit cotton producers to persuade them to adopt the new currency. Cotton, of course, was an essential military item and a large part of north China's economy. Chūgoku rengō junbi ginkō komonshitsu, *Chūgoku rengō junbi ginkō go-nen shi*, 4.

52. Imura, *Jyūgonen sensō jūyō bunken shirizu (17)*, 3.

53. Kubo, "Kōain ni yoru Chūgoku chōsa," 79.

54. Ōyodo, *Miyamoto*, 263.

55. Kubo, "Kōain ni yoru Chūgoku chōsa," 73–103. Unlike former Marxist investigators who primarily worked for Mantetsu, Asia Development Board engineers emphasized factors conducive to industrialization and development as opposed to "semifeudal" conditions.

56. Hensha, "Hajime ni," 10.

57. Mimura, "Technocratic Visions of Empire: The Reform Bureaucrats in Wartime Japan," 242.

58. Furukawa, *Shōwa senchūki*, 19.

59. Yatsugi, *Shōwa dōran shishi*, Vol. 1, 491; Vol. 2, 210–11. Both think tanks gained in importance in the aftermath of the attempted right-wing coup (2.26 Incident) in 1936 and the establishment of the first Konoe cabinet in 1938. The National Policy Research Association consisted mostly of well-established, midlevel or higher people from business, government, academia, and the military, and they had strong connections to the army. The Shōwa Research Association, on the other hand, consisted of younger people from each area, and they had strong connections to the navy. Both were "reformist" and technocratic. See Furukawa, *Shōwa senchūki*, 34–35. For more on the Shōwa Research Association, see Fletcher, *Search for a New Order*.

60. Furukawa, *Shōwa senchūki*, 113–14.

61. For more on the wartime economy's key laws, see Nakamura, "Japanese Wartime Economy."

62. Itō, "Mōri Hideoto ron oboegaki" 235; Nihon hyōron shinsha, *Yōyōko*, 129.

63. Furukawa, *Shōwa senchūki*, 176–77. The Cabinet Planning Board, in its efforts to shape "national morals," clashed with the Imperial Rule Assistance Association, which was the main organization in charge of "spiritual mobilization" and a bastion for the type of irrationalist ideologies that reform bureaucrats attempted to control through their assertion of the importance of science and technology.

64. Gao, *Economic Ideology*, 115–16.

65. Furukawa, *Shōwa senchūki*, 302; Gao, *Economic Ideology*, 77.

66. Kamakura, "Tōsei keizai no hinkon no gennin," 16–18.

67. Kamakura, "Kokumin seikatsu soshiki no kiten," 6.

68. Kamakura, "Kokumin keizai to shieki," 88.

69. Ibid., 86–88; Kamakura, "Kokumin soshiki to tōa kyōdōtai no fukabunsei," 24.

70. Kamakura, "Tōsei keizai no hinkon no gennin," 16–17.

71. Kamakura, " 'Tōa ittai' toshite no seijiryoku," 11.

72. Kamakura, "Kokumin soshiki to tōa kyōdōtai no fukabunsei," 22.

73. Ibid.

74. Kamakura, "Kokumin seikatsu soshiki no kiten," 6.

75. Ibid., 6–7, 10–11.

76. Kamakura, "Handō o kokufuku suru seiji," 10.

77. Mōri, "Keizai shintaisei kōza," pages unnumbered. In this speech, he distinguished between a short-term "wartime economy" and a permanent "national defense economy." This mapped onto his earlier distinction between a "requisition economy" and a "planned economy" in the colonies. He also discussed how parents in Nazi Germany always made their young children stand in moving trains so that they became used to hardship from a very early age.

78. Mōri had even employed quantum mechanics as a metaphor for the society organized by vocation and a strong national defense consciousness. Quantum mechanics went beyond classical atomism by seeing individual particles in terms of a larger, more complicated synthesis. "This is similar to totalitarianism and the managed economy, which is not a logical unification into something uniform and homogenous. It is the synthesis

into a higher standpoint while affirming multiplicity as it is," he said in a dialogue with the Kyoto School philosopher Miki Kiyoshi. He extended this metaphor of quantum theory to the "multiple ethnicities of the Greater East Asian Co-Prosperity Sphere" and its "extremely plural life functions." Miki and Mōri, "Ashita no kagaku Nihon no sōzō," 199. Mōri also argued that production technology was tied to the Japanese nation. For him, technology was "cultural" and the people's creative energies, if integrated into the managed economy and the national life organization, provided a deep pool for technical innovation and development, such as in the strategic area of developing hi-tech synthetic materials that eliminated Japan's dependence on foreign natural resources. On the relationship between production technology and national spirit, see Kamakura, "Gijutsu no kaihō to seiji."

79. Kamei, *Kamei Kanichirō shi danwa sokkiroku*, 37; Furukawa, *Shōwa senchūki*, 119–21.

80. For an overview of the influence of German economic thought in Japan at the time, see Yanagisawa, "Senzen Nihon no tōsei keizai ron." Yanagisawa focuses on the influence of Sombart and Walter Rathenau, both of whom believed that the era of free-market, liberal capitalism was ending and that some degree of state intervention was necessary to alleviate capitalism's ills.

81. Haag, "Othmar Spann and the Quest for a 'True State,'" 237; Haag, "Othmar Spann and the Politics of 'Totality,'" 36.

82. Furukawa, *Shōwa senchūki*, 120; Haag, "Othmar Spann and the Politics of 'Totality,'" 30, 37.

83. Sombart, *Future of Capitalism*.

84. Mimura, *Planning for Empire*, 115.

85. Härd, "German Regulation," 59, 62.

86. Hanel, "Technology and Sciences Under Modern Capitalism," 95–96.

87. Aikawa, *Sangyō gijutsu*, 329.

88. Gassert, "'Without Concessions to Marxist or Communist Thought,'" 224–26; Göttl-Ottlilienfeld, "Fordism," 400–402.

89. Kaneko, "Zentaishugi keizaigaku no ni keikō."

90. Kamei was Mōri's distant relative, and he later married Kamei's daughter. Kamei, *Kamei Kanichirō shi danwa sokkiroku*, 35–36.

91. Oikawa, "Relation Between National Socialism and Social Democracy," 198–99.

92. Kamei, *Kamei Kanichirō shi danwa sokkiroku*, 193.

93. Itō, "Konoe shintō mondai," 144.

94. An example of Kamei's Nazi-inspired work is Kamei, *Natchisu kokubō keizairon*.

95. Itō, "Konoe shintō mondai," 135.

96. Kamei, *Kamei Kanichirō shi danwa sokkiroku*, 4–5, 38, 45–46.

97. "Senji seikatsu sōdanjo to wa," 11.

98. Furukawa, *Shōwa senchūki*, 137.

99. Kobayashi, "Senji seikatsujo no seijiteki igi," 17.

100. "Senji seikatsu sōdanjo to wa," 10.

101. Mōri, "Hitotsu no kotae toshite no seikatsu sōdanjo," 201.

102. Furukawa, *Shōwa senchūki*, 138.

103. Saguchi, "Historical Significance of the Industrial Patriotic Association," 271–72.

104. Furukawa, *Shōwa senchūki*, 314; Saguchi, "Historical Significance of the Industrial Patriotic Association," 281–82.

105. Kashiwabara, "Sangyō hōkoku undō no shin hōshin," 500–502. Furukawa notes that Mōri drafted this document, and it was published under Kashiwabara's name. Furukawa, *Shōwa senchūki*, 315.

106. Mōri, "Tōa keizai no kensetsu ni tsuite, dai ikkō," page unnumbered; Kamakura, "Chūgoku no 'kōsen kenkoku' o hihan su," 7.

107. Mōri, "Tōa keizai no kensetsu ni tsuite, dai ikkō," page unnumbered.

108. Kamakura, "Chūgoku no 'kōsen kenkoku' o hihan su," 7. Mōri apparently studied the "platforms and organizations of the Chinese democratic parties" as a member of Kamei's research group. See Furukawa, "Kakushin kanryō no shisō to kōdō," 19.

109. Kamakura, "'Tōa ittai' toshite no seijiryoku," 10.

110. Mōri, "Dai tōa bunka no igi," page unnumbered.

111. Kamakura, "Tōa kyōseitai kensetsu no sho jōken," 29. Mōri's East Asian nationalism somewhat resembles what Naoki Sakai calls "imperial nationalism." For more on "imperial nationalism" and its continuity with recent formulations of U.S. imperialism and the discipline of Asian Studies, see Sakai, "Subject and Substratum."

112. Mōri, "Dai tōa bunka no igi," page unnumbered.

113. Mōri, "Tōa keizai no kensetsu ni tsuite, dai ni kō," page unnumbered.

114. Kamakura, "Tōa ni okeru bōkyō no igi," 7. Germany was also charged with this "world-historical task," according to Mōri.

115. "Nichi-Man-Shi keizai kensetsu yōkō." October 3, 1940. National Diet Library. http://rnavi.ndl.go.jp/politics/entry/bib00277.php. (Accessed December 13, 2009.)

116. Matsumoto, "Dai niji taisenki no senji taisei kōsō."

117. Furukawa, *Shōwa senchūki*, 271.

118. Adachi, "Dai tōa kensetsu shingikai."

119. Furukawa, *Shōwa senchūki*, 272–73.

120. Herf, *Reactionary Modernism*, 1.

121. Ibid., 2, 3, 8.

122. Kamakura, "Gijutsu no kaihō to seiji."

123. See in particular chapter 5, "Dai tōa kyōeiken minzoku kōsaku to gijutsu taisei," in Kokusaku kenkyūkai, *Dai tōa kyōeiken gijutsu taiseiron*, 30–40.

124. Miki and Mōri, "Ashita no kagaku Nihon no sōzō," 186–87, 199. Mōri, like other right-wing bureaucrats and intellectuals at the time, was influenced by the Nazi geographer Karl Haushofer, who coined the term "space for living" (*Lebensraum*). For example, see Kamakura, "Taiheiyō kūkan no seikaku kakumei," 36

125. The essays in Yamanouchi et al., *Total War and 'Modernization,'* deal with this overall issue of continuity.

Epilogue

1. Quoted in Morris-Suzuki, *Technological Transformation of Japan*, 161.

2. Ibid.

3. Mimura, *Planning for Empire*, 195–200.

4. Ōyodo, *Miyamoto*, 514.

5. Ibid., 519.

6. Satō, *"Motazaru kuni" no shigen ron*, 98. On the Resource Investigation Association, see ibid., 101–35. For a description of the TVA philosophy and program, see Lilienthal, *TVA: Democracy on the March*.

7. Morris-Suzuki, *Technological Transformation of Japan*, 162, 165, 166.

8. Tsutsui, *Manufacturing Ideology*, 136.

9. Ibid., 116–18.

10. Saguchi, "Historical Significance of the Industrial Patriotic Association," 261–87.

11. Partner, *Assembled in Japan*, 107–92.

12. Garon, *Molding Japanese Minds*.

13. Yamanouchi, "Total War and System Integration," 25–26.

14. Quoted in Gotō, "National Land Comprehensive Development Act," 336.

15. Ibid.

16. For more on the TVA in Japan, see Dinmore, "Small Island Nation Poor in Natural Resources," 160–207.

17. Kawamura, "Damu to iu 'kaihatsu pakkeji,'" 79–80; Kodama, "Sakuma damu kensetsu to iu 'gijutsu no shōri,'" 161–63; Machimura, "'Sakuma damu' kenkyū no kadai to hōhō," 1, 15, 18; Yamamoto, "Sakuma damu kensetsu to ryūiki keizaiken no henyō," 44–46.

18. Gotō, "National Land Comprehensive Development Act," 341.

19. Ibid., 342–44.

20. Ibid., 344.

21. Tanaka, *Building a New Japan*.

22. McCormack, *Emptiness of Japanese Affluence*, 25–77. Perhaps the most bizarre example of this type of appeal to technology and public works by Japan's governmental and business leaders was National-Panasonic head Matsushita Kōnosuke's 1977 proposal to build a fifth island in Japan by leveling 20 percent of Japan's mountains and dumping them into the sea. This 200-year project would apparently provide Japan's citizens with new purpose and a sense of national unity, which Matsushita believed was lost after the war. Ibid., 66.

23. Dyer-Witherford, *Cyber-Marx*, 20. "Just-in-time production" links the consumer to the production process by rapidly communicating what the individual consumer wants to the factories involved, which then produce the specific item and deliver it accordingly rather than simply mass-producing different models. This significantly reduces inventory expenses and increases profits.

24. Ibid. See also Morris-Suzuki, *Beyond Computopia*, and Masuda, *Information Society as Post-Industrial Society*.

25. Morris-Suzuki, *Technological Transformation of Japan*, 210–11.

26. McCormack examines the utter failure of Kobe's new high-tech structures to deal with the 1995 Great Hanshin Earthquake. Communications and command centers failed, citizens became trapped and isolated for long periods of time, relief efforts were delayed and unorganized, and building structures were revealed to be substandard. McCormack, *Emptiness of Japanese Affluence*, 7–17.

27. Gotō, "Technology Studies," 345–57.

28. Koschmann, *Revolution and Subjectivity*, 146.

29. Taketani, "Genshiryoku no heiwa-teki riyō," 135.

30. Gotō, "Technology Studies," 347.

31. Ibid., 350–51.

32. Since the 1980s, however, the discipline of Science and Technology Studies has become established in Japan and has generated a lot of work on technology's sociopolitical aspects. For a recent example, see Murata, *Gijutsu no tetsugaku*. At the level of popular culture, a more sophisticated analysis of the dynamics between subjectivity and technology has emerged in *manga* and animation since the 1980s that has questioned common assumptions about the boundaries between the human and nonhuman, animate and inanimate, and life and death through the trope of the cyborg—in ways that explore the multiple possibilities of hybrid subjectivity and rework conventional notions of race, sex, and gender, rather than viewing technology in purely dehumanizing terms. See Brown, *Tokyo Cyberpunk*.

33. Kawamura, "Damu to iu 'kaihatsu pakkeji,'" 83.

34. Nagatsuka, *Kubota*, 327.

35. Nippon Kōei, *Nippon Kōei sanjū go nen shi*, 73.

36. Nagatsuka, *Kubota*, 306–7, 369–70.

37. Ibid., 316–17; Shoenberger, "Japan Aid."

38. Kawamura, "Damu to iu 'kaihatsu pakkeji,'" 75–76.

39. Nitchitsu's founder, Noguchi Jun, died in 1944. George, *Minamata*, 24–25, 32–35.

40. Walker, *Toxic Archipelago*, 159; Molony, *Technology and Investment*, 138.

41. Fukai, *Minamatabyō no seiji keizaigaku*, chapters 1–4. Because Nitchitsu had dominated Minamata's society and economy since the late Meiji era, some have used the term "internal colonialism" to describe the unequal relationship between the company and the region's residents, which formed the basis for the pollution disaster. George, *Minamata*, 40.

42. Machida, "Honma-san no omoide," 76.

43. On the "construction state" and its effects, see Kerr, *Dogs and Demons*, and Woodall, *Japan Under Construction*.

44. Nagatsuka, *Kubota*, 270–76.

45. Nippon Kōei, *Nippon Kōei sanjū go nen shi*, 322–23, 326–27.

46. Aldrich, *Site Fights*, 96; McCormack, "Modernity, Water, and the Environment in Japan," 447.

47. Japanese companies seeking international development project contracts apparently used the Sakuma Dam documentary in their negotiations with national govern-

ments and businesses. Kawamura, "Damu to iu 'kaihatsu pakkeji,'" 75–76. The writer of *The Sun in Kurobe*, Kimoto Shōji, published a nonfiction serial novel in the Mainichi newspaper in 1966 entitled "Hong Kong Water." It was based on the activities of Japanese engineers who were contracted to resolve Hong Kong's water crisis through the construction of a large dam and reservoir. In a striking resemblance to the popular 1940 colonial film *Vow in the Desert* (discussed in Chapter 3), the story included a love affair between the Japanese engineer, Fujisawa Isao, and a Chinese woman, He Lifen, who lost her father during the war with Japan; Fujisawa dying during the construction; and He visiting Fujisawa's grave in Japan to pour Hong Kong water over it. Kimoto, *Hon Kon no mizu*. The book was made into an NHK drama in 1991, and *The Sun in Kurobe* was remade into a Fuji television series in 2009, demonstrating a continuing resonance of the dam-centered technological imaginary in popular culture.

48. McCormack, "Modernity, Water, and the Environment in Japan," 454. Today, control over public works projects has been further consolidated under the 70,000-person-strong Ministry of Land, Infrastructure, Transport and Tourism.

49. Aldrich, *Site Fights*, 95–105.

50. For example, Japan's Fifth Comprehensive National Development Plan (1997) put forth "harmony with nature" and creating a "beautiful national land" as its objective yet simply continued the same developmentalist agenda of funding large-scale projects, this time to cover over much of the ecological damage resulting from previous national plans. McCormack, "Modernity, Water, and the Environment in Japan," 453. A study of the postwar technological imaginary also needs to account for how intellectuals, government officials, and civil society challenged and altered the postwar technological imaginary in ways that did not merely reject it. On Minamata and antipollution movements, see George, *Minamata*. Satō Jin analyzes alternative visions among policymakers and intellectuals toward sustainable natural resource policies. Satō, *"Motazaru kuni" no shigen ron*. Kawakami Yukio, head of the comprehensive planning section in the Ministry of Land, Infrastructure, Transport and Tourism, analyzes the history of national land planning in terms of the continuing clash among planners between visions emphasizing "efficiency" and those prioritizing "balance." Kawakami, *Kokudo keikaku no hensen*. In the wake of the Fukushima nuclear disaster, a whole array of civil society groups have risen in support of sustainable energy, decentralization of power production, open access to medical and environmental data, and democratic accountability of the power industry and national government.

Works Cited

Sources in English

Aldrich, Daniel P. *Site Fights: Divisive Facilities and Civil Society in Japan and the West.* Ithaca, NY: Cornell University Press, 2008.

Aldrich, Daniel P., and Martin Dusinberre. "Hatoko Comes Home: Civil Society and Nuclear Power in Japan." *Journal of Asian Studies* 70, no. 3 (September 2011): 683–705.

Archer, Kevin. "The Limits to the Imagineered City: Sociospatial Polarization in Orlando." *Economic Geography* 73, no. 3 (July 1997): 322–36.

Barnhart, Michael A. *Japan Prepares for Total War: The Search for Economic Security, 1919–1941.* Ithaca, NY: Cornell University Press, 1988.

Barshay, Andrew E. *The Social Sciences in Modern Japan: The Marxian and Modernist Traditions.* Berkeley: University of California Press, 2004.

Baskett, Michael. *The Attractive Empire: Transnational Film Culture in Imperial Japan.* Honolulu: University of Hawaii Press, 2008.

Billington, David, and Donald Jackson. *Big Dams of the New Deal Era.* Norman: University of Oklahoma Press, 2006.

Bix, Herbert. "Rethinking Emperor-System Fascism." *Bulletin of Concerned Asian Scholars* 14 (1982): 20–32.

Bloom, Justin L. *Japan as a Scientific and Technological Superpower.* Potomac, MD: Technology International and Springfield, VA: U.S. Department of Commerce, National Technical Information Service, 1990.

Brown, Steven T. *Tokyo Cyberpunk: Posthumanism in Japanese Visual Culture.* New York: Palgrave Macmillan, 2010.

Buck, Pearl. *The Good Earth.* Reprint. New York: Pocket Books, 2005.

Calichman, Richard F., ed. *Overcoming Modernity: Cultural Identity in Wartime Japan*, trans. Richard Calichman. New York: Columbia University Press, 2008.

Commission on the History of Science and Technology Policy, ed. *Historical Review of Japanese Science and Technology Policy.* Tokyo: Society of Non-Traditional Technology, 1991.

Cusumano, Michael A. "'Scientific Industry': Strategy, Technology, and Entrepreneurship in Prewar Japan." In *Managing Industrial Enterprise*, ed. William D. Wray, 269–315. Cambridge, MA: Council of East Asian Studies/Harvard University, 1989.

Dinmore, Eric G. "A Small Island Nation Poor in Resources: Natural and Human Resource Anxieties in Trans-World War II Japan." Ph.D. dissertation, Princeton University, 2006.

Dower, John W. "The Useful War." *Daedulus* 119, no. 3 (1990): 49–70.

———. *War Without Mercy: Race and Power in the Pacific War*. New York: Pantheon Books, 1986.

Driscoll, Mark. *Absolute Erotic, Absolute Grotesque: The Living, Dead, and Undead in Japan's Imperialism, 1895–1945*. Durham, NC: Duke University Press, 2010.

Duus, Peter, and Daniel I. Okimoto. "Fascism and the History of Prewar Japan: The Failure of a Concept." *Journal of Asian Studies* 39, no. 1 (1979): 65–76.

Dyer-Witherford, Nick. *Cyber-Marx: Cycles and Circuits of Struggle in High-Technology Capitalism*. Urbana: University of Illinois Press, 1999.

Feenberg, Andrew. *Critical Theory of Technology*. New York: Oxford University Press, 1991.

Fletcher, W. Miles. *The Search for a New Order: Intellectuals and Fascism in Prewar Japan*. Chapel Hill: University of North Carolina Press, 1982.

Gao, Bai. *Economic Ideology and Japanese Industrial Policy: Developmentalism from 1935 to 1965*. Cambridge: Cambridge University Press, 2002.

Garon, Sheldon. *Molding Japanese Minds: The State in Everyday Life*. Princeton, NJ: Princeton University Press, 1997.

———. "Rethinking Modernization and Modernity in Japanese History: A Focus on State-Society Relations." *Journal of Asian Studies* 53, no. 2 (May 1994): 346–66.

———. *The State and Labor in Modern Japan*. Berkeley: University of California Press, 1987.

Gassert, Philipp. "'Without Concessions to Marxist or Communist Thought': Fordism in Germany, 1923–1939." In *Transatlantic Images and Perceptions: Germany and America Since 1776*, ed. David Barclay and Elisabeth Glaser-Schmidt, 217–42. Cambridge: Cambridge University Press, 2003.

George, Timothy S. *Minamata: Pollution and the Struggle for Democracy in Postwar Japan*. Cambridge, MA: Harvard University Asia Center, 2002.

Gerow, Aaron. *Visions of Japanese Modernity: Articulations of Cinema, Nation, and Spectatorship, 1895–1925*. Berkeley: University of California Press, 2010.

Gordon, Andrew. *Labor and Imperial Democracy in Prewar Japan*. Berkeley: University of California Press, 1991.

Gorky, Maxim. "Literature and the Soviet Idea." In *Problems of Soviet Literature*, ed. H. G. Scott. Moscow: Cooperative Publishing Society of Foreign Workers in the USSR, 1935.

Gotō, Kunio. "The National Land Comprehensive Development Act." In *A Social History of Science and Technology in Contemporary Japan*. Vol. 3, *High Economic Growth Period, 1960–1969*, ed. Shigeru Nakayama and Kunio Gotō, 333–46. Melbourne: Trans Pacific Press, 2006.

———. "Technology Studies, Technological Innovation and State Monopoly Capitalism." In *A Social History of Technology in Contemporary Japan*. Vol. 2, *Road to Self-Reliance, 1952–1959*, ed. Shigeru Nakayama and Hitoshi Yoshioka, 345–57. Melbourne: Transpacific Press, 2005.

Göttl-Ottlilienfeld, Friedrich von. "Fordism." In *The Weimar Republic Sourcebook*, ed. Anton Kaes, Martin Jay, and Edward Dimendburg, 400–402. Berkeley: University of California Press, 1995.

Griffin, Roger. "The Palingenetic Core of Generic Fascist Ideology." Translated book chapter, 1–16. http://ah.brookes.ac.uk/resources/griffin/coreoffascism.pdf. (Accessed January 14, 2012.)

Grunden, Walter E. *Secret Weapons and World War II: Japan in the Shadow of Big Science.* Lawrence: University Press of Kansas, 2005.

Haag, John. "Othmar Spann and the Politics of 'Totality': Corporatism in Theory and Practice." Ph.D. dissertation, Rice University, 1969.

———. "Othmar Spann and the Quest for a 'True State.'" *Austrian History Yearbook* 12 (1976): 227–50.

Habermas, Jürgen. "Science and Technology as 'Ideology.'" In *Toward a Rational Society: Student Protest, Science, and Politics*, Jürgen Habermas, trans. Jeremy J. Shapiro, 81–122. Boston: Beacon Press, 1970.

Hall, Peter. *Cities of Tomorrow: An Intellectual History of Urban Planning and Design in the Twentieth Century.* 3rd edition. Malden, MA: Blackwell Publishing, 2002.

Hanel, Johannes. "Technology and Sciences Under Modern Capitalism: Sombart's Forgotten Question." In *Werner Sombart (1863–1941): Social Scientist.* Vol. 2, ed. Jürgen Backhaus, 163–178. Marburg: Metropolis, 1996.

Hanes, Jeffrey E. *The City as Subject: Seki Hajime and the Reinvention of Modern Osaka.* Berkeley: University of California Press, 2002.

Hanwell, Norman D., and Kurt Bloch. "Behind the Famine in North China." *Far Eastern Survey* 9, no. 6 (March 13, 1940): 63–68.

Härd, Mikael. "German Regulation: The Integration of Modern Technology into National Culture." In *The Intellectual Appropriation of Technology: Discourses on Modernity, 1900–1939*, ed. Mikael Härd and Andrew Jamison, 33–67. Cambridge, MA: The MIT Press, 1998.

Härd, Mikael, and Andrew Jamison. "Conceptual Framework: Technology Debates as Appropriation Processes." In *The Intellectual Appropriation of Technology: Discourses on Modernity, 1900–1939*, ed. Mikael Härd and Andrew Jamison, 1–15. Cambridge, MA: The MIT Press, 1998.

Harootunian, Harry. *Overcome by Modernity: History, Culture, and Community in Interwar Japan.* Princeton, NJ: Princeton University Press, 2000.

Hatch, Walter F. *Asia's Flying Geese: How Regionalization Shapes Japan?* Ithaca, NY: Cornell University Press, 2010.

Herf, Jeffrey. *Reactionary Modernism: Technology, Culture, and Politics in Weimar and the Third Reich.* Cambridge: Cambridge University Press, 1984.

Hoston, Germaine A. *Marxism and the Crisis of Development in Prewar Japan.* Princeton, NJ: Princeton University Press, 1986.

Hotta, Eri. *Pan-Asianism and Japan's War, 1931–1945.* New York: Palgrave MacMillan, 2007.

Iriye, Akira. "Toward a Cultural Order: The *Hsinmin Hui* (Xinmin Hui)." In *The Chinese and the Japanese: Essays in Political and Cultural Interactions,* ed. Akira Iriye, 254–74. Princeton, NJ: Princeton University Press, 1980.

Johnson, Chalmers. *MITI and the Japanese Miracle: The Growth of Industrial Policy, 1925–1975.* Tokyo: Charles E. Tuttle, 1982.

Ju Zhifen. "Northern Chinese Laborers and Manchukuo." In *Asian Labor in the Wartime Japanese Empire,* ed. Paul Kratoska, 61–80. Singapore: Singapore University Press, 2006.

Kasza, Gregory J. "Fascism from Below? A Comparative Perspective on the Japanese Right, 1931–1936." *Journal of Contemporary History* 19, no. 4 (1984): 607–29.

———. *The State and the Mass Media, 1918–1945.* Berkeley: University of California Press, 1988.

Katada, Saori N. "Why Did Japan Suspend Foreign Aid to China?" *Social Science Japan Journal* 4, no. 1 (2001): 39–58.

Kerr, Alex. *Dogs and Demons: Tales from the Dark Side.* New York: Hill and Wang, 2002.

Kimoto, Takeshi. "Tosaka Jun and the Question of Technology." In *Whither Japanese Philosophy? Reflections Through Other Eyes,* ed. Takahiro Nakajima, 121–40. Tokyo: UTCP, 2009.

———. "Tosaka Jun and the Question of Technology." Paper presented at Association for Asian Studies Annual Meeting, Atlanta, 2008.

Kodama, Fumio. *Analyzing Japanese Advanced High Technologies: The Techno-Paradigm Shift.* London: Pinter Publishers, 1991.

Koschmann, J. Victor. *Revolution and Subjectivity in Postwar Japan.* Chicago: University of Chicago Press, 1996.

———. "Rule by Technology/Technologies of Rule." Unpublished paper, Cornell University, 2002.

———. "The Spirit of Capitalism as Disciplinary Regime: The Postwar Thought of Ōtsuka Hisao." In *Total War and "Modernization,"* ed. Yasushi Yamanouchi, J. Victor Koschmann, and Ryūichi Narita, 97–116. Ithaca, NY: Cornell University East Asia Program, 1998.

Koschmann, J. Victor, and Sven Saaler, eds. *Pan-Asianism in Modern Japanese History: Colonialism, Regionalism, and Borders.* Abingdon: Routledge, 2007.

Kratoska, Paul H., ed. *Asian Labor in the Wartime Japanese Empire: Unknown Histories.* Armonk, NY: M. E. Sharpe, 2005.

Kubo, Tōru. "The Kōain." In *China at War: Regions of China, 1937–1945,* ed. Stephen MacKinnon, Diana Lary, and Ezra Vogel, 51–64. Stanford, CA: Stanford University Press, 2007.

Lary, Diana. "Drowned Earth: The Breaching of the Yellow River Dykes, 1938." *War in History* 8, no. 2 (2001): 191–207.

Layton, Jr., Edwin. *The Revolt of the Engineers: Social Responsibility and the American Engineering Profession*. Cleveland, OH: The Press of Case Western Reserve University, 1971.

Lilienthal, David. *TVA: Democracy on the March*. New York: Harper and Brothers, 1944.

Lo, Ming-cheng M. *Doctors Within Borders: Profession, Ethnicity, and Modernity in Colonial Taiwan*. Berkeley: University of California Press, 2002.

Maier, Charles S. "Society as Factory." In *In Search of Stability: Explorations in Historical Political Economy*, 19–69. Cambridge: Cambridge University Press, 1987.

Marcuse, Herbert. *One Dimensional Man*. New York: Beacon Press, 1968.

Maruyama, Masao. "The Ideology and Dynamics of Japanese Fascism." In *Thought and Behaviour in Modern Japanese Politics*, ed. Ivan Morris, 25–83. Oxford: Oxford University Press, 1969.

Marx, Karl. "Theses on Feuerbach." In *Writings on the Young Marx on Philosophy and Society*, ed. Lloyd D. Easton and Kurt Guddat. Garden City, NY: Doubleday, 1967.

Marx, Leo. "The Idea of 'Technology' and Post-Modern Pessimism." In *Does Technology Drive History? The Dilemma of Technological Determinism*, ed. Merritt Roe Smith and Leo Marx, 237–58. Cambridge, MA: The MIT Press, 1994.

Masuda, Yoneji. *The Information Society as Post-Industrial Society*. Tokyo: Institute for the Information Society, 1980.

Matsusaka, Y. Tak. *The Making of Japanese Manchuria, 1904–1932*. Cambridge, MA: Harvard University Press, 2001.

McCormack, Gavan. *The Emptiness of Japanese Affluence*. Armonk, NY: M. E. Sharpe, 2001.

———. "Modernity, Water, and the Environment in Japan." In *A Companion to Japanese History*, ed. William Tsutsui, 443–59. Oxford: Wiley-Blackwell, 2009.

———. "Nineteen-Thirties Japan: Fascism?" *Bulletin of Concerned Asian Scholars* 14 (1982): 2–19.

Mendl, Wolf, ed. *Japan and Southeast Asia*. Vol. 1. London: Routledge, 2001.

Mimura, Janis. *Planning for Empire: Reform Bureaucrats and the Japanese Wartime State*. Ithaca, NY: Cornell University Press, 2011.

———. "Technocratic Visions of Empire: Technology Bureaucrats and the 'New Order for Science-Technology.'" In *The Japanese Empire in East Asia and Its Postwar Legacy*, ed. Harald Fuess, 97–118. Munich: Iudicium-Verlag, 1998.

———. "Technocratic Visions of Empire: The Reform Bureaucrats in Wartime Japan." Ph.D. dissertation, University of California, Berkeley, 2002.

Mitchell, Timothy. *Rule of Experts: Egypt, Techno-Politics, Modernity*. Berkeley: University of California Press, 2002.

Mizuno, Hiromi. *Science for the Empire: Scientific Nationalism in Modern Japan*. Stanford, CA: Stanford University Press, 2009.

Molony, Barbara. *Technology and Investment: The Prewar Japanese Chemical Industry*. Cambridge, MA: Council on East Asian Studies at Harvard University, 1990.

Moore, Aaron S. "Para-Existential Forces of Invention: Nakai Masakazu's Theory of Technology and Critique of Capitalism." *Positions East Asia Culture Critique* 17, no. 1 (March 2009): 127–57.

———. "The Yalu River Era of Developing Asia: Japanese Expertise, Colonial Power, and the Construction of Sup'ung Dam." *Journal of Asian Studies* 72, no. 1 (2013): 1–25.

Morley, James W. *Dilemmas of Growth in Prewar Japan.* Princeton, NJ: Princeton University Press, 1974.

———. "Introduction: Choice and Consequence." In *Dilemmas of Growth in Prewar Japan*, ed. James W. Morley, 3–30. Princeton, NJ: Princeton University Press, 1974.

Morris-Suzuki, Tessa. *Beyond Computopia: Information, Automation and Democracy in Japan.* London: Kegan Paul International, 1988.

———. *The Technological Transformation of Japan: From the Seventeenth to the Twenty-First Century.* Cambridge: Cambridge University Press, 1994.

Nakagane, Katsuji. "Manchukuo and Economic Development." In *The Interwar Economy of Japan: Colonialism, Depression, and Recovery, 1910–1940*, ed. Michael Smitka, 73–97. London: Routledge, 1998.

Nakamura, Miri. "Making Bodily Differences: Mechanized Bodies in Hirabayashi Hatsunosuke's 'Robot' and Early Shōwa Robot Literature." *Japan Forum* 19, no. 2 (July 2007): 169–90.

Nakamura, Takafusa. "The Japanese Wartime Economy as a 'Planned Economy.'" In *Japan's War Economy*, ed. Erich Pauer, 9–22. London: Routledge, 2002.

———. *The Postwar Japanese Economy: Its Development and Structure*, trans. Jacqueline Kaminski. Tokyo: University of Tokyo Press, 1981.

Noble, David F. *America by Design: Science, Technology, and the Rise of Corporate Capitalism.* Oxford: Oxford University Press, 1977.

Nornes, Markus Abé. *Japanese Documentary Film: The Meiji Era Through Hiroshima.* Minneapolis: University of Minnesota Press, 2003.

Oikawa, Eijiro. "The Relation Between National Socialism and Social Democracy in the Formation of the International Policy of the Shakai Taishuto." In *Nationalism and Internationalism in Imperial Japan: Autonomy, Asian Brotherhood, or World Citizenship?* ed. Dick Stegewerns, 197–228. London: Routledge Curzon, 2003.

Park, Hyun Ok. *Two Dreams in One Bed: Empire, Social Life, and the Origins of the North Korean Revolution in Manchuria.* Durham, NC: Duke University Press, 2005.

Partner, Simon. *Assembled in Japan.* Berkeley: University of California Press, 2000.

Pauer, Erich. "Japan's Technical Mobilization." In *Japan's War Economy*, ed. Erich Pauer, 39–64. London: Routledge, 1999.

Perrins, Robert J. "Doctors, Diseases, and Development: Engineering Colonial Public Health in Southern Manchuria, 1905–1926." In *Building a Modern Japan: Science, Technology, and Medicine in the Meiji Era and Beyond*, ed. Morris Low, 102–32. New York: Palgrave, 2005.

Pietz, David. "Controlling the Waters in Twentieth-Century China: The Nationalist State and the Huai River." In *A History of Water.* Vol. 3, *Water Control and River Biographies*, ed. Terje Tvedt and Eva Jakobsson, 92–119. London: I. B. Tauris, 2006.

———. *Engineering the State: The Huai River and Reconstruction in Nationalist China, 1927–1937.* London: Routledge, 2002.

Rassweiler, Anne D. *The Generation of Power: The History of Dneprostroi*. Oxford: Oxford University Press, 1988.

Reuss, Martin. "Seeing Like an Engineer: Water Projects and the Mediation of the Incommensurable." *Technology and Culture* 49, no. 3 (2008): 531–46.

Reynolds, E. Bruce. *Japan in the Fascist Era*. New York: Palgrave Macmillan, 2004.

Robertson, Andrew. "Mobilizing for War, Engineering the Peace: The State, the Shop Floor and the Engineer, 1935–1960." Ph.D. dissertation, Harvard University, 2000.

Robertson, Jennifer. "Robo Sapiens Japanicus: Humanoid Robots in the Posthuman Family." *Critical Asian Studies* 39, no. 3 (2007): 369–98.

Rubinstein, Modest. "Relations of Science, Technology, and Economics Under Capitalism, and in the Soviet Union." In *Science at the Crossroads: Papers Presented to the International Congress of the History of Science and Technology Held in London from June 29th to July 3rd, 1931 by the Delegates of the U.S.S.R*, 2nd ed., ed. Nikolai Bukharin et al. London, UK: Frank Cass and Company, 1971 [1931].

Saguchi, Kazurō. "The Historical Significance of the Industrial Patriotic Association: Labor Relations in the Total War State." In *Total War and "Modernization,"* ed. Yasushi Yamanouchi, J. Victor Koschmann, and Ryūichi Narita, 261–88. Ithaca, NY: Cornell University East Asia Program, 1998.

Sakai, Naoki. "Subject and Substratum: On Japanese Imperial Nationalism." *Cultural Studies* 14, no. 3 (2000): 462–530.

———. *Translation and Subjectivity: On Japan and Cultural Nationalism*. Minneapolis: University of Minnesota Press, 1997.

Samuels, Richard J. *"Rich Nation, Strong Army": National Security and the Technological Transformation of Japan*. Ithaca, NY: Cornell University Press, 1994.

Sand, Jordan. *House and Home in Modern Japan*. Cambridge, MA: Harvard University Asia Center, 2003.

Schatzberg, Eric. "*Technik* Comes to America: Changing Meanings of Technology Before 1930." *Technology and Culture* 47, no. 3 (July 2006): 486–512.

Schmid, Andre. "Colonialism and the 'Korea Problem' in the Historiography of Modern Japan: A Review Article." *Journal of Asian Studies* 59, no. 4 (November 2000): 951–76.

Scott, Howard. *Introduction to Technocracy*. London: J. Lane, The Bodley Head, 1933.

Scott, James C. *Seeing Like a State: How Certain Schemes to Improve the Human Condition Have Failed*. New Haven, CT: Yale University Press, 1999.

Sewell, Bill. "Rethinking the Modern in Japanese History: Modernity in the Service of the Prewar Japanese Empire." *Japan Review* 16 (2004): 313–58.

Shoenberger, Karl. "Japan Aid: The Give and Take." *Los Angeles Times*, June 9, 1992.

Siegelbaum, Lewis H. *Stakhanovism and the Politics of Productivity in the USSR, 1935–1941*. Cambridge: Cambridge University Press, 1988.

Silverberg, Miriam. *Erotic Grotesque Nonsense: The Mass Culture of Modern Japanese Times*. Berkeley: University of California Press, 2006.

Skya, Walter A. *Japan's Holy War: The Ideology of Radical Shintō Ultranationalism*. Durham, NC: Duke University Press, 2009.

Smith, Henry D. *Japan's First Student Radicals*. Cambridge, MA: Harvard University Press, 1972.

Smith, Kerry. *A Time of Crisis: Japan, the Great Depression, and Rural Revitalization.* Cambridge, MA: Harvard University Asia Center, 2001.

Sombart, Werner. *The Future of Capitalism*. Berlin-Charlottenburg: Bucholz and Weisswange, 1932.

Spaulding, Robert M. "The Bureaucracy as a Political Force, 1920–1945." In *Dilemmas of Growth in Prewar Japan*, ed. James W. Morley, 33–80. Princeton, NJ: Princeton University Press, 1974.

———. "Japan's New Bureaucrats." In *Crisis Politics in Prewar Japan*, ed. George M. Wilson, 51–70. Tokyo: Sophia University Press, 1970.

Steinhoff, Patricia G. *Tenkō: Ideology and Societal Integration in Prewar Japan*. New York: Garland, 1991.

Sugiyama, Mitsunobu. "The World Conception of Japanese Social Science: The Kōza Faction, the Ōtsuka School, and the Uno School of Economics." In *New Asian Marxisms*, ed. Tani Barlow, 205–46. Durham, NC: Duke University Press, 2002.

Tanaka, Kakuei. *Building a New Japan: A Plan for Remodeling the Japanese Archipelago*, trans. Simul International. Tokyo: Simul Press, 1973.

Tanaka, Stefan. *Japan's Orient: Rendering Pasts into History*. Berkeley: University of California Press, 1993.

Tanaka, Toshiyuki. *Hidden Horrors: Japanese War Crimes in World War II*. Boulder, CO: Westview Press, 1996.

Tansman, Alan. "Introduction: The Culture of Japanese Fascism." In *The Culture of Japanese Fascism*, ed. Alan Tansman, 1–28. Durham, NC: Duke University Press, 2009.

Tatsuno, Sheridan. *The Technopolis Strategy: Japan, High Technology, and the Control of the Twenty-First Century*. New York: Prentice Hall, 1986.

Tsurumi, Shunsuke. *An Intellectual History of Wartime Japan, 1931–1945*. London: Kegan Paul International, 1986.

Tsutsui, William M. *Manufacturing Ideology: Scientific Management in Twentieth-Century Japan*. Princeton, NJ: Princeton University Press, 1998.

Tucker, David V. "Building 'Our' Manchukuo: Japanese City Planning, Architecture, and Nation-Building in Occupied Northeast China." Ph.D. dissertation, University of Iowa, 1999.

———. "Labor Policy and the Construction Industry in Manchukuo: Systems of Recruitment, Management, and Control." In *Asian Labor in the Wartime Japanese Empire: Unknown Histories*, ed. Paul Kratoska, 21–57. Armonk, NY: M. E. Sharpe, 2005.

Veblen, Thorstein. *The Engineers and the Price System*. Kitchener, ON: Batoche Books, 2001.

Walker, Brett L. *Toxic Archipelago: A History of Industrial Disease in Japan*. Seattle: University of Washington Press, 2010.

Watanabe, Kazuko. "Militarism, Colonialism, and the Trafficking of Women: 'Comfort Women' Forced into Sexual Labor for Japanese Soldiers." *Critical Asian Studies* 26, no. 4 (1994): 3–17.

Weber, Max. *The Protestant Ethic and the Spirit of Capitalism*, trans. Talcott Parsons. London: Routledge, 1992.

Weiner, Susan B. "Bureaucracy and Politics in the 1930s: The Career of Gotō Kunio." Ph.D. dissertation, Harvard University, 1984.

Weisenfeld, Gennifer. *MAVO: Japanese Artists and the Avant-Garde, 1905–1931*. Berkeley: University of California Press, 2002.

White, Richard. *The Organic Machine: The Remaking of the Columbia River*. New York: Hill and Wang, 1995.

Williams, Peter, and David Wallace. *Unit 731: Japan's Secret Biological Warfare in World War II*. New York: Free Press, 1989.

Woodall, Brian. *Japan Under Construction: Corruption, Politics and Public Works*. Berkeley: University of California Press, 1996.

Yamanouchi, Yasushi. "Total War and System Integration: A Methodological Introduction." In *Total War and "Modernization,"* ed. Yasushi Yamanouchi, J. Victor Koschmann, and Ryūichi Narita, 1–39. Ithaca, NY: Cornell University East Asia Program, 1998.

Yang, Daqing. "Japanese Colonial Infrastructure in Northeast Asia: Realities, Fantasies, Legacies." In *Korea at the Center*, ed. Charles K. Armstrong, Samuel S. Kim, and Stephen Kotkin, 90–107. Armonk, NY: M. E. Sharpe, 2006.

———. *Technology of Empire: Telecommunications and Japanese Expansion in Asia, 1883–1945*. Cambridge, MA: Harvard University Asia Center, 2011.

Young, Louise. *Japan's Total Empire: Manchuria and the Culture of Wartime Japanese Imperialism*. Berkeley: University of California Press, 1998.

Zaiki, Masumi, and Tōgō Tsukahara. "Meteorology on the Southern Frontier of Japan's Empire: Ogasawara Kazuo at Taihoku University." *East Asian Science, Technology and Society: An International Journal* 1, no. 2 (2007): 183–203.

Sources in Japanese

Adachi Hiroaki. "Dai tōa kensetsu shingikai to keizai 'kensetsu kōsō'—'Dai tōa kensetsu kihon hōsaku' no keisei o megutte." *Shien* 65, no. 1 (2004): 58–81.

Aikawa Haruki. *Bunka eigaron*. Tokyo: Kasumigaseki shobō, 1944.

———. "'Dai niji sangyō kakumei' setsu no hihan—Gijutsushi no hōhō ni yosete." *Kagakushi kenkyū*, no. 25 (1953): 12–16.

———. *Gendai gijutsuron*. Tokyo: Mikasa shobō, 1940.

———. *Gijutsu no seisaku to riron*. Tokyo: Kigensha, 1942.

———. *Gijutsu oyobi ginō kanri: Taryō seisan he no tankan*. Tokyo: Tōyō shokan, 1944.

———. *Gijutsuron*. Reprint. Tokyo: Kyūzansha, 1990 [1935].

———. *Gijutsuron nyūmon*. Tokyo: Mikasa shobō, 1942.

———. "Hitachi seisakujo Hitachi kōjō: Kōjō kengakuki—Hatsumei hōshōsei no moderu." *Gijutsu hyōron* (September 1941): 40–42.

———. *Sangyō gijutsu*. Tokyo: Hakuyōsha, 1942.

———. "Sobieto-teki ningen no keisei." *Sekai hyōron* 5, no. 3 (1953): 8–16.

———. *Tōna Ajia no shigen to gijutsu*. Tokyo: Mikasa shobō, 1944.

———. "Zaiso minshū undō no ichi kessan—Sono seika to jiko hihan." In *Aikawa Haruki shōden*, ed. Chūō kōron jigyō shuppan, 271–97. Tokyo: Yanashisada, 1979.

Akaeda Masayoshi. "'Nessa no chikai' no omoide." In Kōyūkai shōshi henshū iinkai, ed., *Kohō banri: Kahoku kensetsu shōshi [KB]*, 141–43.

Aki Kōichi. "Chūgoku no kasen ni manabu." In *KB*, 99–102.

Akigusa Hisao. "Kasen." In *KB*, 66–71.

Akinaga Tsukizō, Hoshino Naoki, Mōri Hideoto, Shiina Etsusaburō, Takahashi Kōjun, et al. "Manshūkoku keizai no genchi zadankai." *Tōyō keizai shinpō* (October 24, 1936): 26–39.

Antō shi kōsho, ed. *Yakushin Antō*. Shinkyō: Manshū jijō annaijo, 1939.

Asano Tatsuo. "Konnichi no tatakai." *Bunka eiga* 2, no. 6 (June 1942): 66–71.

———. "Konnichi no tatakai: Seisaku hōkoku." *Bunka eiga* 2, no. 6 (June 1942): 30–31.

"Chikoku no daihon: Chisui jigyō." In *Manshūkoku gensei kōtoku go-nen han*, ed. Manshūkoku tsūshinsha, 33–34. Reprint. Tokyo: Kuresu shuppan, 2000 [1938].

"Chisui o kanete hatsuden tekichi o tosa." *Manshū nippō*. February 5, 1935.

Chōsen sōtokufu, ed., *Ōryokkō kaihatsu iinkai kankei [OKIK]*. Keijō: Chōsen sōtokufu. Adachi Tōru Papers (copy), Office of Dr. Hirose Teizō, Fukuoka University.

Chōsen sōtokufu teishinkyoku. *Chōsen suiryoku chōsasho dai ikkan (sōron)*. Keijō (Seoul): Chōsen sōtokufu teishinkyoku, 1930.

Chūgoku rengō junbi ginkō komonshitsu, ed. *Chūgoku rengō junbi ginkō go-nen shi*. Beijing: Chūgoku rengō junbi ginkō komonshitsu, 1944.

Chūgokujin kyōsei renkō Nishimatsu kensetsu saiban o shien suru kai, ed. *Senzen no "Suihō" kara "Yasuno" no ima he—Nishimatsu kensetsu no sensō sekinin [NKSS]*. Hiroshima: Chūgokujin kyōsei renkō Nishimatsu kensetsu saiban o shien suru kai, 1999.

Chūō kōron jigyō shuppan, ed. *Aikawa Haruki shōden*. Tokyo: Yanashisada, 1979.

Chūō Nikkan kyōkai, ed. *Chōsen denki jigyōshi*. Tokyo: Chūō Nikkan kyōkai, 1981.

Eda Kenji, Kai Gaku Shi, and Matsumura Takao, eds. *Mantetsu rōdōshi no kenkyū*. Tokyo: Nihon keizai hyōronsha, 2002.

Emori Yasuhei. "Kensetsu sōsho rekidai gikan ni tsuite." In *KB*, 11–15.

Fukai Junichi. *Minamatabyō no seiji keizaigaku: Sangyōshi-teki haikei to gyōsei sekinin*. Tokyo: Keisō shobō, 1999.

Furukawa Takahisa. "Kakushin kanryō no shisō to kōdō, 1935–1945." *Shigaku zasshi* 99, no. 3 (1990): 1–38.

———. *Shōwa senchūki no sōgō kokusaku kikan*. Tokyo: Yoshikawa kōbunkan, 1992.

Furumi Tadayuki. *Wasureenu Manshūkoku*. Tokyo: Keizai ōraisha, 1978.

Gaimubu iin. "Suibotsuchi jūmin no shori." (1938?). In *OKIK*, 227–28.

Harada Kiyoshi. *Suihō hatsudenjo kōji taikan*. Amagasaki: Doken bunkasha, 1942.

Haraguchi Chūjiro. "Manshū no kasen ni tsuite (sono ichi)." *Manshū gijutsu kyōkaishi* 14, no. 101 (1937): 416–23.

———. "Ryōga no kaishū ni tsuite 1." *Manshū no gijutsu* 14, no. 101 (1937): 249–52.

———. "Ryōga no kaishū ni tsuite 2." *Manshū no gijutsu* 14, no. 104 (1937): 287–94.

———. "Ryōga no kaishū ni tsuite (kan)" *Manshū no gijutsu* 15, no. 112 (1938): 389–94.

Haraguchi Chūjirō no yokogao kankōkai, ed. *Haraguchi Chūjirō no yokogao.* Tokyo: Haraguchi Chūjirō no yokogao kankōkai, 1966.

Hata Ikuhiko. *Kanryō no kenkyū.* Tokyo: Kōdansha, 1983.

Hazama-gumi hyakunenshi hensan iinkai, ed. *Hazama-gumi hyakunen shi, 1889–1945.* Tokyo: Hazama-gumi, 1989.

Heianhokudō iin. "Keibi oyobi hōan torishimari taisaku." (1938?). In *OKIK*, 232–41.

———. "Tochi kaishū taisaku." (1938?). In *OKIK*, 216–26.

Hensha. "Hajime ni." In *Kōain to senji Chūgoku chōsa*, ed. Honjō Hisako, Uchiyama Masao, and Kubo Tōru, 1–20. Tokyo: Iwanami shoten, 2002.

Henshū iin Ogawa. "Soshiki no hensen to nisseki shokuin no posuto." In *KB*, 7–9.

Hirao Katsu. "Omoide." In *KB*, 265–67.

Hirose Teizō. "Chōsen ni okeru tochi shūyō rei—1910–1920 nendai o chūshin ni." *Niigata kokusai jōhō daigaku jōhō bunka gakubu kiyō* 2 (March 1999): 1–22.

———. "Chōsen sōtokufu no doboku kanryō." In *Nihon no Chōsen Taiwan shihai to shokuminchi kanryō*, ed. Matsuda Toshihiko and Yamada Atsushi, 260–332. Tokyo: Shibunkaku, 2009.

———. "Gunju keiki to denryoku kensetsu kōji." In *Doboku: Sangyō no shakaishi 12*, ed. Tamaoki Motoi, 135–64. Tokyo: Nihon keizai hyōronsha, 1993.

———. "'Manshūkoku' ni okeru Suihō damu no kensetsu." *Niigata kokusai jōhō daigaku jōhō bunka gakubu kiyō* 6 (March 2003): 1–25.

———. "Shokuminchiki Chōsen ni okeru Suihō hatsudenjo kensetsu to ryūbatsu mondai." *Niigata kokusai jōhō daigaku jōhō bunka gakubu kiyō* 1 (March 1998): 39–58.

———. "Suihō hatsudenjo kensetsu ni yoru suibotsuchi mondai—Chōsen o chūshin ni." *Chōsen gakuhō* 139 (June 1991): 1–35.

Hiroshige Tetsu. *Kagaku no shakaishi: Kindai Nihon no kagaku gijutsu.* 2 vols. Tokyo: Chūō kōronsha, 1979.

"Hokushi kensetsu sōsho ni kinin shitaru Miura Shichirō hakase to sono ichigyō." *Kōji gahō* 14, no. 7 (July 1938): 33–36.

"Hokushi sangyō gokanen keikaku an tekiyō." December 1941. (Accessed September 3, 2009.)

Hokushina hōmengun sanbōchō. "Pekin seikō shin shigai kensetsu keikaku ni kan suru ken." *Rikushi mitsu dai nikki*, no. 72 (1939). www.jacar.go.jp. (Accessed July 23, 2009.)

"Hōman damu manninkō." http://www.ac.auone-net.jp/~miyosi/db3.htm. (Accessed July 27, 2012.)

Honma Mihoko, ed. *Honma Norio o shinonde [HNS].* Tokyo: Honma Mihoko, 1977.

Honma Norio. "Manshūkoku suiryoku denki jigyō ni tsuite." *Kōgyō kokusaku* 2 (August 1939): 28–38.

"Honma Norio ni taisuru—tōjitsu no shitsugi ōtō sokkiroku." *Kōgyō kokusaku* 2 (August 1939): 38–42.

"Honma-san suiden o kataru." *Dengyō* 10, no. 109 (1944): 14–19.

Hori Kazuo. "'Manshūkoku' ni okeru denryokugyō to tōsei seisaku." *Rekishigaku kenkyū*, no. 564 (February 1987): 13–30, 58.

Horii Kōichi, "Shinminkai to Kahoku senryō seisaku (chū)," *Chūgoku kenkyū geppō* 47, no. 1 (1993): 1–13.

Hōseikyoku. *Kōa gijutsu iinkai kansei o sadamu*, August 25, 1939. www.jacar.go.jp. (Accessed September 17, 2010.)

Hoshino Naoki. "Honma Norio-kun to Hōman damu." In *HNS*, 45–55.

Hyōya Sahei. "Ōryokkō mokuzaikai no tenbō," *Shingishū shōkō kaigijo geppō* 138 (January 1941): 16–18.

Igarashi Shinsaku. "Ryūga chisui kōji ni tsuite." *Doboku Manshū* 3, no. 1 (1943): 2–16.

Iida Kenichi. *Gijutsu: Ichigo no jiten*. Tokyo: Sanseidō, 1995.

Imamura Taihei. *Kiroku eigaron*. Kyoto: Daiichi geibunsha, 1940.

Imura Tetsuo, ed. *Jyūgonen sensō jūyō bunken shirizu (17): Kōain kankō tosho zasshi mokuroku*. Tokyo: Fuji shuppan, 1994.

Inose Yasuo. "Pekin Tōseikō." In *KB*, 115–18.

Ishimoto Itsuo, Kubo Kichizō, Shiina Etsusaburō, et al. "Nichi-Man-Shi o tsūzuru keikaku keizai no sai ginmi zadankai." *Chōsa shūhō* (February 9, 1939): 1–10.

Itō Takashi. "Mōri Hideoto ron oboegaki." In Itō Takashi, *Shōwaki no seiji (zoku)*. Tokyo: Yamakawa shuppankai, 1993.

———. "Shōwa jūsannen Konoe shintō mondai kenkyū oboegakisho." In *"Konoe shintaisei" no kenkyū*, ed. Nihon seiji gakkai. Tokyo: Iwanami shoten, 1972.

Iwamiya Noboru. "Sainan no omoide." In *KB*, 155–56.

"Kahoku no sangyō kaihatsu." *Tōa shinpō*, July 9, 1940. In Miyamoto Takenosuke, *Miyamoto Takenosuke nikki, 1937–1941*. Nagoya: Denki tsūshin kyōkai tōkai shibu, 1971. Inserted between pages 86 and 87 in the year 1940 section.

Kajimura Hideki. "Sen kyūhaku sanjū nendai Manshū ni okeru kōnichi tōsō ni taisuru Nihon teikokushugi no sho hōsaku—'Zaiman Chōsenjin mondai' to kanren shite." *Nihonshi kenkyū*, no. 94 (November 1967): 25–55, 95.

Kamakura Ichirō. "Chūgoku no 'kōsen kenkoku' o hihan su." *Kaibō jidai* (February 1939): 4–15.

———. "Chūshōteki bukka to gutaiteki bukka." *Kaibō jidai* (April 1940): 4–35.

———. "Gijutsu no kaihō to seiji." *Kaibō jidai* (September 1939): 4–8.

———. "Handō o kokufuku suru seiji: Kigen nisen roppyaku hyaku-nen sengen." *Kaibō jidai* (January 1940): 4–12.

———. "Jiken dai yonki wa seiji o tenkai su." *Kaibō jidai* (December 1938): 70–77.

———. "Kokumin keizai to shieki." *Kaibō jidai* (May 1939): 83–89.

———. "Kokumin seikatsu soshiki no kiten." *Kaibō jidai* (November 1939): 4–11.

———. "Kokumin soshiki to tōa kyōdōtai no fukabunsei." *Kaibō jidai* (January 1939): 22–28.

———. "Nihon kokumin keizai no keisei to seiji: Hō toshite no tōa no shin chitsujo." *Kaibō jidai* (April 1939): 25–32.

———. "Taiheiyō kūkan no seikaku kakumei—Sekai seiji to tōa kyōeiken no hensei." *Chūō kōron* (November 1940): 34–42.

———. "'Tōa ittai' toshite no seijiryoku." *Kaibō jidai* (November 1938): 6–11.

———. "Tōa kyōdōtai to gijutsu no kakumei." *Kaibō jidai* (March 1939): 4–12.

———. "Tōa kyōseitai kensetsu no sho jōken." *Kaibō jidai* (October 1938): 23–29.

———. "Tōa ni okeru bōkyō no igi." *Kaibō jidai* (June 1939): 4–10.

———. "Tōsei keizai no hinkon no gennin: Shizenryoku ka soshikiryoku ka." *Kaibō jidai* (December 1939): 16–21.

Kamei Kanichirō. *Kamei Kanichirō shi danwa sokkiroku.* Tokyo: Nihon kindai shiryō kenkyūkai, 1970.

———. *Natchisu kokubō keizairon.* Tokyo: Tōyō keizai shuppanbu, 1939.

Kan Chae-On. *Chōsen ni okeru Nitchitsu kontserun.* Tokyo: Fuji shuppan, 1985.

Kaneko Hiroshi. "Zentaishugi keizaigaku no ni keikō: Gottoru to Shupānu." *Kokumin keizai zasshi* 65, no. 2 (August 1938): 35–48.

KA-sei. "Miura Shichirō o mukaete." *Dōro no kairyō* 21, no. 3 (March 1939): 143–45.

Kashiwabara Hyōtaro. "Sangyō hōkoku undō no shin hōshin." In *Shiryō Nihon gendaishi.* Vol. 7, ed. Kanda Fumihito, 500–502. Tokyo: Ōtsuki shoten, 1981.

Kawahara Hiroshi. *Shōwa seiji shisō kenkyū.* Tokyo: Waseda daigaku shuppanbu, 1979.

Kawai Kazuo. "Dai niji suiryoku chōsa to Chōsen sōtokufu kanryō no suiryoku ninshiki." In *Nihon no Chōsen Taiwan shihai to shokuminchi kanryō,* ed. Matsuda Toshihiko and Yamada Atsushi, 303–32. Tokyo: Shibunkaku, 2009.

Kawakami Yukio. *Kokudo keikaku no hensen: Koritsu to kōhei no keikaku shisō.* Tokyo: Kajima shuppankai, 2008.

Kawamura Masami. "Damu to iu 'kaihatsu pakkeji.'" In *Kaihatsu no kūkan, kaihatsu no jikan: Sakuma damu to chiiki shakai no han seiki,* ed. Machimura Takashi, 73–92. Tokyo: Tokyo daigaku shuppankai, 2006.

Keimukyoku iin, "Keibi oyobi hōan torishimari taisaku" (1939?). In *OKIK,* 130–36.

"Kensetsu no uta." In *KB,* 137.

Kensetsu sōsho Pekin-shi kensetsu kōteikyoku. *Pekin-shi tōseikō shin shigaichi chizu.* 1940.

Kensetsu sōsho toshi kyoku. *Pekin toshi keikaku yōzu.* Beijing: Kensetsu sōsho toshi kyoku, 1939.

Kimoto Shōji. *Hon Kon no mizu.* Tokyo: Kōdansha, 1972.

Kitamura Sukehiro. "Kodoku no tensai." In *KB,* 326–27.

Kitsurin kōteikyoku. "Shōkakō entei hatsuden kōji gaikyō." *Manshū no gijutsu* 17, no. 134 (1940): 208–14.

Kitsurin shōkō kōkai, ed. *Kitsurin shōkō kōkai jigyō hōkokusho.* Jilin: Kitsurin shōkō kōkai, 1940.

———. *Kitsurin shōkō kōkai jigyō hōkokusho.* Jilin: Kitsurin shōkō kōkai, 1941.

———. *Suiryoku hatsuden ni tomonai kagaku kōgyō toshika suru Kitsurin.* Jilin: Kitsurin shōkō kōkai, 1938.

Kōain. *Dokuryū nyūkai genka shori hōsaku* (September 1941).

———. *Pekin toshi keikaku gaiyō.* Tokyo: Kōain, 1941.

Kobayashi Gorō. "Senji seikatsujo no seiji-teki igi." *Kokumin hyōron* 12, no. 7 (July 1940): 16–20.

Kobayashi Hideo. *Dai tōa kyōeiken no keisei to hōkai.* Tokyo: Ochanomizu shobō, 1975.

———. "Kahoku senryō seisaku no tenkai katei: Kō shokutaku han no kessei to katsudō o chūshin ni." *Komazawa daigaku keizaigaku ronshū* 9, no. 3 (1977): 191–203.

———. *Mantetsu—"Chi no shūdan" no tanjō to shi*. Tokyo: Yoshikawa kōbunkan, 1996.

———. "Shokuminchi keiei no tokushitsu." In *Iwanami kōza kindai Nihon to shokuminchi: Shokuminchi to sangyōka*. Vol. 3, ed. Ōe Shinobu, Asada Kyōji, Mitani Daiichirō, Gotō Kenichi, Kobayashi Hideo, Takasaki Sōji, Wakabayashi Masahiro, and Kawamura Minato, 3–26. Tokyo: Iwanami shoten, 1993.

Kodaira Keima, ed. *Shingishū shōkō yōran Shōwa jūshichi nenkan*. Reprint. Seoul: Keijin bunkasha, 1989 [1942].

Kodama Reon. "Sakuma damu kensetsu to iu 'gijutsu no shōri.'" In *Kaihatsu no kūkan, kaihatsu no jikan: Sakuma damu to chiiki shakai no han seiki*, ed. Machimura Takashi, 153–70. Tokyo: Tokyo daigaku shuppankai, 2006.

"Kōga shori yōkō (an)—kyokumitsu." In Kashiwabara Hyōtarō monjo, Document 458. [Date unknown.] Kensei shiryōshitsu. National Diet Library, Tokyo.

Kohiro Yoshio. "Hotei sekōjo, jimukanjo tōji no omoide." In *KB*, 194–95.

Kokusaku kenkyūkai. *Dai tōa kyōeiken gijutsu taiseiron*. Tokyo: Nihon hyōronsha, 1945.

Komagome Takeshi. *Shokuminchi teikoku Nihon no bunka tōgō*. Tokyo: Iwanami shoten, 1996.

Kōshinkai zaika gyōseki kiroku henshū iinkai, ed. *Ōdo no gunzō*. Tokyo: Kōshinkai, 1983.

Koshizawa Akira. "Daitōkō no keikaku to kensetsu (1937–1945)—Manshū ni okeru mikan no dai kibo kaihatsu purojekuto." *Nihon dobokushi kenkyū happyō ronbunshū* 6 (1986): 223–34.

———. *Harupin no toshi keikaku*. Tokyo: Chikuma shobō, 2004.

———. "Nihon senryōka no Pekin toshi keikaku, 1937–1945." In *Dai gokai Nihon doboku shi kenkyū happyō ronbunshū*, ed. Doboku gakkai Nihon doboku shi kenkyū iinkai. Tokyo: Doboku gakkai, 1986: 265–76.

———. "Nitchū sensō ni okeru senryōchi toshi keikaku ni tsuite." *Toshi keikaku bessatsu*, no. 8 (1979): 385–90.

———. "Sangyō kiban no kōchiku to toshi keikaku: Taiwan, Manshū, Chūgoku no toshi keikaku." In *Kindai Nihon to shokuminchi 3: Shokuminchika to sangyōka*, ed. Ōe Shinobu, Asada Kyōji, Mitani Daiichirō, Gotō Kenichi, Kobayashi Hideo, Takasaki Sōji, Wakabayashi Masahiro, and Kawamura Minato, 183–242. Tokyo: Iwanami shoten, 1992–1993.

———. *Shokuminchi Manshū no toshi keikaku*. Tokyo: Ajia keizai kenkyūjo, 1978.

Kōtsūbu daijin kanbō shiryōka. *Kotsūbu yōran*. Shinkyō: Kōtsūbu daijin kanbō shiryōka, 1944.

Kōtsūbu ryōga chisui chōsasho, ed. *Ryōga suikei shiryō*. 3 vols. Shinkyō: Kōtsūbu ryōga chisui chōsasho, 1938–1940.

Kōtsubu Shōbu chisui kōteisho. *Ryūga chisui kōji no taiyō*. Zhangwu: Kōtsubu Shōbu chisui kōteisho, 1939.

Kōyūkai shōshi henshū iinkai, ed. *Kohō banri: Kahoku kensetsu shōshi* [*KB*]. Tokyo: Kōyūkai, 1977.

Kubo Kenji. "Kōjō to bunka eiga." *Bunka eiga* (May 1942): 34–37.

Kubo Tōru. "Kōain ni yoru Chūgoku chōsa." In *Kōain to senji Chūgoku chōsa*, ed. Honjō Hisako, Uchiyama Masao, and Kubo Tōru, 74–103. Tokyo: Iwanami shoten, 2002.

Kubota Yutaka. "Watakushi no rirekisho." In *Watakushi no rirekisho*. Vol. 27, ed. Nihon keizai shinbunsha, 241–321. Tokyo: Nihon keizai shinbunsha, 1966.

Kuga Tokuhei. *Beikoku ni okeru kōentei shisatsu hōkokusho*. Tokyo: Denki kasen shinpōsha, 1937.

Kuroda Shigeharu. "Futōkō 'Daitōkō' ga mondai ni naru made." In *Daitōkō futō jōkyō shisatsu kōen*, ed. Antō shōkō kōkai, 7–19. Andong: Antō shōkō kōkai, 1939.

Kuwabara Mamoru. "Kawa kaihatsu no shisō." *Oruta* 3 (1992): 33–45.

Kuwano Hiroshi. *Senji tsūka kōsaku shi ron: Nitchū tsūka sen no bunseki*. Tokyo: Hōsei daigaku shuppan kyoku, 1965.

Kuzuoka Masao. "Shinkyō kōgakuin." In *Haraguchi Chūjirō no yokogao*, ed. Haraguchi Chūjirō no yokogao kankōkai, 145–46. Tokyo: Haraguchi Chūjirō no yokogao kankōkai, 1966.

Machida Yoshitomo. "Honma-san no omoide." In *HNS*, 72–79.

Machimura Takashi. "'Sakuma damu' kenkyū no kadai to hōhō." In *Kaihatsu no kūkan, kaihatsu no jikan: Sakuma damu to chiiki shakai no han seiki*, ed. Machimura Takashi, 1–28. Tokyo: Tokyo daigaku shuppankai, 2006.

"Manshū dengyō shi" henshū iinkai, ed. *Manshū dengyō shi* [MDS]. Tokyo: Manshū dengyōkai, 1976.

Manshū denki kyōkai. *Shōkakō suiryoku hatsuden keikaku gaiyō*. Dalian: Manshū denki kyōkai, 1940.

Manshū jijō annaijo, ed. *Daitōkō to hōkō Tōhendō*. Shinkyō: Manshū jijō annaijo, 1940.

———. *Kitsurin jijō*. Shinkyō: Manshū jijō annaijo, 1941.

Manshūkoku shi hensan kankōkai, ed. *Manshūkoku shi (Kakuron)*. Tokyo: Manmō dōhō engokai, 1970–71.

Manshūkoku suiryoku denki kensetsu kyoku. *Shōkakō dai ichi hatsudenjo kōji shashinchō dai ichi go*. Shinkyō: Manshū denki kyōkai, 1940.

Manshūkoku tsūshinsha kaisha. "Shōkakō hatsudenjo kensetsu no kushin o kiku zadankai." *Manshū denki kyōkai kaihō*, no. 28 (1943): 29–37.

Matsumoto Toshirō. "Dai niji taisenki no senji taisei kōsō ritsuan no ugoki." *Okayama daigaku keizaigakukai zasshi* 25, no. 1–2 (1993): 99–123.

Matsumoto Tōtarō. *Kitsurin ringyōshi*. Jilin: Kitsurin mokuzai dōgyō kumiai, 1940.

Matsuura Shigeki. "Konkureeto damu ni miru senzen no sekō gijutsu." *Dobokushi kenkyū*, no. 18 (1998): 569–78.

———. *Senzen no kokudo seibi seisaku*. Tokyo: Nihon keizai hyōronsha, 2000.

Miki Kiyoshi. "Gijutsu tetsugaku." In Vol. 7 of *Miki Kiyoshi zenshū*. 19 vols, ed. Ōuchi Hyōei, 197–330. Tokyo: Iwanami Shoten, 1966–68.

Miki Kiyoshi and Mōri Hideoto. "Ashita no kagaku Nihon no sōzō." *Kagakushugi kōgyō* (January 1941): 186–207.

Mikuriya Takashi. *Seisaku no sōgō to kenryoku: Nihon seiji no senzen to sengo.* Tokyo: To-kyo daigaku shuppankai, 1996.

Minami Manshū tetsudō kabushiki kaisha chōsabu. *Shina toshi fudōsan kankō chōsa shiryō: Sainan ni kansuru hōkokusho.* Dalian: Mantetsu chōsabu, 1943.

———. *Tōko ni kansuru hōkokusho.* Dalian: Mantetsu chōsabu, 1942.

Minami Manshū tetsudō keizai chōsakai. *Manshū chisui hōsaku.* Dalian: Minami Manshū tetsudō keizai chōsakai, 1935.

Miura Shichirō. "Hokushi ni okeru doboku jigyō no gaiyō." *Doboku gakkai shi* 24, no. 12 (1938): 1385–88.

Miyamoto Takenosuke. *Gendai gijutsu no kadai.* Tokyo: Iwanami shoten, 1940.

———. *Gijutsu to kokusaku.* Tokyo: Kagakushugi kōgyōsha, 1940.

———. *Gijutsusha no michi.* Tokyo: Kagakushugi kōgyōsha, 1939.

———. "Jogen." In Miyamoto Takenosuke, *Tairiku kensetsu no kadai,* 1–2. Tokyo: Iwa-nami shoten, 1941.

———. "Kōa gijutsu no kihon seikaku." In Miyamoto Takenosuke, *Gendai gijutsu no kadai,* 111–29. Tokyo: Iwanami shoten, 1940.

———. "Kōa gijutsu no konpon genri." In Miyamoto Takenosuke, *Tairiku kensentsu no kadai,* 141–54. Tokyo: Iwanami shoten, 1941.

———. "Kōa gijutsu no mittsu no seikaku." In Miyamoto Takenosuke, *Tairiku kensetsu no kadai,* 177–83. Tokyo: Iwanami shoten, 1941.

———. "Kōa gijutsu no shidō seishin." In Miyamoto Takenosuke, *Gijutsu to kokusaku,* 9–14. Tokyo: Iwanami shoten, 1940.

———. "Kōa kensetsu no suiri mondai." In Miyamoto Takenosuke, *Tairiku kensetsu no kadai,* 103–14. Tokyo: Iwanami shoten, 1941.

———. "Manmō mondai to gijutsuka." *Kōjin* (March 1932): 29–33.

———. *Miyamoto Takenosuke nikki, 1937–1941.* Nagoya: Denki tsūshin kyōkai tōkai shibu, 1971.

———. "Nanman oyobi hokushi zakkan." In Miyamoto Takenosuke, *Gijutsusha no michi,* 280–340. Tokyo: Kagakushugi kōgyōsha, 1939.

———. "Nikka shinzen to gijutsu-teki teikei." In Miyamoto Takenosuke, *Gijutsu to kokusaku,* 77–86. Tokyo: Iwanami shoten, 1940.

———. "Ō Chō Mei to Kō Yū Nin." In Miyamoto Takenosuke, *Gijutsusha no michi,* 119–22. Tokyo: Kagakushugi kōgyōsha, 1939.

———. "Shin tōa kensetsu to gijutsu no shimei." In Miyamoto Takenosuke, *Tairiku kensetsu no kadai,* 184–202. Tokyo: Iwanami shoten, 1941.

———. "Shina kaihatsu to gijutsu." *Doboku gakkai shi* 24, no. 7 (1938): 711–13.

———. "Shina Manshū bekken ki." In Miyamoto Takenosuke, *Gijutsusha no michi,* 223–79. Tokyo: Kagakushugi kōgyōsha, 1939.

———. "Tairiku hatten to gijutsu." In Miyamoto Takenosuke, *Gijutsu to kokusaku,* 106–21. Tokyo: Iwanami shoten, 1940.

———. *Tairiku kensetsu no kadai.* Tokyo: Iwanami shoten, 1941.

———. "Tekunokurashī no kenkyū." In Miyamoto Takenosuke, *Gijutsu to kokusaku,* 25–54. Tokyo: Iwanami shoten, 1940.

———. "Tōhoku shinkō keikaku hihan." In Miyamoto Takenosuke, *Gijutsu to koku-saku*, 60–72. Tokyo: Iwanami shoten, 1940.

Mōri Hideoto. "Dai tōa bunka no igi." *Mōri Hideoto kankei bunsho*. Document 222. [Date unknown.] Kensei shiryōshitsu. National Diet Library.

———. "Dai tōa sensō o tsūjite." *Mōri Hideoto kankei bunsho*. Document 213. 1942 or 1943. Kensei shiryōshitsu. National Diet Library.

———. "Haihin kaishū roku." *Kagakushugi kōgyō* (December 1939): 162–64.

———. "Hitotsu no kotae toshite no seikatsu sōdanjo." In *Senji seiji keizai shiryō*. Vol. 1, ed. Kokusaku kenkyūkai, 8–9. Tokyo: Hara shobō, 1982.

———. "Keizai shintaisei kōza." *Mōri Hideoto kankei bunsho*. Document 221. [Date unknown.] Kensei shiryōshitsu. National Diet Library.

———. "Seisan keizai no konpon rinen." *Kagakushugi kōgyō* (October 1940): 22–27.

———. "Shina no sangyō kaihatsu." *Tōyō* (August 1939): 73–83.

———. "Tai shi keizai gijutsu no sōzō." *Keizai jōhō* (June 1939): 97–105.

———. "Tōa keizai kensetsu ni tsuite, dai ikkō." *Mōri Hideoto kankei bunsho*. Document 218. September 1944. Kensei shiryōshitsu. National Diet Library.

Mōri Hideoto, Sakomizu Hisatsune, Kashiwabara Hyōtarō, and Minobe Yōji. "Kakushin kanryō: Shintaisei o kataru zadankai." *Jitsugyō no Nihon* (January 1941): 52–67.

Morikawa Kakuzō. *Natchisu Doitsu no kaibō*. Tokyo: Koronasha, 1940.

Murata Junichi. *Gijutsu no tetsugaku*. Tokyo: Iwanami shoten, 2009.

Nagatsuka Riichi. *Kubota Yutaka*. Tokyo: Denki jōhōsha, 1966.

Naimukyoku iin. "Tochi kaishū taisaku." (1941?). In *OKIK*, 29–32.

Nakai Masakazu. "Bigaku nyūmon." In *Nakai Masakazu zenshū 3: Gendai geijutsu no kūkan*, ed. Kuno Osamu, 1–139. Tokyo: Bijutsu shuppansha, 1981.

Nakamura Aijirō. "Kahoku no omoide no ki." In *KB*, 247.

Nakamura Seiji. *Gijutsuron ronsōshi*. 2 vols. Tokyo: Aoki shoten, 1975.

Nakamura Takafusa. *Senji Nihon no Kahoku keizai shihai*. Tokyo: Yamakawa shuppan-sha, 1983.

Nakayama Shigyō. *Yakushin Kitsurin no dai kōgyō to kankō jigyō ni tsuite*. Jilin: Kitsurin shōkō kōkai, 1939.

Nan Longrui. "'Manshūkoku' ni okeru Hōman suiryoku hatsudenjo no kensetsu to sengo no saiken." *Ajia keizai* 48, no. 5 (2007): 2–20.

———. "'Manshūkoku' ni okeru Nihon no shokuminchi tōchi to sengo Tōhoku chiiki no saiken—Shokuminchi tōchi no nimensei to rekishi no renzoku danzetsu ni kans-uru kenkyū." Ph.D. dissertation, Tsukuba University, 2007.

Nan Sō Sui. "Gakuchō kakka." In *Haraguchi Chūjirō no yokogao*, ed. Haraguchi Chūjirō no yokogao kankōkai, 147–49. Tokyo: Haraguchi Chūjirō no yokogao kankōkai, 1966.

Naoki Rintarō. "Daitōkō futō jōkyō shisatsu kōen." In *Daitōkō futō jōkyō shisatsu kōen*, ed. Antō shōkō kōkai, 1–6. Andong: Antō shōkō kōkai, 1939.

———. *Gijutsu seikatsu yori*. Tokyo: Tokyodō, 1918.

———. "Manshū no suiryoku denki jigyō." *Suiri to doboku* 10, no. 11 (1938): 2–9.

———. "Sōsa ni Manshū he." *Kōjin* (January 1934): 3.

Nessa no chikai. Videocassette. Directed by Watanabe Kunio. 1940. Tokyo: Tōhō kabu-shiki kaisha, 1990s.

"Nichi-Man-Shi keizai kensetsu yōkō." October 3, 1940. National Diet Library. http://rnavi.ndl.go.jp/politics/entry/bib00277.php. (Accessed December 13, 2009.)

Nihon hyōron shinsha, ed. *Yōyōko: Minobe Yōji tsuitōroku.* Tokyo: Nihon hyōron shin-sha, 1954.

Nippon Kōei. *Nippon Kōei sanjū go nen shi.* Tokyo: Nippon Kōei, 1981.

Nōrinkyoku iin. "Ryūbatsu taisaku" (1938?). In *OKIK,* 246–53.

———. "Ryūbatsu taisaku" (1939?). In *OKIK,* 141–47.

———. "Ryūbatsu taisaku" (1941?). In *OKIK,* 24–28.

Oida Hiromi. "Shiryō: 'Shūdan' buraku ni tsuite." In *NKSS,* 86–91.

Ōkochi Masatoshi. "Hokushi no kōgyō." *Kagakushugi kōgyō* (August 1938): 68–76.

———. *Shihonshugi kōgyō to kagakushugi kōgyō.* Tokyo: Kagakushugi kōgyōsha, 1938.

Ōmura Miyoji. "Zai tairiku hōjin no jūtaku mondai ni tsuite." *Kenchiku zasshi* 56, no. 691 (1942): 756–63.

Orisaka Rigorō. "Pekin shi kōcho kōmukyoku." In *KB,* 38–41.

"Ōryokkō guchi ni okeru Sen-man chikukō to toshi keikaku." *Shokugin chōsa geppō* (February 1940).

"Ōryokkō suiden suibotsuchi happyō saru." *Shokugin chōsa geppō,* no. 10 (March 1939).

Ōyodo Shōichi. *Gijutsu kanryō no seiji sankaku: Nihon no kagaku gijutsu gyōsei no maku hiraki.* Tokyo: Chūkō shinsha, 1997.

———. *Kindai Nihon no kōgyō rikkokuka to kokumin keisei: Gijutsusha undō ni okeru kōgyō kyōiku mondai no tenkai.* Tokyo: Suzusawa shoten, 2009.

———. *Miyamoto Takenosuke to kagaku gijutsu gyōsei.* Tokyo: Tōkai daigaku shup-pankai, 1989.

Pekin Nihon shōkō kaigisho. *Pekin Nihon shōkō meikan.* Beijing: Pekin Nihon shōkō kaigisho, 1942.

Rikugunshō shinbunhan. *Kokubō no hongi to sono kyōka no teishō.* Tokyo: Rikugunshō shinbunhan, 1934.

"Rikushi mitsuden gojūgo-go." February 13, 1939 in *Rikushi dai nikki,* no. 13 (1939). www.jacar.go.jp. (Accessed March 2, 2010.)

Ryōga chisui chōsajo. "Ryōga suiri shiken no kinkyō." *Kensetsu* 4, no. 6 (1939): 24–29.

Ryōga chisui keikaku shingikai gijiroku. January 1938.

"Ryūbatsu mondai masu masu jūdaika—Suiden gawa no seii ni kitai," *Shingishū shōkōkaigisho geppō* 133 (August 1940): 10–13.

Saigusa Hiroto. "Gijutsu no shisō." In *Saigusa Hiroto chosakushū 7,* 117–234. Tokyo: Chūo kōronsha, 1972–77.

Satō Jin. *"Motazaru kuni" no shigen ron: Jizoku kanō na kokudo o meguru mō hitotsu no chi.* Tokyo: Tokyo daigaku shuppankai, 2011.

Satō Kurō. "Shinkyō no yūjintachi." In *Haraguchi Chūjirō no yokogao,* ed. Haraguchi Chūjirō no yokogao kankōkai, 135–37. Tokyo: Haraguchi Chūjirō no yokogao kankōkai, 1966.

Satō Takeo. "Hokushi ni okeru hōjin jūtaku mondai no bekken." *Kenchiku zasshi* 55, no. 678 (1941): 728–37.

Satō Yoshimasa and Yamazaki Hiroshi. "Dōro." In *KB*, 53–66.

Sawai Minoru. "Kagaku gijutsu shintaisei kōsō no tenkai to gijutsuin no tanjō." *Osaka daigaku keizaigaku* 41, nos. 2–3 (1991): 367–95.

"Sekai dai ni no Suihō damu kankō." *Shashin shūhō*, no. 194 (November 1941): 6–7.

"Senji seikatsu sōdanjo to wa." *Shashin shūhō* (June 12, 1940): 10–11.

Shibata Yoshimasa. "Kōain to Chūgoku senryōchi gyōsei." In *Kōain to senji Chūgoku chōsa*, ed. Honjō Hisako, Uchiyama Masao, and Kubo Tōru, 22–46. Tokyo: Iwanami shoten, 2002.

Shinohara Takeshi. "Chōki kensetsu kokumin sai soshiki to 'gijutsu' no igi no kakujū." *Gijutsu Nihon*, no. 191 (1938).

———. "Kagaku no sōgōka to gijutsu no sōgōka." *Kōgyō kokusaku* 1, no. 7 (1938): 21–29.

Shiobara Saburō. "Kahoku ni okeru toshi kensetsu." *Toshi kōron* 27, no. 6 (1944).

———. "Kahoku ni okeru toshi kensetsu." *Toshi kōron* 27, no. 8 (1944).

———. "Toshi." In *KB*, 107–15.

———. *Toshi keikaku ten to sen*. Tokyo: Shiobara toshi keikaku konsorutanto, 1971.

Son Jung Mok. *Nihon tōchika no Chōsen toshi keikakushi kenkyū*, trans. Ichioka Miyuki, Nishigaki Yasuhiko, and Lee Jong Ji. Tokyo: Kashiwa shobō, 2004.

Sosiroda Saburō. "Hokushi manmō ni okeru Nihonjin no jyūkyo kenchiku kōzō chōsa hōkoku." *Kenchiku zasshi* 55, no. 676 (July 1941): 32–38.

Sugai Shirō, ed. *Kokudo keikaku no keika to kadai*. Tokyo: Tameido, 1975.

"Suihō damu chikuzō to Ōryokkō ryūbatsu mondai." *Shingishū shōkōkaigisho geppō*, no. 128 (January 1940): 25–33.

"Suihō damu chikuzō to tsūbatsu shisetsu mondai." *Shingishū shōkōkaigisho geppō*, no. 122 (June 1939): 9–11.

"Suihō damu shōjo hatsu sōden." *Keijō nippō*. August 26, 1941.

Suiryoku denki kensetsu kyoku. *Manshūkoku suiryoku dengen chōsa nenpō*. Shinkyō: Suiryoku denki kensetsu kyoku, 1939.

Sunaga Noritake. "Manshū ni okeru denryoku jigyō." *Rikkyō keizaigaku kenkyū* 59, no. 2 (2005): 67–100.

Suzuki Takeo. "Chōsen ni okeru toshi keikaku to kokudo keikaku." *Chōsen* (February 1940): 1–15.

Tabuchi Jurō. *Aru doboku gishi no han jijoden*. Nagoya: Chūbu keizai rengōkai, 1962.

Tada Reikichi. *Nanpō kagaku kikō*. Tokyo: Kagakushugi kōgyōsha, 1943.

Tagami. "Nidai suiden kōji o miru 2." *Manshū nichi nichi shinbun*. September 22, 1938.

Taketani Mitsuo. "Genshiryoku no heiwa-teki riyō." In *Taketani Mitsuo chosakushū 3: Sensō to kagaku*, ed. Taketani Mitsuo, 129–55. Tokyo: Keisō shobō, 1968–70.

Tamaoki Shōji. "Chōsen no suiryoku ni tsuite." June 22, 1960, interview, transcript and digitized audio at Research Institute for Oriental Cultures. Gakushūin University, Tokyo, Japan.

Tanaka Shunsuke. "Ōryokkō suiden kaisha no yōchi kaishū nado no jigyō suikō ni taisuru jimotomin no ikō." In *Ōryokkō suiden shisetsu ni kan suru suibotsuchi taisaku kankei shorui*, ed. Chōsen sōtokufu shiseikyoku chihōka, 391–96. Keijō: Chōsen sōtokufu, 1939.

Taniguchi Saburō. *Tairiku no kyokusen*. Tokyo: Zen Nihon kensetsu gijutsu kyōkai, 1950.

"Tekunokurashī ni kansuru bunken." *Kōjin*, no. 134 (June 1933): 47.

"Tekunokurashii zadankai." *Kōjin*, no. 134 (June 1933): 23–45.

Tōa mondai chōsakai, ed. *Saishin Shina yōjin den*. Osaka: Asahi shinbunsha, 1941.

Tochika keikaku mata. "Daitōkō toshi keikaku no naiyō." In *Daitōkō 2*, ed. Daitōkō kensetsukyoku, 74–83. Andong: Daitōkō kensetsukyoku, 1942.

Tomii Masanori, Iizuka Wataru, Yoshida Tadashi, and Kuwano Hisao. "Pekin ni okeru kyū Nihonjin jūtakuchi ni kansuru kenkyū: Pekin bōzan kabushiki kaisha no san jūtakuchi o jirei toshite." In *Nihon kenchiku gakkai taikai gakujutsu kōen gaiyō shū* (August 1992): 5–8.

Uchida Hiroshi, Nagaoka Jirō, Ishikawa Gorō, Mori Kyōji, and Sasaki Jirō, eds. *Hōman damu: Shōkakō entei hatsuden kōji jitsuroku*. Tokyo: Taihō kensetsu, 1979.

Ueno Mitsuo. "Ishizu unga no kaisō." http://www.shin-nihon.net/main/forum/unga. htm. (Accessed October 27, 2008.)

Umezawa Mitsukuni. "Soshū no omoide." In *KB*, 165–67.

Unikowski. "Manshūkoku no kaihatsu to chisui mondai." In *Ryōga to shōkakō*, ed. Manshū jijō annaijo, 205–15. Tokyo: Nichiman jitsugyō kyōkai, 1935.

Wen Zhonglin. "Genba de no shōgen—Manninkō to kakurijo 'taiheibō' seki o nozomu genchi de: Wen Zhonglin." In *NKSS*, 51–52.

Yamada Hiroyoshi. "Chihō keikaku ni tsuite." In *Daitōkō 2*, ed. Daitōkō kensetsukyoku, 2–7. Andong: Daitōkō kensetsukyoku, 1942.

Yamamoto Masao. "Hōman chosuichi ni yoru kōtoku jyū nendo kōsui chōsetsu seika hōkoku." *Doboku Manshū* 4, no. 2 (April 1944): 16–20.

———. "Honma moto suiden kyokucho o shinobu." In *HNS*, 135–47.

Yamamoto Tadahito. "Sakuma damu kensetsu to ryūiki keizaiken no henyō." In *Kaihatsu no kūkan, kaihatsu no jikan: Sakuma damu to chiiki shakai no han seiki*, ed. Machimura Takashi, 29–50. Tokyo: Tokyo daigaku shuppankai, 2006.

Yamasaki Keiichi. "Sen kyūhaku sanjūhachi nen no Kahoku toshi keikaku no omoide." In *KB*, 297–98.

Yanagisawa Osamu. "Senzen Nihon no tōsei keizai ron to Doitsu keizai shisō: Shihonshugi no tenka shūsei o megutte." *Shisō*, no. 921 (2001): 120–44.

Yatsugi Kazuo. *Shōwa dōran shishi*. 3 vols. Tokyo: Keizai ōraisha, 1971–78.

Yokoyama Mitsuo. "Daitōkō toshi kensetsu jigyō to enchi." *Toshi kōron* 23, no. 8 (August 1940): 111–16.

Yoneda Masafumi. "Daitōkō kensetsu jigyō ni tsuite." *Shinkyō keizai kihō* 1, no. 4 (December 1941): 16–19.

———. "Ozawa Kyūtarō naimu gishi to naru ki." In *Ozawa Kyūtarō*, ed. Ozawa Kyūtarō kinen jigyō iinkai, 107–8. Tokyo: Ozawa Kyūtarō kinen jigyō iinkai, 1968.

Sources in Korean

Kang Ik-sŏn. "Hŏmsan chullyŏng, munjŏn oktap ilcho e sugukhwa rani hanbat'ang kkum iroda—Oo widaehal son kwahak ŭi him iyŏ." *Mansŏn ilbo*. January 28, 1940.

———. "Paeksŏl ŭl kŏdŏch'ago paesuryŏng ŭl nŏmŏsŏni Hwangbuk en odumach'a pangul sori yoranhago chŏgi chigetkun ŭi susimga hŭnggyŏpta." *Mansŏn ilbo*, January 27, 1940.

———. "Parabononi, simsan yugok—Chŏnch'ang chungsŏkkwang chidae—Sangasansŏng e hoeŏk to saerowŏra." *Mansŏn ilbo*, January 29, 1940.

———. "Sumolchi rŭl ttŏnanŭn tongp'o yŏ kiri pongnok ŭl nurira—natsŏn kojang to chŏngdŭlmyŏn nae kohyang toenŭn kŏt ŭl." *Mansŏn ilbo*, February 1, 1940.

Index

Page numbers in italic indicate tables and figures.

The authorized representative in the EU for product safety and compliance is:
Mare Nostrum Group
B.V Doelen 72
4831 GR Breda
The Netherlands

www.ingramcontent.com/pod-product-compliance
Lightning Source LLC
Chambersburg PA
CBHW020336270326
41926CB00007B/202